QUANTA, ALTERITY, AND LOVE

Quantum Physics and Philosophy

QUANTA, ALTERITY, AND LOVE
VOLUME III

MARK DREHER

MixerMuse, LLC

Edition 1.10 – May 24, 2026

Volume 3

ISBN: 9798995130024

For permission requests, please contact the author at:

Mark Dreher

mdreher@MixerMuse.com

Disclaimer for External Links:

This book contains links to external websites provided for informational purposes only. The author and publisher do not endorse, approve, or assume responsibility for the accuracy or content of any third-party site.

First published by MixerMuse, LLC.

mixermuse.com

Formatted with Vellum

CONTENTS

QUANTA, ALTERITY, AND LOVE-VOLUME 3

APPENDICES

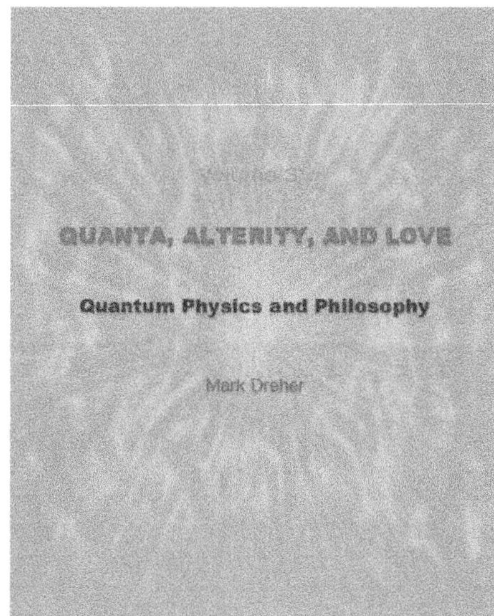
QUANTA, ALTERITY, AND LOVE
Quantum Physics and Philosophy
Mark Dreher

PHILOSOPHICAL REFLECTIONS ON PHYSICS

After Physics

What lies after the conclusion of physics in this volume is not a continuation of physics by other means, nor a speculative ethics derived from scientific indeterminacy. To do so would be to repeat the very violence this work resists: the reduction of alterity to knowledge, of responsibility to explanation, of the Other to a function of the Same. Ethics does not arise from the quantum. Responsibility is not encoded in wave functions, probabilities, or informational fields. And yet—having passed through the undoing of classical certainty, we are no longer permitted to return to ethics as if nothing has happened.

The question is no longer *whether* classical metaphysics and its reduction into 'common-sense' can ground responsibility. That question has already been answered in the negative. The question now is how responsibility appears **when grounding fails**, when justification collapses, and when no appeal to nature, reason, God, or systems can absolve us of the demand placed upon us by the Other. That discussion starts here. The first and second volumes of this work examined how twentieth-century physics brought Western thought to the brink of a "spookiness" which has the effect of loosening the grip of our history and language so the gap created may bring us to an unaccounted for exteriority. Quantum mechanics dismantles the metaphysical assurances of classical science and its underpinnings in metaphysics, exposing the fragility of concepts such as causality, locality, objectivity, and presence. In doing so, it forced a reconsideration of what it means to speak of reality at all. The first and second volumes ended at an edge.

They did not end with a theory completed, a system secured, or a worldview stabilized. They ended with a rupture—an exposure produced by the failure of classical metaphysics and science to close the world in upon itself. Quantum physics, taken seriously rather than metaphorically, disclosed not a new ontology but the inadequacy of ontology itself as a final ground. What emerged was not a replacement metaphysics, but a clearing, an opening, in which certainty, mastery, and causal closure no longer reign. This second volume continues after this opening. What follows is not a replacement of science by ethics, but an exposure of the point at which science—precisely in its most advanced forms—can no longer tell us how to live, decide, or answer for how we face the other.

Introduction: From Ontological Failure to Ethical Demand

Quantum physics did not teach us how to live. But it has taught us something no less consequential: that the world does not submit to total comprehension, that reality exceeds representation, and that the dream of a closed, self-sufficient order—whether scientific, political, or moral—is untenable. What was once secured by causality, locality, and determinism has been displaced by superposition, entanglement, and irreducible relationality. The universe is no longer a stage populated by objects; it is a field of relations in which observation, interaction, and description are inseparable from what appears. Yet the temptation remains strong to convert this disruption into a new metaphysics—to speak of "quantum ethics," "informational morality," or a physics-based responsibility. This volume refuses that move. Ethics does not arise where explanation succeeds. It arises where explanation fails.

The collapse of classical metaphysics does not give us moral content; it gives us exposure. It reveals that the desire for ethical grounding—like the desire for ontological closure—has often been a desire to escape responsibility rather than assume it. To ground ethics in being, reason, nature, or law is to seek shelter from the irreducible demand that comes from outside all systems. What quantum physics has inadvertently accomplished is the removal of that shelter. In the absence of a closed ontology, responsibility can no longer be deferred to systems, structures, or necessities. There is no longer a "because" sufficient to absolve action. The ethical demand emerges not as a rule to be followed, but as a disturbance—an interruption that cannot be anticipated, calculated, or mastered. This volume turns toward that interruption.

Alterity Beyond Knowledge

The ethical relation does not begin with understanding. It begins with encounter. The Other—the stranger, the widow, the orphan—does not present themselves as 'object' to be known or a problem to be solved. The Other arrives as excess: that which resists assimilation into my categories, my narratives, my histories, and my systems of meaning. Unlike quantum indeterminacy, which remains within the domain of description, ethical alterity is **irreducible**. It does not resolve with better models or finer measurements. Where physics encounters limits, ethics encounters obligation. This distinction is decisive. The exteriority opened by quantum physics is not the exteriority of the ethical relation. The former reveals the instability of objects; the latter exposes the instability of the self. Responsibility does not arise from uncertainty about the world but from vulnerability before the Other. It is not grounded in what *is*, but in what *demands*. This demand does not ask permission. It does not wait for justification. It does not consult ontology. **It precedes freedom.**

Method: A Different Kind of Rigor

Because ethics does not arise from explanation, it cannot be treated with the tools of classical theory alone. This volume therefore adopts a different rigor—one that resists system-building, resists totalization, and remains vigilant against the temptation to convert alterity into concept. Philosophy here will be read *against itself*, particularly where it has sought mastery under the guise of reason. Thinkers of alterity, responsibility, and exteriority will be engaged not as authorities but as interlocutors—voices that help articulate what cannot be secured. Theology, phenomenology, and political thought will appear not as foundations but as sites of tension, failure, and demand. The goal is not coherence but fidelity: fidelity to the fact that responsibility cannot be deduced, calculated, or resolved.

What Is at Stake

We live in a world increasingly governed by systems—technological, political, economic, and informational—that promise optimization, prediction, and control. Artificial intelligence, algorithmic governance, and techno-scientific rationality extend the classical dream of mastery even as the sciences that enabled them undermine their metaphysical foundations. The danger is not that these systems are powerful. The danger is that they provide alibis. When action can be attributed to systems, responsibility dissolves. When harm is justified by necessity, ethics disappears. When the Other becomes data, risk, or abstraction, violence becomes administrable. This volume insists that responsibility cannot be automated. It cannot be delegated. It cannot be optimized. It can only be borne.

A Few Vocabulary Reminders

In the first volume I developed some important terms that will again find further refinements in this volume. Here are some reminders:

chaos - the Greek word *χάος*. In Greek it means a gap, yawn, breach, separation, or void. Let's be very clear about what chaos is **not** in the context of these volumes. Chaos is not disorder. Order and disorder belong to a conventional historical hierarchy of binary oppositions. Together they form a totalizing oppositional unity to each other setting up a bipolar opposition wherein all degrees of order and disorder are fully accounted for. There is no excess or alterity which can exceed our lived symbiotic identities of order and disorder.

genesis - in these volumes referring to multiple 'origins' not as metaphysics' history dictates but as emergence from chaos, rhizomic

phusis - ancient linguistic root is a cognate from the Proto-Indo-European language family meaning "to grow"

chaogenesis - chaos growing as what we now think of as beginnings, as genesis or origin

rhizo-phusis - rhizome emergent at one point in the ancient Greek word *phusis*

exformation - Here the prefix '*ex*' in ancient Greek means 'governs the genitive' or governs genesis, origin, motion, cause and effect, etc. However, the oddity of 'what governs genesis' knows nothing about 'origin'. In this case, '*ex*' also carries with it the more modern sense of externality or in our case radical externality.

ONE MORE IMPORTANT NOTE: Much of current literature is only available on-line or in very restrictive academic journals. This means many of the reader's accessible references are hyperlinks on-line. When I moved the eBook to these volumes there were many hyperlinks which did not move correctly because printed books cannot support hyperlinks. At the very end of this volume in "About the Author" section, the last item is a printed address for a free, downloadable .pdf file on my web site. This file will have all the links which can be clicked on to see the referenced material in your browser.

PHUSIS: TOWARDS RADICAL EXTERIORITY AS PHILOSOPHY

I

PHILOSOPHICAL REFLECTIONS ON PHYSICS

In previous sections I have tried to show how quantum physics has introduced a radically other orientation to anything we think of as 'reality'. These findings are uncomfortable as follows,

- They rest on the confidences we have in science.
- We have no easy way to make sense of these ideas.

I have also tried to put these disturbing truths into a social and historical context to show how in many different avenues our social, political, cultural, and environmental securities are being eroded from within by sociological externalities which cannot be reconciled to our basic commonsense presumptions of the 'real world'. The reactionary way back to safe classicisms is mutually assured self-destruction. Extremists and authoritarian reactions want to reinstate worldviews which are being denuded of their confidences not from the outside but from within themselves. First, we need to recognize that we can no longer insist 'reality' adhere to centuries of history which established them. Second, we need to let externality, the insufficiency of everything we thought we knew as 'reality' simply speak without reacting emotionally with insecurity, fear, anger, or retreat. We need to let go of having to know to simply listen and learn. What we must let go of will mean we will not have all the security of a definite future as this is always being hollowed out from the inside. Our responsibility is not made in the choice of our freedom or adapting to a history. Our responsibilities speak from exteriority in the saying of the Other (Autrui). Here, history and choice end. Each of us has no freedom to choose standing in the responsibility we face. Levinas understands responsibility to go far below the narratives in which we have the freedom to choose. We desire to retreat away from radical alterity and towards the depletion of self into the retreat of ego. In a way, I see Levinas asking us to become more like little children, starting yet again on the path of superposition-like birth forging towards unimaginable 'unfamiliars'. Levinas positions Ethics and responsibility beneath the classical tales we tell ourselves in common sense. He shows us how the abyss as chaos, classical thought in terms of the other of order, is excess - radical alterity as responsibility for the he or she who faces us and how this brings us towards a new future with vastly different repercussions for our daily lives.

The quandary we find ourselves in from physics and philosophy has been elaborated through this entire text so far. This was important so that we could feel ourselves towards a predicament which has no easy answer. Rather than retreat into fading realities of the past we need to look towards the future, chaos where the dark incestuous and abstract polarities of war and peace, mind, and body, subject and object, finite and infinite, static and dynamic, changeless and becoming, order and disorder, same and other, Being and nothing can no longer congeal into a semblance of a world. Chaos is the haze our future holds. Diminishing security is the violent demise we will face looking backwards to the past. We have little time to take, to survive as a species. We face extinction from without, the environment, and from within, nuclear annihilation in the Machiavellian despotism of war. Retreating into the past will only hasten catastrophe. First, we must begin to understand why answers from the past are insufficient and 'unable to be able'. This has been the quest so far in these volumes. Now, we can look forward to the void which faces us.

At the least we can say from chaos observation forces branching. Branching can no longer be determined from epistemology and ontology, from the neutrality of essence and Being. We cannot create a fantastical new Nietzschean future based on the nobility of ascending morality. The history of ancient Greece from Hesiod and Heraclitus onward has brought us to where we are. Affirming heroically Nietzsche's infinite recurrence of the same will only bring on our current dilemma with no time left to replay it. Only chaos remains and faces us. In the face of the radical exteriority of chaos, there is no choice as our classical narratives tell us. Suicide is not a choice for relief or cessation. Suicide only addresses, is a resolution of, is cast from, classical narratives of escape, death, and nothingness. The threat or consolation of the extinction of death is based in the narrative of absolutes: spacetime, 'thingness', mechanism, and the metaphysic of nothingness. As we have seen, 'reality' is not necessarily what it seems from the cause-and-effect history of absolute time and space. What we must face is not knowing absolutely. Ideas as historical narratives are not the 'be all in all' we are conditioned to believe. Radical exteriority cannot be answered or measured by history, by spacetime, even by death. The absolute finality of the death of me as an existing subject in spacetime arrives with contingent certainty in the way sleep arrives. However, no such certainty arrives from the wave function. For Heidegger, death is the possibility of the impossibility of death, of 'me' in classicism's time and space which can no longer *be* that we must authentically face. For Levinas, death is the impossibility of all our possibility of 'me' in Being, in classicism's narratives of time and space, that we face every day. This impossibility is the 'me' of self-origin, of historic narratives of certainty which becomes 'me'. How did this possibility even come about? Isn't it from classically spun narratives which situate us in their 'realities'? Hasn't the discussion on quantum physics at least shaken some of our unfounded confidences? Why must we play these skeletal poker games of transactional confidences in our allusions of certainties? Do we have any such thing as choice? Quantum physics is telling us that we must face an abyss where knowledge is no longer absolute and spooky excess leaves us 'unable to be able' as our lived classical narratives inform us. Even the quantum wave function tells us that we are not masters of our destiny. Our branching into classicism finds no absolute, no essence, no 'reality' but only abyssal contingency. This section will explore the radical externality which remains for each of us and humanity. Spookily, spacetime emerges from exformation as the Big Bang's Big Wave Function. Exformation is chaos in terms of spacetime but nevertheless spacetime is emergent from exformation in/with the observer effect. What we face in chaos is not spacetime but radical exteriority.

Echoes from the self-love of Narcissus are retreat from externality which faces us. The gaze of Orpheus can never *be*, or enter into, his love's presence. Self rises from simmering and glistening exteriority, gaze which cannot found its own completion, gaze which cannot found origin. Hades promises certainty, bringing narcissistic self, self-love, to presence. From the underworld's becoming to the heights of the totality of Concept cycling endlessly through presences and absences only to fall back into the hazy swamp of uncertainty. Orpheus' desperate, silent, yearnings for his self-love's completion through 'same and bad faith other' fall back perpetually and tragically into the underworld with only an uncertain "goodbye". Just as Orpheus looks backward his love flees from his sight leaving only a yawning gap. In the agony of death, Narcissus hears an echo of his name spoken by the other, his love's last goodbye. Only a flower remains of his body as an uncertain promise. We have been entranced by the echo of our own history, culture, and language. We will die but not in certainty, in spooky uncertainty. Entangled from radical externality, an other we never knew in all our effacements, we die in a shroud of haze we *know* as 'certainty'. However, can we see with a child's wonder as newness which takes no notice of dying in echoes of ourselves, our dying histories consuming us as narratives. Exformation as *phusis* brings us towards proximity to a radical exteriority we face once again for the first time.

ON MY USE OF ANALOGY AND THE QUANTUM WAVE FUNCTION

For this major section, I want to reiterate why I use a highly unorthodox approach to Levinas in analogies of the quantum wave function. This volume is not written for professional philosophers. I am doing the best I can to make Levinas relatable to folks that have never seriously studied philosophy. From the previous discussion, I simply find analogies of the quantum wave function as perhaps an easier way to get an idea of where Levinas is coming from. Just as there is a danger in reading the quantum wave function from the lens of classicism, there is a danger in reading Levinas as yet another metaphysician of history. That would be a mistake and a misunderstanding. However, anyone not steeped in the history of philosophy will find that much of his writing is inaccessible. Do I believe that these analogies between the quantum wave function and certain ideas of Levinas are more than an analogy? Well, I do believe that science and philosophy are coming together in ways which have not been possible, at least since Roman times. It has become apparent since the beginning of the 20th century that what underlies all our classical assumptions and truths has a very spooky underbelly. In fact, quantum physics is so spooky that it suspends basic assumptions from our entire philosophical history over a yawning gap. It is difficult for us to even find easily accessible words to begin to grasp what has come to the fore recently in our history. As Bohr told us, "*Anyone who is not shocked by quantum theory has not understood it.*" However, quantum physics is just beginning to scratch the surface of its far-reaching effects on the macro classical world we think of as 'real'. Even the problem of consciousness is still many years from having a good understanding of how quantum physics might account for it. The question of the other, ourselves, humanity, and its spooky rootedness as the branching of the observer effect may be hundreds of years away (if Homo sapiens can survive that long). So, since none of us are going to be around that long we can only extrapolate from what we know now. We know and have the spooky feeling that radical exteriority accounts for all we think of as 'real', as 'true', and especially as common sense. It seems that the other and others and the phenomenon which calls itself 'me' is far stranger than history has led us to believe. But analogies, like intuition, are just that. I feel very awkward about making equivalences between Levinas and quantum physics. Any associations I make in this regard are pure speculation in an attempt to try to present Levinas' in a more accessible fashion to non-philosophers. If any philosophers or physicists ever read this volume (which is highly unlikely), I have no doubt they will simply write it off as idiotic. And it may well be idiotic, but it is really only me doing the best I can to articulate a lifetime of inquiry and a hope which may be of use to someone else.

ON THE END OF SENSIBILITY

Modernity and British Enlightenment was based on sensibility. The thinking machine was the marriage of Descartes and modernity along with the Latin history of substance. Substance became the universal thing-like matter that composed all reality. However, reality is not merely a proper or authentic compilation of each animal's (inclusive of human) biological sense data. Biological senses are typically thought of as sight, smell, touch, taste, and hearing. The sensors are electrically connected to some kind of conduit like the central nervous system for humans. The conduit provides informational input and output to a processing unit of some type. For humans our brain and our central nervous system synthesizes and unifies electrical sensory neurons into a cohesive narrative which either initiates some behavior, chooses inaction, or simply goes unnoticed. The outcome of this processing determines whether a motor response is required and, if required, what kind of motor response is initiated. However, all this is much more complicated than it seems. Let's look at a few of these complications.

The sense of touch is believed to be the first human sense to develop.

> Touch is thought to be the first sense that humans develop, according to the Stanford Encyclopedia of Philosophy. Touch consists of several distinct sensations communicated to the brain through specialized neurons in the skin. Pressure, temperature, light touch, vibration, pain and other sensations are all part of the touch sense and are all attributed to different receptors in the skin.
>
> Touch isn't just a sense used to interact with the world; it also seems to be very important to a human's well-being. For example, touch has been found to convey compassion from one human to another, according to a study published by the University of California, Berkeley.
>
> Touch can also influence how humans make decisions. Texture can be associated with abstract concepts, and touching something with a texture can influence the decisions a person makes, according to six studies by psychologists at Harvard University and Yale University, published in the June 24, 2010, issue of the journal Science.
>
> "Those tactile sensations are not just changing general orientation or putting people in a good mood," said Joshua Ackerman, an assistant professor of marketing at the Massachusetts Institute of Technology. "They have a specific tie to certain abstract meanings." *

This article shows how the other traditional senses of the world are not so abstract, modernistic, and detached but more complexly intertwined with how we live, experience, feel life. Multiple specialized receptors give us a sense of touch. Different receptors detect pressure, temperature, light touch, vibration, pain, and other receptors all combine into the categorical notion of touch. There is no simple touch receptor. There is a heterogeneity, a plurality, of receptors we abstractly assign to the simple word touch. This complex heterogeneity applies to all the common words of sight, smell, touch, taste, and hearing all arouse kinesthetically, affectively, and sensually.

The assumption of mechanism ignores the observer's role in detecting complexity, where the

* "The Five (and More) Senses." *Live Science*. Last modified February 12, 2026.

Ackerman, Joshua M., Christopher C. Nocera, and John A. Bargh. "Incidental Haptic Sensations Influence Social Judgments and Decisions." *Science* 328, no. 5986 (2010): 1712-15.

See Bradford, Alina (23 October 2017). "The Five (and More) Senses". Live Science. Retrieved 2021-06-16.

system's collective behavior gives rise to structural modifications and new hierarchical arrangements, emergence and novelty which cannot remain confined in all our narratives. Similarly, in physics, emphasizing the constituent parts at the expense of the whole physical system can lead to oversimplified models and experiments that cannot be extrapolated to accurately describe complex physical systems. Emerging systems in physics are examples of complex systems where new and complicated properties emerge in response to collective properties and not to the sum of properties from individual parts. As we saw earlier, complexity in biology is increasingly looking towards quantum physics for answers. Problems of emergent properties in biology, learning in biological, error detection in organisms, quantum tunneling in viruses, and quantum cooling in biological systems, etc.

> Fascinating combinations of physics and biology can be understood already now. We have identified a large number of interconnects between quantum physics and the life sciences and the status of present experimental skills is great. But the complexity of living systems and high-dimensional Hilbert spaces is even greater.
>
> When we talk about quantum information, the discussion always circles around exponential speed-up. But in living systems any improvement by a few percent might already make the difference in the survival of the fittest. Therefore, even if coherence or entanglement in living systems were limited to very short time intervals and very small regions in space—and all physics experiments up to now confirm this view—simple quantum phenomena might possibly result in a benefit and give life the edge to survive.
>
> We still have to learn about the relevance and evolutionary advantage of quantum physics in photosynthesis, the sense of smell, or the magnetic orientation of bird. We still do not know whether quantum entanglement is useful on the molecular level under ambient conditions, whether quantum information processing could possibly be implemented in organic systems. We still do not fully understand and appreciate the philosophical implications of the quantum-to-classical transition even under laboratory conditions.
>
> We thus conclude that the investigation of quantum coherence and entanglement in biological systems is timely and important. And it will need even more visions, further refined theories, and above all—a significantly broadened basis in carefully worked out and interdisciplinary experiments. *
>
> Aristotle wrote of accidental properties. We might now think of these as emergent.
>
> "A second way one might characterize the autonomy of emergent entities is through realization by base properties that does not depend on any specific realization."
>
> "Almost all accounts of emergence suppose that emergents are not just distinct from, but also distinctively efficacious as compared to, their bases. Weak emergentists typically deny that emergents have any fundamentally new powers, on grounds that such new powers would be,

* Vedral, Vlatko. "Living in a Quantum World." *Scientific American* 304, no. 6 (2011): 38-43.
Arndt, Markus, Thomas Juffmann, and Vlatko Vedral. "Quantum Physics Meets Biology." *HFSP Journal* 3, no. 6 (2009): 386-400.
See Quantum physics meets biology
And Emergent Properties (Stanford)

either directly or indirectly, powers to produce certain basic physical effects, violating the causal closure of the fundamental physical realm. They maintain, however, that distinctive efficacy does not require having a new power. Distinctive efficacy might be understood in terms of distinctive counterfactual patterns over time, yoked to a counterfactual account of causation (LePore & Loewer 1987 and 1989); difference-making or "proportionality" considerations in accounting for macroscopic as against microscopic effects (Yablo 1992); or more generally through the proposal that special science laws track comparatively abstract levels of causal grain (Antony & Levine 1997, Wilson 2010)...

Strong emergentists, by contrast, standardly take emergents to introduce fundamentally novel causal powers—powers that their lower-level physical bases do not have. These are often taken to be powers of the high-level features themselves, and are directed "downwardly" at the structures from which they emerge (as well as "horizontally" in contributing to emergent features of the system at subsequent times). But some propose that high-level structural features induce novel powers in component entities; others suggest that systemic "transformations" occur in which parts either lose their identity when caught up in emergent wholes or have their behavior transformed in virtue of such embeddedness. Details concerning these proposals are provided in section 4.2 below." *

Where and how do these heterogeneous sources get taken up as a word, or better, a thought which we naively take as final and explanatory? For modernity our history, language, culture, and individual choices form a network for making 'sense' from 'sensibility'. Let's observe how even in stating, "making sense from sensibility" we are really understanding something by simply repeating it in the words 'sense' and 'sensibility'. We are not logically processing all our heterogeneous sources of information like a digital, Boolean based computer program would in order to make 'sense' of our world. The most common repetition of this kind gives us redundant and repetitive certainty that we know something (e.g. reality) and is simply summarized by the verb 'to be'. I pointed out this circularity previously when discussing how we are hindered, rendered awkwardly, in even **being** 'able to be able' to think quantum uncertainty without thinking the '*is*' from the context of quantum uncertainty which actually contests the very notion of '*is*' or the generality named by 'to be'.

Physics has proven in many ingenious and sophisticated experiments that the state of the water-tops is not just mathematical uncertainty—it **is** the uncertainty of all possibilities occurring at once in 'reality'. Let's notice the awkwardness in juxtaposing what quantum physics is telling us and what the language I used in the last sentence is telling us. The use of the words 'reality' and 'is' in writing it—***is*** *the uncertainty of all possibilities occurring at once in 'reality'—**is** exactly what a major thesis of these volumes is all about.* Are we not awkwardly doubling, tripling down again by subsequently and repeatedly writing again—***is*** *exactly what a major thesis of* these volumes *are all about*? This is not an accidental coincidence. Let's notice a kind of circularity in our use of the word '**is**' which is repeated in both statements. **'Is' is** the present tense, third person singular of the verb 'to be'. This awkwardness

* O'Connor, Timothy, and Hong Yu Wong. "Emergent Properties." *Stanford Encyclopedia of Philosophy*. Last modified September 30, 2020.

See Emergent Properties (Stanford)

And Quantum tunnelling in the context of SARS-CoV-2 infection

And Quantum physics meets biology

built into the structure of Western language [perhaps language as such] is a circularity fundamentally built into language which we automatically and unawares simply assume. We name this circularity 'Being'. We use it all the time as if we all know what it means. We think of being in all its grammatical forms as a kind of filler word which means nothing but somehow needs to be used to make sense of our sentences. I already discussed this in the section on Martin Heidegger. For now, let's notice that there is an awkwardness in language to even explicate, in linguistic terms, what mathematical, experimental, empirically proven evidence has told us over the last century about "the dominant trait of quantum physics". The circularity we are seeing is not from quantum physics. It is from language which is the product of history. The notion of 'Being' is the fundamental bedrock upon which our Western language is built on. It is so submerged in language that we use it all the time as if it were an invisible 'given', a forgotten inheritance of our history.

Quantum physics has exposed many such limitations such as:

- uncertainty which cannot find any 'reality' which undermines it.
- exformation which has nothing to do with cause and effect.
- the wave and/or particle in which are both true.
- the wave function which updates information or branches off 'many-worlds' in some spooky way. *

Eidetic memory is often used to mean a visual memory identically, an acuity for identity. Eidetic is from the Greek word εἶδος (eîdos, meaning form or idea). † Identity is an important idea in philosophy. The idea of identity **is** the basis for any kind of comparison. To recognize degrees of similarity or dissimilarity, knowingly or not, we must utilize some idea of identity. This **is** the binary in logic (*logos*). We inherently (or inheritly) know 'difference' by what something **is.** **'Is' is** the double play, the repetition of the same. If we think A = A we think identity. This formulation is a statement of logic. This is called a tautology in philosophy, something which **is** or must **be** necessarily true. It **is** epistemologically certain apart from anything to do with phenomena. We could also think A **is** A. This **is** thought of as repetition of phenomenon. In this case the notion of identity **is** thought phenomenally as two phenomenally different instances which are the same in every way. This **is** the ontological interpretation of identity. Let's notice the words bolded in the previous sentence. In logic **'is'** (or 'to be') **is** called the copula. The copula **is** taken as what simply and innocuously connects premises and conclusions in logic.

Socrates **was** a man.

All men **are** mortal.

Therefore, Socrates **was** mortal. ‡

Note that '*was*' and '*are*' are simply forms of the verb 'to be'. What shows itself in our language in my digression, is an inescapable repetition built into thinking itself. For Sartre this repetition has "No Exit' or what I have called the 'no exit of thinking'. Husserl's philosophy wanted to step back and look at experience without attaching any particular truth or existence value to it. He called this the eidetic

* Dreher, Mark Randall. *Quanta, Alterity and Love*. Denver: MixerMuse, LLC, 2026.

† See eidetic (Wiktionary)

‡ "Aristotle." Wikipedia. Last modified January 12, 2026.

reduction. This involved not theorizing about what **is** the case in any way. It also meant not simply taking, for example, classical science's findings (e.g. absolute time and space) as relevant. Husserl wanted to get back to the phenomenon itself in perception and experience. This is where lived time, lived space, lived body and lifeworld all found relevance in Husserl. I understand Husserl's intent here was to avoid the '**is**' which already comes with interpretations (which philosophers have subsequently found difficult or problematic for Husserl). Husserl was Heidegger's mentor. Heidegger accepted the hermeneutical circle of language. By this, he means how our interpretations incessantly recycle upon themselves. Heidegger wanted to phenomenologically investigate the question of Being and beings which were necessarily circular. And then came Levinas. Husserl and Heidegger were Levinas' mentors. However, for Levinas, Heidegger was not deepening the question of Being but doubling down on the totality and violence of Being. He also understood Husserl as preserving the "avatars of interpretation". Here Levinas explains his issues with Hegel, Husserl, Heidegger, and much of philosophy from the ancient Greeks very compactly,

> The position of the subject in the philosophy issued from Husserl — existence, axiological emotion, practical intentionality, thought of Being and even man as a sign, or man as guardian of Being — preserves, across all the avatars of the interpretation, the theoretical sense of signification insofar as openness, manifestation, phenomenality, appearing remain the proper event, the *Ereignis*, the "appropriating," of *esse*. *

Levinas in thinking of radical exteriority wants us to ask ourselves about:

- What cannot be reappropriated?
- What cannot once again repeat the mantra of the same?
- What cannot inevitably reduce the Other (Autrui) to the same?
- What cannot reduce the Other (Autrui) to a circularity?
- What cannot reduce the Other (Autrui) to a hegemony built from 'to be', the logic of being and nothing, that must always miss and erase any trace of the he or she which faces us? †

The awkwardness to think quantum physics will not once again be solved, resolved, discovered in 'to be'. I plan to work through all this and more in the rest of this volume. For now, let's start with philosophical history, the soil from which externality in quantum physics arises and which also makes it so difficult.

Towards an Excess

In order to understand the links between quantum physics and philosophy we need to look at our most basic common assumptions about 'reality'. We all inherit vague but practical notions about what is

* Emmanuel Levinas, *Otherwise than Being: Or Beyond Essence*, translated by Alphonso Lingis (Pittsburgh: Duquesne University Press, 1998), 24.

† Mark Randall Dreher, *Quanta, Alterity and Love* (MixerMuse, LLC, 2026), 1.7.

real. When we are born, we have no such notions. We are born in a quantum superposition-like fog which slowly 'becomes', branches into, probable certainties of our surroundings. We develop a sense of what is real. Reality conveys practicality and repetition. These perceptions slip in unnoticed from our language, our culture, and a history which was never exclusively of the kind as 'my personal history'.

My history is always veiled in the uncertainty of birth and the certain decay into the oblivion of death. Yet for many of us common narratives span our lifetime orienting us in a way known as 'local realism'. Local realism is the foundation of modernism. Local realism is best illustrated as the metaphysics of mechanism. A car has parts located in close proximity such as cylinders, fuel injectors, transmission, tires, etc., all made complete by an orderly arrangement which makes the car a practical method for us to get around. The local arrangement of these parts persists through time, at least until the car gets old and starts to break down. The car does not pop into existence when we need it and disappears from reality when we leave it in our garage. Locality in this case is the practical arrangement of the car's parts which enable transportation. Realism implies the car will continue to exist when we are not looking at it. What is local realism?

> Measurements of entangled particles contradict at least one of the following two assumptions:
>
> 1. Realism: Objects have properties that exist regardless of whether anyone is observing them. Observation merely reveals properties that the objects had all along.
>
> 2. Locality: The measurement of one object can't affect the measurement of another object that is arbitrarily far away.
>
> The combination of the two is called local realism: the assumption that objects have definite properties, independent of our knowledge of them, and independent of measurements performed on other objects. Local realism is deeply embedded in our common sense. When I measure the length of my left foot, I determine the length it already had, without affecting the length of my right foot. And yet this common-sense claim would be exactly wrong if my feet were entangled particles. *

Common sense serves us well when it comes to practical activities because our human world has already been tailored in a way which makes a certain kind of practicality a benefit. History, language, and culture tell us something very important about the kind of being we are. We are historical beings. We overlay our existence in the world with a historically acquired grid which does two things:

It informs us how to succeed or fail in day-to-day life.
It also shapes our world from generation to generation.

Narratives which tailor 'reality' and practicality have a symbiotic relationship. A symbiotic relationship is of the kind Empedocles told us about in which love and strife incessantly feed off each other as if swirling in a circular vortex. This symbiotic relationship reveals a unique capacity for being human. However, this capacity dampens our ability to deal with change. History is the record of change. It chronicles lived 'reality'. The problem with our tendency to conserve a particular narrative

---

* Albert, David Z. "Quantum Mechanics and Reality," *Scientific American* 270, no. 2 (February 1994): 63.
Albert, David Z. "Quantum Mechanics and Reality." *Scientific American* 270, no. 5 (1994): 58–67.

of history is a source of violence spawned by resistance to change. We resist change for many reasons. We are not comfortable with change. Change challenges established power structures. However, change is perhaps the only real constant over time. Evolution is all about change, emergence, adaptation. When our worldviews are undermined, there is a very natural tendency to resist the onslaught of angst which we fear will result in chaos. Chaos is perceived as a threat to cultural mores. It upsets the conservatism of day-to-day practicality. However, history is the story of the violent desire to conserve and to resist the inevitable chaos which wears the face of change. When we stop and reflect on the historic prohibition of chaos, we certainly see that chaos in terms of disorder and order form a binary duality. Change, perturbation, introduces chaos into a system. History is the chronicle of change brought on when chaos upsets order. Conserving order is a tale we tell ourselves to repeat and prolong tragic and violent legacies. Who would even think that order itself could be the source of horrific histories? And who would think that chaos might bring value to the legacy of the survival of Homo sapiens?

Let's remember in classical mechanics, chaos arises from nonlinearity and extreme sensitivity to initial conditions. Tiny differences in initial states can lead to vastly different outcomes over time. Quantum chaos investigates how chaotic classical systems can be described using quantum theory. The central question is: What is the relationship between quantum mechanics and classical chaos? The correspondence principle suggests that classical mechanics emerges as the limit of quantum mechanics when Planck's constant approaches zero. But how does quantum chaos underlie classical chaos? Quantum systems exhibiting classical chaos often display:

- Spectral level repulsion: Energy levels avoid clustering too closely together.
- Dynamical localization: In time evolution, certain states remain localized despite chaotic classical behavior.
- Enhanced stationary wave intensities: Quantum features concentrate in regions where classical dynamics have unstable trajectories. [*]

Coherence refers to a quantum system's "quantumness." When a system loses coherence due to interactions with its environment (quantum to classical transition), quantum chaos signatures may be suppressed. This loss of coherence affects the system's ability to serve as quantum information or be manipulated. While quantum chaos primarily focuses on physical systems, its principles could potentially impact biological processes. Adaptation in evolution involves genetic changes over generations. Quantum effects at the molecular level might influence mutation rates, genetic stability, and phenotypic variation. [†]

Stuart Kauffman, a complex-systems researcher, proposed the concept of autocatalytic sets in the 1980s and 1990s. These sets consist of molecules that reproduce collectively. Kauffman suggested that if a diverse chemical soup of polymers existed, these autocatalytic sets would spontaneously emerge as a phase transition—akin to the shift from solid to liquid. These sets function holistically, mutually catalyzing the formation of all their molecular members. His inspiration came from advances in

---

* Giacomo Casati, Boris V. Chirikov, and Italo Guarneri, "Energy-Level Statistics of Chaotic Quantum Systems," *Physical Review Letters* 54, no. 13 (April 1985): 1350.

† See Quantum chaos (Wikipedia)
And Quantum Chaos (Springer Link)
And Finding coherence in quantum chaos

network mathematics, where phase transitions occur as connectivity increases. Kauffman's key insight is motivated by what he calls "the nonergodic world," which includes objects more complex than atoms. While simple atoms can exist in all their possible states over a reasonable time, interactions among atoms to form molecules lead to an astronomical number of possible states. Only a tiny fraction of moderately complex proteins (e.g., 200 amino acids long) have emerged throughout the Universe's history. Given these limitations, Kauffman explores how existing complexity ever comes into being. He introduces concepts like closure, where processes are linked in closed cycles, driving each other forward. Quantum laws, discovered relatively recently in the timescale of human evolution, manifest at different organizational levels: inert matter, living organisms, and cultural artifacts. These quantum laws hint at the nature of the fundamental constituents of our world and their role in evolution. Quantum nonseparability plays a crucial role in understanding emergence. T. E. Humphreys proposed the notion of "fusion," where quantum nonseparability contributes to emergent phenomena.

> In systems with broken spatial symmetry, directed transport of particles can be observed in the absence of any time-averaged forces or gradients, if the system is kept away from thermal equilibrium. The underlying principle of such "ratchets" has found applications in particle separation and in the modelling of molecular motors in living systems, and has also attracted much interest from a fundamental point of view. In particular, chaos in classical ratchets, and quantum effects such as tunnelling, have each been found to have a strong effect on direction and magnitude of the particle current. Here we are interested in the question how classically chaotic ratchets would behave in the quantum regime. Can chaotic behaviour be observed in quantum ratchets, or, alternatively, could ratchet effects be generated from quantum chaotic behaviour? From the point of view of experimentalists, we discuss these questions with a focus on mesoscopic electron devices, and point to possible topics for theoretical investigations. *
>
> We present a view of the evolving reality based on our understanding of the recently discovered (in the time scale of human evolution) quantum laws, how they manifest at the different organizational levels of inert matter, living organisms and cultural artifacts, and what they possibly imply regarding the nature of the stuff the world is made of. What emerges is a pancognitivist framework in which the notion of quantumness interlaces with multiple aspects of our world, including the appearance of the complex life forms on the surface of our planet. †

This perspective suggests that emergence cannot be reduced solely to brain states or mental substances. Emergence requires irreducible levels of explanation in physics, chemistry, and biology. This pattern of emergence offers a fresh approach to the problem of consciousness—one that transcends mere brain states or mental substances. The intersection of quantum physics and emergence is opening up new ways of understanding biology and life. ‡

---

* Heiner Linke, "Quantum Chaotic Ratchets," *Applied Physics A* 75, no. 2 (August 2002): 167.
See Chaos in Quantum Ratchets

† Ignazio Licata and Leonardo Chiatti, "Quantumness and Pancognitivism," in *The Quantum Self: A New Philosophy of Quantum Mechanics*, edited by Ignazio Licata (Cham: Springer, 2024), 213.
See Quantum Perspectives on Evolution

‡ See The new physics needed to probe the origins of life
And Quantum Perspectives on Evolution
And Emergence and Quantum Mechanics

Athens would have never had democracy without the chaotic disruption of tyrannical order of ancient Greek clans. Athenian democracy happened because of the political changes made by Cleisthenes, who broke up the unlimited power of the nobility by organizing citizens into ten groups based on where they lived, rather than on their wealth.

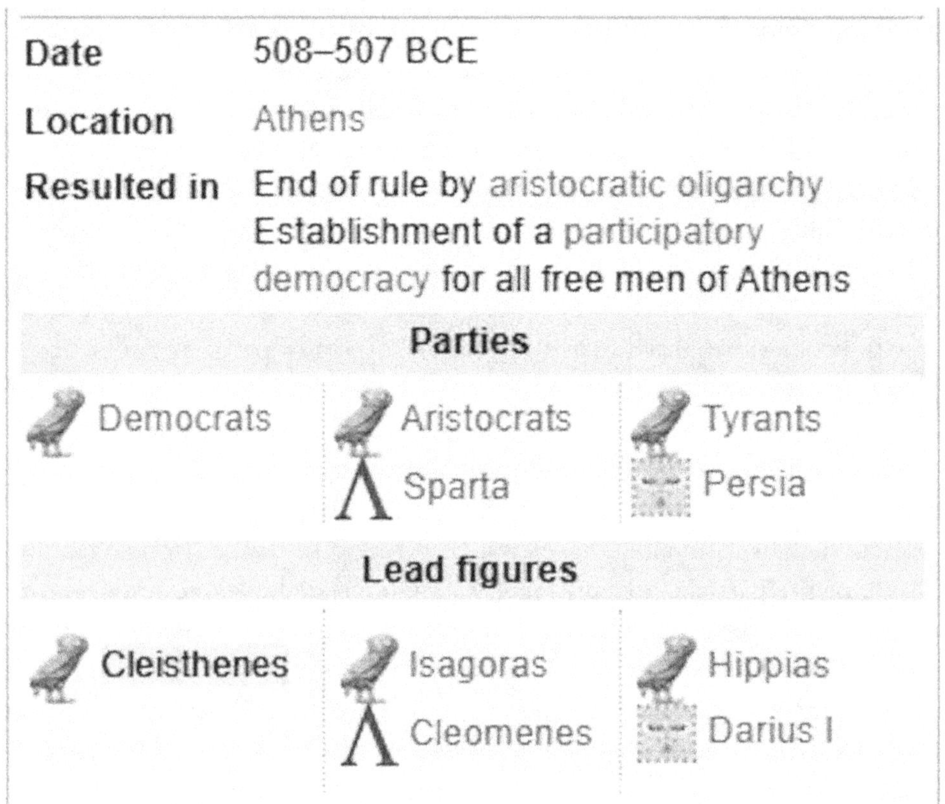

Date	508–507 BCE
Location	Athens
Resulted in	End of rule by aristocratic oligarchy Establishment of a participatory democracy for all free men of Athens

Parties		
Democrats	Aristocrats Sparta	Tyrants Persia
Lead figures		
Cleisthenes	Isagoras Cleomenes	Hippias Darius I

[*]

This new democratic power structure was created to forestall strife between the traditional clans, which had led to the tyranny in the first place. The tyranny had been a terrible and bloody failure, and even the Spartans acknowledged that a moderate form of democracy would be preferable. The goal of Athenian democracy was that all citizens should have equal political rights and the ability to fully participate in either the council or the Assembly. [†] Later, chaos coming on the heels of the breakdown in feudal systems in Europe ushering in the notion of democracy. During the French Revolution in 1789, the feudal system was officially abolished. This marked a crucial turning point in European history, as it dismantled the traditional feudal structure and paved the way for more modern forms of

---

And Mind and Emergence: From Quantum to Consciousness

* "Athenian Revolution Infobox," *Wikipedia*, last modified February 17, 2024.
Image based on "Athenian Revolution" by Wikipedia contributors, licensed under CC BY-SA 4.0.
See Athenian Revolution - Wikipedia

† See Athenian democracy - Wikipedia
And Athenian Revolution - Wikipedia
And The Final End of Athenian Democracy (pbs.org)
And Democracy of the Ancient Athens | Short history website

governance. In England, the remnants of feudalism persisted until they were abolished by Parliament in 1645. After the Restoration, Charles II further solidified the end of feudal bonds in 1660. While the system lingered on in parts of Central and Eastern Europe until the 1850s, significant reforms gradually eroded its influence. For instance, slavery in Romania was abolished in 1856, and Russia finally abolished serfdom in 1861. * Later, the United States fell into the chaos of its Civil War in which incompatibilities in the idea of democracy and slavery were brought to the fore. And beginning in the 20th century with relativity, quantum mechanics, phenomenology, analytic philosophy with its many linguistic turns, chaotically and controversially leading to postmodernism concurrently ushered in the idea that democracy was not sustainable when gender, ethnicity, and sexual orientation were exempted.

What does this tell us about ourselves and cultural disruptions? More importantly, let's observe that values are conserved in the historic play of reductive binary dualisms in which, for example, order is good, and chaos is bad. But the virtue of conservation can always be reappropriated in common narratives to cover over any excess to its worldview. This is how virtue always gets reassigned in the play of binary dualisms. We see this in the United States in the frenetic and violent gasp of the anti-woke conservatives who believe they are maintaining an illusory order which is deemed virtuous. The need to conserve narratives as values until the chaos of disorder perturbs the system and order can be reestablished as the new order. History has taught us this reductive scheme in which binary dualisms can never allow an excess by continual reappropriation. This is why those of us still caught up in the wake of classicism and modernism have lost the ancient sense of the externality of chaos. Even more, a sense of logic in common sense has historically given values the veneer of unquestionable absolute truth.

One important point to make here is that chaos does not necessarily include value judgements. Chaos is an agent of change, but change is not always considered good. Adaptation can go either way to increase survivability or decrease survivability. It seems to me that Nietzsche's entire passion was centered on chaos, no matter what the outcome or the pre-ordained values associated with them. For Nietzsche, the heroic Greek affirmation of tragedy (or worse eternal reoccurrence of the same) elevated the individual human and societal humans. For Levinas, what elevates humanity is Ethics based on the radical alterity of the Other (Autrui).

As was previously mentioned, in the early part of the 20th century when relativity was revolutionizing physics, phenomenology was revolutionizing philosophy. It is important to note here that an exhaustive study of the history of physics and philosophy would show us that this transition was not an abrupt change. I spent a little time discussing this transition for philosophy in my recent post on George Orwell and Emanuel Levinas. See George Orwell and Emmanuel Levinas Introspective: Socialism and the Other (https://www.mixermuse.com/blog/capitalism-and-marxism/george-orwell-and-emmanuel-levinas-introspective-socialism-and-the-other/). The point here is that chaos had already intervened in history, and always intervenes in history, to deplete the meaning bestowing practicality of existing paradigms, the *mythos* of the 19th century mechanistic worldview. The depletion of modernity, our worldview, is asymptotic, a diminishing limit. Those of us whose practicality is established and maintained under worldview, can only ever after approach a smaller and smaller significance derived in that lifeworld. This was brilliantly exhibited in 1936 with Chaplin's "Modern Times".

* See A History of Europe
And feudalism
And Feudalism (Wikipedia)
And Feudalism and Knights in Medieval Europe

Modernity has exhausted itself under its norms, its values, its practicality. It no longer functions well as when an old car starts breaking down. We, as Homo sapiens, must adapt or face extinction, if for no other reason than climate change. Or, in a frenetic reaction to the impending doom of an epoch, we can forcefully impose a veneer of the old order...like a Nazi version of "Groundhog Day". Modernity is simply out of answers. Its naïve optimism of George Jetson, space age, push button, efficiency has exhausted itself. This apparent practicality becomes more evident by dysfunctional politics and social upheavals. In its death grip, the same old tired answers bring on the rise of authoritarianism and violence around the world. The climate is dying for humans but looking great for cock roaches. When an age dies, cataclysmic events spell human tragedy or species extinction. If we cannot find a way forward, the climate we have poisoned will silence the war of all against all unless nukes beat us to it.

Since recorded history we have seen these kinds of transitions prompted by the breakdown of societal practicality. Ancient Greece was aware of this recurring movement as *peras* and *apeiron*, order and chaos. We should not be surprised that the 21st century has once again stared into this chaotic abyss which Nietzsche tells us always stares back at us. My personal studies of philosophy and science have shown me the necessity of facing the 'yawning gap'. The abyss gaped at us in the 20th century response to the breakdown of classical physics and philosophical modernity beginning with the wonder of relativity, quantum physics, and phenomenology. We have yet to fully feel its effects as we would rather sleep, and dream of days gone by. What constitutes the inflexibility of worldview and the necessity of chaos? How does practicality cease to function and lose its hold on societal norms? Is there an externality which looks us in the face and exhausts our death grip on dying realties?

When our historicity becomes absolute, it becomes a meta of *phusis*. It becomes 'reality'. It no longer merely orients our practicality. It takes on the garb of the sacred. Here sacred means it loses the capacity to be questioned. This is not as much from choice but more as a desperate fanaticism towards divination. The invisible gods of yesteryear speak in our day-to-day practicalities where our motions feel more and more like cardboard dreams. Existence takes on a plasticity, but it comes on so gradually the resignation of mere transactional gamesmanship becomes reality. Before we can allow reflective thought to come to the fore we are already immersed in the surety of our intentions. Practicality is not mere reality, it is practice. This practice veils itself in simply 'what is', as if the matter is self-enclosed, as if it is simple tautology or what is necessarily true. This absolute reduction has more than intriguing parallels to the breakdowns of 'absolutes' in classic science such as Newton's absolute time and space. While practicality suggests a well-trod confidence in action it also imprisons action into a context which must forever remain oblivious. But chaos, change, persists. Postmodernity cast shadows of the failures of our epoch against our cave walls in conjunction with nagging questions in physics of any such absolute of 'reality' and even going all the way back to Plato's Forms. Quantum physics places uncertainty at the fore of the void left by the failure of such classical notions as 'absolute'.

> In the tradition of classical physics, the observer is a disembodied mind external to nature, and the objective essence of nature is revealed in the mathematical intuition of its structure. In quantum mechanics, however, the observer is not a disembodied mind, nor does the mathematical intuition represent an 'objective' presence but the probability of a presence. In quantum mechanics, the observer is a human consciousness embodied in instruments that serve as a bodily extension of the observer's embodiment. The empirical observations of quantum physics consequently involve a bilateral relation between the observer's enhanced

> embodied consciousness (on the subject side) and what is observed (on the object side), each having its place and context within the lifeworld of Nature as culture. The concepts and judgments of quantum physics consequently are contextualized subjectively and objectively by the 'natural world' as structured by science. *

In quantum physics and phenomenology both subjectivity and objectivity cannot be taken ontologically as something 'real' or 'existent' from which we, the observer, can stand in a privileged position independent from 'my observational effect' in which objectivity thrives. Classic physics along with classic metaphysics were in firm agreement that such a perspective as purely objective was not only possible but practical. For Husserl the lifeworld was the phenomenal experience of our lived world. Our lifeworld comes to us as beliefs, emotions, and as unions of culture, language, and history. We can never merely be an absolute passive *cogito* which is removed from our immersion in experience. This notion of a god's-eye view is itself a lived immersion in phenomena not some underlying classic 'reality'. At the end of histories and epochs, everyday practicality is the submergence of the god's-eyed artifice in the intentional task at hand from which the context of classical history becomes determinate. However, the confluence of relativity, quantum physics, and phenomenology has interrupted historic narrative as the collapse of absolute time and space into the relativity of spacetime referential frames. But concurrently with relativity quantum physics obliterated millennia of foundations and underpinnings of classical certainties. It turns out the 'reality' itself was conjured up by history like aether. Let's step back and take a look at how perceptions on a universal and personal level converge into 'me now'.

I have already illustrated relativity and how the speed of light affects time and space under acceleration. But acceleration is gravity. There is no difference. If I am standing on a weight scale accelerating in space and I weigh myself to be the same weight that I am on earth, this means the acceleration I am undergoing is the same in space as it is on earth. We think of gravity as a force we feel on earth and that space has no gravity. This would mean that gravity is the same on the earth and in space. What is the 'same thing' is acceleration. Gravity is not really a force like the three fundamental forces of electromagnetism and the weak and strong nuclear forces. The perception of gravity as a force which always acts locally is not correct. Gravity is really nothing other than the geometric curvature of spacetime. Our lived experience of gravity is weight. Our perception of gravity as a force is shown to be a historic narrative not what 'gravity' actually is.

We looked at how entangled systems give rise to notions of near and far in terms of more or less entangled. We have seen how spacetime emerges from entanglement. Remember how distance can be determined by the amount of entanglement in a system? Our body is an entangled system. Our lived experience of 'my body' is the concrete embodiment of this scientific abstraction. Our bodies are more densely entangled particles, so we 'feel' locality. Locality is an abstract historic narrative which we think explains this feeling. But locality does not exist as we abstractly think it does. For modernism the sense organ of the eye and light determines what we see. But interestingly, even as far back as modernism's early voice in Descartes the sense perception of seeing was contested by what could not be doubted –

* Patrick A. Heelan, "The Cognitive and Ontological Role of Technology in the Life-World of Quantum Mechanics," in *The Quantum Self: A New Philosophy of Quantum Mechanics*, edited by Ignazio Licata (Cham: Springer, 2024), 29.
See Consciousness, Quantum Physics, and Hermeneutical Phenomenology (.pdf)

thinking. But gradually seeing and all the senses became detached as Frankensteinian body parts into a universe of dead, inanimate, neutral 'things', matter.

In the English play Lingua, the five senses engage in combat for superiority. Ultimately, Sight is declared the victor by *Communis Sensus*, the vice regent of Queen Psyche. German Jesuit Jacob Balde also explored a similar competition in his work Urania victrix. Dutch art witnessed the emergence of allegorical paintings depicting the five senses, created by artists like Rembrandt and van Bijlert. Notably, a collaboration between Jan Brueghel the elder and Peter Paul Rubens produced a famous series of five paintings, one of which—Sight (1617)—captures diverse ideas about vision. In this painting, a young Venus sits amidst visual splendor, viewing a small painting of Jesus restoring sight to the blind. The excess of visual elements—gold coins, a peacock, and a flower arrangement—hints at the vanity of beauty and visual pleasure. Astronomical and navigational instruments scattered throughout emphasize the commercial and scientific value of optics. Animals like monkeys, parrots, and dogs remind us of shared sensations. Extreme empiricism challenged ideas like mind, self-consciousness, self, etc. as mere idealism, asserting that sensory experiences were fundamental. Diderot provocatively stated, "*if you want me to believe in God, I would have to touch Him*," emphasizing the importance of tactile experience. Empiricism broadly emphasized the role of sensory experience in forming concepts and acquiring knowledge. [*] Seeing was irrevocably changed by 20th century relativity.

In relativity, the inertial frame is just like any other frame with the exception that we can abstractly and randomly make it our non-accelerated frame from which we view the spacetime of other frames. This means if we have two clocks running precisely together. We can place one clock in our inertial frame and put the other clock in another frame. Not only are times different for us from my lived frame observing another person's lived frame but practicality, already by narrated history, informs us that there is a simultaneity which we take as common for both me and the other. From relativity seeing is relative to frames of reference where clocks and lengths can and must change even on the smallest scales but yet appear to stay the same on larger scales. On small scales the mass of our bodies causes each one of us to have separate 'clocks' which are not synchronized. If we live on top a high rise or stand on a mountain, we age more slowly than those below on the ground where gravity is stronger. Sometimes we feel our own lived time as slow if we are bored. Sometimes we feel our own lived time as fast if we are having fun. Those kinds of feelings tend to get discounted as 'merely subjective', in which history's 'common sense' always prevails before we even think about it. Actually, lived spacetime and relativity's spacetime from which I observe the other are diachronous. They are two extremely different types of time.

Modernist notions have us believe that all our clocks are synchronized, and the senses are the gold standard of this historically abstracted narrative. Relativity's frames, on a larger scale, change the practicalities we live by. Our lives are not a synchronized succession of events as we pass through the day. We start work at nine o'clock. We do not expect that all our synchronized clocks will run differently, causing everyone arriving at nine o'clock to start work at different times. However, this effect happens on the smallest scale even between me and you, me and the world. We are just acclimated practically to understanding reality by historic habituation, so it all appears homogeneous, the same, to us. The

---

* See Senses, Early Modern Theories of the
And The Senses in Philosophy and Science: From the nobility of sight to the materialism of touch
And Empiricism, Early Modern
And The Senses: Classic and Contemporary Philosophical Perspectives

twentieth century's spacetime has not yet become a part of our commonsense narrative. Even more, our history colors us as all the same 'things'. Our lived spacetime and relativity's inertial spacetime frame has no absolute reference. And all our common senses get much more exasperated by exformation. Absolute is an abstraction which we take as certain. But there is no absolute spacetime and there are no absolute others who are the same as me, existing subjects in a world of objects - all taken as absolute, as what is, as common sense. 'Myself' as a randomly chosen inertial frame is relative from the smallest to largest scales of different clocks and sizes in a spooky sea of non-local, indeterminate uncertainty which has some odd connection to what I observe. History's abstract subjectivity is simply the frame we choose as a given of common sense. It is important to realize that even in our passivity we make a choice to consider all of this the 'same'. But beneath the historic narrative of freedom to choose we face the Other (Autrui) in which responsibility cannot be chosen only oriented to as retreat.

Gravitational waves propagate through the universe. Stronger ones are caused by cataclysmic events like when two blackholes combine into one. These waves wash over us changing spacetime at the speed of light. These changes cause everything in our experience to grow larger and smaller as waves and simultaneously time to go faster and slower (including all clocks) but we never feel the difference. We feel like everything just plods along as usual. The entire fabric of space time stretches and contracts as if we were in the surf on the beach. Quantum jitter is a fact of quantum particles. It never stops. The jitter is so small it is impossible to measure on smaller particles. However, it can be measured on larger particles. We never feel it. The vacuum of space is a tight knit system of entangled particles jittering and, like bubbles that wash up on the ocean's shore, they combine to make larger bubbles of energy and disperse quickly. As mentioned earlier these virtual particles can combine to make lightning under the right circumstances and many other events we can measure. Vacuum jitter and endlessly pulsating energy below Planck levels is too small to measure directly. And this jitter is not just for the 'vacuum', this is the fabric of all spacetime. Vacuum jitter is real in its effects, but it has a hazy relationship with what we call reality. Even our biological eyes always see the past. It takes minimally about 120 milliseconds for our brain to register what the eyes see. We never 'see' the future or the present. It is always unseen. On top of that, for entanglement, an effect can come before a cause. This is Alice in Wonderland, but we call it normal, everyday, practical reality. Perhaps Alice saw more clearly than us. These are the circumstances we have found ourselves in when the 20th century shattering our naiveté.

On Transactionalism

Throughout this series of volumes I have referred to transactionalism as gamesmanship. I want to clarify what I mean by transactionalism based on previous comments. Transactionalism is the economy of echo. Echo arises from narratives of history based on lived embodiment. Lived embodiment is phenomenological. In physics vernacular lived embodiment is the observer effect of branching into classical narratives as 'myself', as us, as narratives of reality, existence, and being. Therefore, transactionalism appeals to the self as already narrated by being-in-the-world. There is a necessity imposed by this history for acting out of self-interest as Adam Smith would tell us. But pure self-interest has played itself out in modernism's mechanics of the self as thinking machine. The proof of this is shown in the unabashed pride in acquiring capital. The ideal of capital acquisition has even usurped the notion of religion and democracy as the opiate of the people. While many folks regard themselves as unabashed capitalist in the United States, most folks will never reap the idealized rewards of their devotion to self-interest. Instead, all their lives they will act out their self-interest in petty gamesmanship. The game of Trump-styled, self-interest long ago was contrived as the latest and most ingenious 'public relations' stunt. This stunt was rooted in modernity's alienation of the self in the mechanism of the thinking machine of the senses. It was inevitable that marketing would become the tool of perception as reality. And technology has and is increasingly become the Matrix-styled pods where human labor and energy are harvested. In the Matrix,

> Humans are tricked into existing in a fake world that they believe to be real while their energy is being used and converted into power. Their energy is used to fuel the machines that exist in the real world, and keeping them in the pods keeps humans easy to manage and easy to harvest energy from. *

Just to be clear, Communism was also the just another 'public interest' campaign of the powerful few who idealized themselves as the elite creators of the type fictionalized by Ayn Rand. The answer to a successful public relations campaign is not yet another public relations campaign. The problem arises from our unique capacity as historical embodiment. Lived embodiment historically interpreted imposes a survival instinct on all of us. Death as the final ruin of the thinking machine is the mechanism for lived dreams. Dreams of immortality haunt our lived day-to-day with what Heidegger calls being-towards-death. History has evolved and afforded us an obsession with mortality which imposes a necessity to act-as-if pure self-interest, echoing ourselves back to ourselves, will bring Eurydice back from Hades only to find that death awaits our blank stares.

From the narrative of the historic abyss on which live the hallucinations of our own echo habitually becomes practice. And, by the way, this does not necessarily involve having any explicit comprehension of the subject matter of these volumes. It involves a different kind of day-to-day practical choice which many folks can and do find their way towards intuitively with practical responsibility to the Other (Autrui) which does not totalize the Other (Autrui). Quantum physics has told us in no uncertain terms that all our historic narrations which found us as commonsense and day-to-day practicalities face an extreme excess. This is not the religious myth of a transcendent God. This is not the hallucination of unrestrained self-interest. This is the abyss we face where all we see is chaos. This is not the horror of disorder or its symbiotic dream of order, logic culminating in Concept. We face radical alterity not as the horror of death but as the face of the Other (Autrui) with whom we are indebted. The scope of this indebtedness will be explored in this major section of this volume. We will find that the self, 'my' self, is not my invention, my origin as perpetual echo, but emerges from an embodied and lived withdrawal fleeing from exteriority which cannot come under my dominion or control. This is not death. This is the tree of life.

The Dying Epoch: Abyssal Limit Gathering Origin to Neutrality

If exformation is prior to time and space, then we can see that it might satisfy Aristotle's requirements

---

* *The Matrix*, directed by Lana Wachowski and Lilly Wachowski (Warner Bros., 1999). See Why Humans Are In Pods In The Matrix.

for *telos* gathering the limit, culmination, final cause which is uncaused. Aristotle states*, "*For the final cause is not only "the good for something," but also "the good which is the end of some action"*, the *end for some action* gathering as *telos*. If exformation is uncaused as a radical exteriority not the product of the endless chain of causes and effects in spacetime, it cannot be thought of in terms of origin. 'Origin' in this sense is thought of as the first term in an endless series of cause and effect. It may be a helpful conceptual tool to abstractly in conformity to historic narrative to think of exformation in terms of 'the first cause' from which spacetime is emergent (perhaps, vis a vis the gravitational transformation of mass into the singularity's creation of virtual particles resulting in what we might call the Big Bang). However, the emergence of spacetime from exformation is a limited thought experiment in terms of 'kind' which we are pointing towards in thinking of exformation. Exformation is not a cause among other causes. It is not a species of a genus. It is not an element in the set of all causes. While spacetime may be emergent from exformation, exformation is oblivious to spacetime. The relationship is not one of reciprocity in which spacetime emerges from exformation and exformation is the dialectical determinate of spacetime. There is no higher order which transforms spacetime and exformation into a higher unity. Exformation is not a negative of spacetime or vice versa. Exformation is not merely the 'good for something' as in the emergence of spacetime. It is also the 'end as in culmination [*telos*] of some action' with regard to spacetime. *Telos* is the branching into classicism's narratives. As for *telos,* exformation cannot be recovered from within spacetime. It cannot be understood from the history of classic physics or even strictly from Einstein's relativity. In other words, exformation is a extreme disjuncture, a yawning gap, which is anarchic.

Exformation does not recognize form or order. It is not one 'kind' of form and order. Neither is it a negative of form and order. It cannot be thought to exist or not exist. It can only remain an uncertain haze. From the perspective of spacetime it is all information without order, without time or cause and effect. It is all possibilities instantaneously which cannot be actualized as such except as random amounts of uncertainty or chaos in the anarchic [without origin] act of observation. This is what is meant by radical exteriority. As radical exteriority exformation is the *telos*, culmination or fulfillment of not just some action but all actuality. However, exformation does not recognize or observe this end vis a vis actuality, as objectivity. This abrupt disjunction can only be recognized as such by an attentive observer. What is meant here by 'attentive observer' is one that can recognize how extremely odd the notion of exformation is. Even the greatest thinkers of the 20th century such as Einstein, Bohr, Heisenberg, and Schrödinger who understood the ramifications of exformation were troubled by its historic incongruity. Niels Bohr, one of the most important founders of quantum physics told us that "Anyone who is not shocked by quantum theory has not understood it". †

In my estimation, perhaps exformation in Aristotle's sense as 'the good for something', as anything that is something, inclusive of personal and historic narratives, must be emergent from entangled virtual quantum foam popping in and out of existence which may rise from bare (virtual) existence to energy levels which can be, appropriated as 'reality' and detected and bound to wave/particles [wavicles] together with the spooky observer effect. However as discussed it makes no sense to think of exformation in reductionary terms of origin as it has a hazy exteriority in which reciprocity is irrelevant, the

* See NASA Quantum Computer Mission Boldly Goes From Hilbert Space to Outer Space

† See Beyond Weird: Why Everything You Thought You Knew about Quantum Physics Is Different.

spooky wave function which never 'sees' (for lack of a better word) spacetime and historic narrative yet makes spacetime and historic narrative possible. Aristotle states,

> In the latter sense [**the end of some action**] it applies to immovable things, although in the former [the good **for something**] it does not; and it causes motion as being an object of love, whereas all other things cause motion because they are themselves in motion. Now if a thing is moved, it can be otherwise than it is. Therefore, if the actuality of "the heaven" is primary locomotion, then in so far as "the heaven" is moved, in this respect at least it is possible for it to be otherwise, i.e. in respect of place, even if not of substantiality. But since there is something—X —which moves while being itself unmoved, existing actually, X cannot be otherwise in any respect. For the primary kind of change is locomotion, and of locomotion circular locomotion; and this is the motion which X induces. Thus, X is necessarily existent; and qua necessary it is good and is in this sense a first principle. For the necessary has all these meanings: that which is by constraint because it is contrary to impulse; and that without which excellence is impossible; and that which cannot be otherwise, but is absolutely necessary. [*]

There are no "*movable things*" in the quantum wave function. Motion, force, energy, entropy, etc. are all movable things which all have to do with spacetime, things which can be moved, can have motion. Everything in motion can change implicating temporality. For temporality, entity, being, existence can be other than it is, by its very isness. If "*the heaven*", spacetime, is thought of as primary entity, being, existence its place can change (e.g., from Big Bang to now to heat death of the universe). If something can move as spacetime and all at the same time, simultaneity, cannot be moved (the quantum wave function) then there is an 'existence' to it (e.g., the observer effect) and is pure actuality while concurrently without regard to change, temporality, it moves all, branching of many-worlds, while itself is never moved (e.g., ~~universal~~[†] wave function as simultaneity and all possibilities (potential). The good in Aristotle's sense is contrary to impulse, momentum, and without which excellence is impossible. Excellence cannot be other than itself and as such is "*absolutely necessary*". As excellence the good is otherwise than Being, *ousai*, awareness and beyond essence. Aristotle deems this the good. And Plato deems this the good beyond Being. This is the first principle but also an excess of first principles, *aperion* -> unlimited, without form, void, chaos as privation and its excess - therefore, "*the good*".

The quantum wave function is unmoved, does not recognize classical branching of the observer effect. Yet, it is absolutely necessary for movement, actuality, everything which 'is' and has locomotion. It seems to me that Aristotle had one of the best explanations for the quantum wave function millennia before there was even the possibility of such an idea. I call this intuition. In Aristotle's latter sense as *"the good which is the end of some action"* exformation is the culmination of action which cannot be thought in terms of the same kind or dialectical necessity of, actuality. It only becomes actual in experimental observation. Apart from observation it cannot be thought in terms of 'actual'. This does not mean that entanglement only happens when we observe it. Nor does it mean the contrary,

---

[*] Aristotle, *Metaphysics*, translated by Hugh Tredennick (Cambridge, MA: Harvard University Press, 1933), 1072b. See Aristotle, Metaphysics 12.1071b through Aristotle, Metaphysics 12.1072b

[†] Jacques Derrida is the primary postmodern philosopher associated with "erasure" or overstriking (sous rature) words like "universal," "truth," or "being." This technique signals that a word is necessary for language yet inadequate or deceptive, challenging the stability of traditional philosophical concepts.

that it happens spontaneously without observation. It is not a binary totality which fits into historic categories. Pure subjectivity or pure objectivity is preposterous. Yet, it is responsible for everything that is anything, every transition from one quantum system's state to another. However, the essential impossibility of determination of the wave function apart from experimental observation casts an impenetrable mist upon the banks of 'reality'...like it or not. Exformation as "**the** characteristic trait of quantum mechanics" remains purely **contingent** potentiality, as the sum of all possibilities which may become actual. Even this actuality is still infected with fundamental indeterminateness. In terms of Aristotle, motion cannot be exformation. Exformation in this sense is Aristotle's unmoved mover. Exformation has no dependence on spacetime. Exformation cannot actually exist. The first principle is necessary for locomotion, for classicism's narratives. In Aristotle's notions the unmoved mover is a primary being, *ousai*, awareness and beyond essence. Locomotion is an accidental being. The good is necessary as *telos*, fulfillment, completion, excellence. We need to rid ourselves of this modernist holdover in thinking of the wave function as pertaining only to 'matter', dead inanimate stuff, because it results in our lived life and experiences as branching narratives which is neither personal nor collective but both and exceeds both making actual narratives of who we *are*.

As we have seen, exformation is before space and time, cause, and effect, and is pre-originary. Origin achieves its necessity, its demand and requirement, through the unfolding of space and time. Origin is an after-effect, a narrative of exformation theory. The questions the predecessors of ancient Greece could never resolve was, what is the origin of 'origin'? This is because it must be an endless recursive question, "What came before X" or the well-known chicken-egg scenario. In effect, current physics tells us origin loses all meaning with exformation. Exformation does not have any relationship to origin as is commonly conceived in a classical framework. Therefore, it is anarchic by definition, literally meaning no origin in Greek. It does not recognize the relativity of time and space. It has no 'origin' in any of the historic and philosophical ways we have of understanding origin. Aristotle wrote, "matter exists potentially, because it may attain to the form; but when it exists actually, it is then in the form." (Metaphysics 1050a15. Greek: *ἔτι ἡ ὕλη ἔστι δυνάμει ὅτι ἔλθοι ἂν εἰς τὸ εἶδος: ὅταν δέ γε ἐνεργείᾳ ᾖ, τότε ἐν τῷ εἴδει ἐστίν*). For Aristotle, potentiality or possibility does not exist in any sense of the word. Actuality is existence with regard to motion, change, activity or for our case the relativity of time and space, cause and effect, form and order.

What Hath Quanta Wrought?

Ideally, information theory would be the record down to the most infinitesimal detail of everything that has ever happened. Information in this sense is classical information. Classical information is what we call history. That is, information is a factual recount of everything which happened to explain how transitions from one state to another occurred. The universe is a giant, inconceivable repository of all information which disappears from complexity into the limit of chaos. Even blackholes are believed to retain all information, a record of all the history of what went in them, at the 'inner horizon' just below the event horizon of the blackhole.* But all this information is chaotic exformation given by the mathematical allegory of a singularity. Ideally and classically, we could recount the entire record of cause and

* See Black holes store information as holograms at the event horizon, says Stephen Hawking.
And The Most Famous Paradox in Physics Nears Its End

effect since the Big Bang. This would not just be the account of physics, biology, and the natural sciences but also the histories of sociologies, psychologies, anthropologies, ethnologies of languages and customs, etc. In philosophy this school was called positivism as it believed everything could be logically or mathematically including morality and ethical judgements. In 1898 Lester Ward wrote,

> "The most important thing to determine was the natural order in which the sciences stand—not how they can be made to stand, but how they must stand, irrespective of the wishes of any one. ... This Comte accomplished by taking as the criterion of the position of each the degree of what he called "positivity," which is simply the degree to which the phenomena can be exactly determined. This, as may be readily seen, is also a measure of their relative complexity, since the exactness of a science is in inverse proportion to its complexity. The degree of exactness or positivity is, moreover, that to which it can be subjected to mathematical demonstration, and therefore mathematics, which is not itself a concrete science, is the general gauge by which the position of every science is to be determined. Generalizing thus, Comte found that there were five great groups of phenomena of equal classificatory value but of successively decreasing positivity. To these he gave the names astronomy, physics, chemistry, biology, and sociology." *

In his early years, Ludwig Wittgenstein was highly influenced by positivism when he wrote the *Tractatus Logico-Philosophicus* (Latin for '*Logical-Philosophical Treatise*') between 1914 and 1918. Wittgenstein wrote,

> We can say straight away: Instead of: this proposition has such and such a sense: this proposition represents such and such a situation. It portrays it logically. Only in this way can the proposition be true or false: It can only agree or disagree with reality by being a picture of a situation" †

And later wrote,

> The great problem round which everything I write turns is: Is there an order in the world a priori, and if so what does it consist in?" ‡

While physics in the 20th century was making all these startling discoveries philosophy was reflecting its own abstraction—shattering findings which we, as common-sense folk, have yet to even have a hint of. All of this changed with Husserl and his student Heidegger.

---

* Lester Frank Ward, "The Place of Sociology among the Sciences," *American Journal of Sociology* 1, no. 1 (July 1895): 18.

† Ludwig Wittgenstein, *Tractatus Logico-Philosophicus*, translated by C. K. Ogden (London: Kegan Paul, Trench, Trubner & Co., 1922), 4.03.

‡ Ludwig Wittgenstein, *Notebooks, 1914–1916*, edited by G. H. von Wright and G. E. M. Anscombe, translated by G. E. M. Anscombe, 2nd ed. (Oxford: Basil Blackwell, 1979), 53.

See Wittgenstein,Tolstoy and the Folly of Logical Positivism

> While quantum mechanics was taking shape in Göttingen, Leipzig, and Munich, where Heisenberg studied and worked, the phenomenology of Edmund Husserl was academically prominent. Later, Heisenberg was also one of a circle of professional scientific intellectuals who met regularly during the summer with Martin Heidegger in the Black Forest. In Germany, phenomenology and hermeneutics were *wissenschaftlich* approaches to psychology, art, literature, music, and natural science, establishing them both as 'scientific' and 'philosophical,' on a par academically with the role of logic and analysis in the contemporary USA. In addition to Husserl and Heidegger, Dilthey's work on history, Maurice Merleau-Ponty's work on perception, and Hans-Georg Gadamer's work on literature and art also inspired a strong current in European academic culture before and after the war. These currents of thought also inspired Michael Polanyi, who was Wigner's scientific mentor. *

The early 20th century was the last gasp of modernity's sensing machine which passively observed absolute time and space as God's eye of pure unfolding 'reality'. In modernity God's eye is my eye. My eye is locked in a prison of our lostness to historic narrative. What still remains to be seen is that seeing the other simply as my narrative of the other is precisely where my narrative fails. Another way to think of this is the practical, cultural breakdown in observation which has lost itself in its own myopia, its own covered over assumptions of a certain pre-narrated cultural motif. The resistance to what is coming and has already started around the world is a violent desperation to return to a dying 'reality'. The feeling of necessity drives this fanaticism. It is because we are not aware of the historic pathways of common sense which no longer 'work'. We intuit this socially but in passivity have not taken the time and effort to think through what is dying in our social mores. These breakdowns in 'seeing' which cannot yet see is a lens which we can also see in Einstein's own dogged determination to reject the 'spooky' action of quantum entanglement in order to find a local explanation.

> In a phenomenological analysis, it means first, that the laboratory with its measuring instruments belongs to 'the extended body' of the observer, and thus, that the observer is 'the embodied human consciousness so extended.' Consequently, the observer lives in and through the laboratory measuring instruments as oriented towards practice; this is the channel of his/her 'noetic' intentionality. Under such circumstances, the measure-number becomes just a coded messenger, and like the photons received by the eyes in seeing, the message delivered by the measure numbers is interpreted within the context of the measurement. In vision, it is not the photons which are 'observed' by the culturally prepared viewer, but the illuminated objects; so also in measurement, the measure-numbers are not what are ' observed ' by the culturally prepared 'observer' but what they point to beyond themselves, namely, the presence of micro-systems in the laboratory situation, unaccompanied, however, by any intuitive description of them in the space-time of the laboratory. †

---

* Patrick A. Heelan, "The Cognitive and Ontological Role of Technology in the Life-World of Quantum Mechanics," in *The Quantum Self: A New Philosophy of Quantum Mechanics*, edited by Ignazio Licata (Cham: Springer, 2024), 28.
See Consciousness, Quantum Physics, and Hermeneutical Phenomenology (.pdf)

† Patrick A. Heelan, "The Cognitive and Ontological Role of Technology in the Life-World of Quantum Mechanics," in *The Quantum Self: A New Philosophy of Quantum Mechanics*, edited by Ignazio Licata (Cham: Springer, 2024), 28.
See Consciousness, Quantum Physics, and Hermeneutical Phenomenology (.pdf)

The phenomenal experience of the breakdown of praxis, the inability to be able, holds the promise and threat of totem and taboo. Lived experience in action concretizes reward and punishment. It sets an expectation of a condition which minimizes the angst of lack of origin and uncertainty of death, the oblivion of birth and death. It anesthetizes the lived experience of birth and death which brings chaos to the fore. In thinking of chaos, we are bringing the lived experience of exteriority to the fore. In Western history enlightenment and modernity culminated in reducing humans to mechanism, to sensing, enlightened thinking machines failed to achieve its goal of educated, critical thinking machines. The twentieth century saw this failure in the face of two world wars, human suffering, oppression, and vast inequities in wealth distribution. It also brings to the immanent fore the threat of nuclear holocaust and the environmental extinction of the human species. Certainly, Enlightenment ushered in a fuller realization of democratic states. However, we have seen recently that social acceptance of the consequences of democracy has generated anti-democratic movements all over the world wanting to supplant democracy with authoritarian governments. This is clearly evidenced in the human extermination given by the continuing inability to overcome such authoritarian brutalities as fascism and communism. It is also shown in cultural prescriptions as ethnic bias, gender bias, and sexual preference biases. All of these failures demonstrate how power interests have found ways to exacerbate the most anti-democratic and anti-Enlightenment tendencies with the highly effective, scientific marketed motto that 'perception is reality'. In this case 'perception' is not the observer but the determinations of the observer of the observed.

When democracy devolves into a bourgeois strategy of protectionism it is no longer democratic but a pretense for the conservation of power structures. Democracy as the government of educated and enlightened citizens stops short of the promise of capitalism when it transitions to a tool of wealth interests. Technology combined with marketing has intensified the potency of vested and invested perceptions over reality. Quantum physics' notions of uncertainty and observation and relativity's elevation of the inertial (observational) frame of reference co-arise with these economic politico-societal changes. This is not a coincidence. It is another sign of the breakdown of a worldview which can no longer function in its disappearance as practicality. The 20th century spread of democracies all over the world tells us that the seeds of Enlightenment and subsequent modernity may no longer be merely Western. Democracy has had worldwide effects both in its rise and in its demise. Many advocates of capitalism, democracy, and modernity adhere to an incremental approach of eventually fulfilling mechanistic promises. There is one immediate reason why this notion can only fail - climate change.

NASA tells us,

> Earth's climate has changed throughout history. Just in the last 800,000 years, there have been eight cycles of ice ages and warmer periods, with the end of the last ice age about 11,700 years ago marking the beginning of the modern climate era — and of human civilization. Most of these climate changes are attributed tovery small variations in Earth's orbit (http://climate.nasa.gov/blog/2949/why-milankovitch-orbital-cycles-cant-explain-earths-current-warming/) that change the amount of solar energy our planet receives.
>
> The current warming trend is different because it is clearly the result of human activities since the mid-1800s, and is proceeding at a rate not seen over many recent millennia.[1] It is undeniable that human activities have produced the atmospheric gasses that have trapped more of the Sun's energy in the Earth system. This extra energy has warmed the atmosphere, ocean,

> and land, and widespread and rapid changes in the atmosphere, ocean, cryosphere, and biosphere have occurred.
>
> Earth-orbiting satellites and new technologies have helped scientists see the big picture, collecting many different types of information about our planet and its climate all over the world. These data, collected over many years, reveal the signs and patterns of a changing climate.
>
> Scientists demonstrated the heat-trapping nature of carbon dioxide and other gasses in the mid-19th century.[2] Many of the science instruments NASA uses to study our climate focus on how these gasses affect the movement of infrared radiation through the atmosphere. From the measured impacts of increases in these gases, there is no question that increased greenhouse gas levels warm Earth in response.
>
> Ice cores drawn from Greenland, Antarctica, and tropical mountain glaciers show that Earth's climate responds to changes in greenhouse gas levels. Ancient evidence can also be found in tree rings, ocean sediments, coral reefs, and layers of sedimentary rocks. This ancient, or paleoclimate, evidence reveals that current warming is occurring roughly 10 times faster than the average rate of warming after an ice age. Carbon dioxide from human activities is increasing about 250 times faster than it did from natural sources after the last Ice Age. *

We no longer have the time for the human species to possibly and incrementally or better phantasmically reap the diminishing rewards of the promise of Enlightenment strained through the decline of modernism. It is imperative that we rise to new challenges of modernism's breakdown and the reactionary movement back to a failing worldview. Modernism failed to root out the old face of force and violence in order to perpetually reinstall the old guard. The result of modernism's basis in the senses and self-interest is technology. Heidegger tells us that technology sets up the environment as 'standing reserve' waiting to be used. Self-interest is focused on short term gain. Pure self-interest has proven itself to be ineffective in dealing with the urgent and pending planetary apocalypse of climate change. What can we gleam from the remains of our day? How can we rise to new possibilities given the fractures of classic and enlightened 'reality'? What does this breakdown show us phenomenologically about ourselves, our societal norms of being with each other which must inevitably face the void as threat? The breakdown of order and the advent of chaos, loss of origin, implores us to lean forward into the abyss which stares back at us.

As we have seen, the classical world was a historic narrative which never really existed in phenomenon in spite of all the attempts to make it so. The *a priori* world of origin and order was an abstraction of history. As we know from quantum mechanics, the best we can do with information about physics is statistical probability. Even in classical physics probability is mathematically related to the statistical amount of uncertainty. Uncertainty is calculated by Shannon entropy calculated in bits. Uncertainty and entropy are related. Shannon entropy can range from absolute certainty about classical information to random uncertainty or maximal entropy. Quantum systems measure information very differently. Quantum information is stored as a 'qubit'. The qubit is the amount of information

* NASA Science, "Evidence," Climate Change: Vital Signs of the Planet, October 23, 2024, https://science.nasa.gov/climate-change/evidence/.

See EVIDENCE: How Do We Know Climate Change Is Real?

that can be stored in the simplest quantum system like the spin property of an electron. The point here is that absolute determinism and certainty does not exist except as a historic narrative and abstraction of science and philosophy. The end of these notions in quantum physics and postmodernism started in the late 19th century and in philosophy came to fruition in the 20th century with phenomenology.

Historically, philosophy and science have roughly followed with similar paradigm shifts. A case and point of this being the 20th century's shift in classic physics to relativity and quantum mechanics along with the 20th century's shift in philosophy with analytic philosophy's transition from positivism to the 'linguistic turn' where hermeneutics play a central role. For the continental tradition, philosophy shifted from Kantian categories of understanding and German idealism and the reaction to these movements in existentialism to phenomenology. Einstein's relativity and Husserl's phenomenology were in lock step in the 20th century. As I have mentioned, classic physics looked at time and space as absolutely linear, local, and absolutely distinct from each other. Relativity looked at relativistic frames and how spacetime is curved where time and space does not conform to our classical idea of them but distorts classical linearities in space and time as length contraction and time dilation which is not merely visceral – it is non-classical 'reality', an oxymoron of sorts. Quantum physics upends classical notions of locality. Modernity since Descartes looked for absolutes in human existence, nature, and history. Phenomenology looks at how phenomenon shows itself in cultural and personal lived frames of reference.

Classical time is abstract as absolute linear now moments proceeding from a past to a future. Phenomenology starts with how phenomenon shows itself to us apart from these philosophical and metaphysical abstractions. Phenomenology looks at how phenomenon shows itself in cultural and personal lived frames of reference. Lived temporalities have a kind of stretch from a past to a future, retention and protention, perhaps reminiscent of relative frames of reference which prohibit any kind of universal temporal synchronicity. More modern digital cultures can experience time in shorter time increments measured in the experience of digital devices. Ancient cultures may utilize agrarian time as longer planting and harvesting or positions of the planets or sun. In my lived time I may feel time passing quickly when I am having fun or slowly when I am having anxiety.

Space in classical terms was linear units of distance which again turns out to be an abstraction. In phenomenology, space for individuals is experienced in terms of regions. We bring different regions closer or further away depending on our experience. For example, we may be wearing glasses while looking at a cup of coffee in our hand. Even though the glasses on our face are closer, we are inhabiting the region of the coffee, we bring it near in our attention. We also have cultural sacred spaces or memorial places. Marketing is an example of how regions of affectivity can play more into consumer purchasing than logical judgements of quality or even price of a product. Marketing understands lived experience as more central to who we are than logic and truth.

There seems to some similarities in quantum physics in which spatiality is shaped by exformation density which in wave function collapse is not merely the abstract idea of bare distance and extension but also in some spooky way participates in historic *narratives* and self-consciousness more akin to lived space, living intentionality, lived regionality, where bringing regions far and near have more to do with lived settings. Perhaps we could think of this case as the region of a theatric stage, of a play. For an audience, the stage of the play is 'closer' than the seats we are sitting in. It seems to me there is an element of entanglement here in which linear, abstract distance falls short - is simply abstract.

Additionally, contrary to positivism, empirical certainty obtained through the sense finds history as language which is spookily interpretive. There is no pure classical understanding of history which can be divorced from interpretation. Hermeneutics is the study of how history gets interpreted. In Greek mythology Hermes was the playful, mischievous trickster which bought messages from the gods to humans with a certain amount of embellishment. So, while we have a sense of objective truth, that sense can never be detached from interpretation and experience just as classic physics can never reach its ideal in its inventive abstraction we simply think of as *reality*. *Reality* can only exist as an abstraction which gives us some information but can never reach the ideal of a full account of all information. As quantum mechanics shows us the record of information must always remain inherently statistical and probable.

But for what purpose? This asks the ancient philosophical question of origin which I wrote about extensively in my philosophy series on this blog. [*] The ancient answer of Hesiod to origin is chaos or 'yawning gap'. In contemporary language, chaos indicates the negation of knowledge. Contrarily, the yawning gap for the ancient Greeks indicates an incompleteness of knowledge. An excess of knowledge which cannot be measured by knowledge. Contemporaneously, this is more like what we mean by the term infinity which indicates an overflowing of itself. Even 'infinity' has a tendency to be abducted by the assumed knowledge of neutrality. I prefer Levinas' phrase radical exteriority or radical alterity.

In one form or another, the issue of the ultimate beginning has engaged philosophers and theologians in nearly every culture. It is entwined with a grand set of concerns, one famously encapsulated in an 1897 painting by Paul Gauguin: *D'ou venons-nous? Que sommes-nous? Ou allons-nous?* Where do we come from? What are we? Where are we going? The artwork depicts the cycle of birth, life and death--origin, identity and destiny for each individual--and these personal concerns connect directly to cosmic ones.

[†]

We can trace our lineage back through the generations, back through our animal ancestors, to early forms of life and proto-life, to the elements synthesized in the primordial universe, to the amorphous

---

[*] See Hesiod
And The Origin

[†] Paul Gauguin, *Where Do We Come From? What Are We? Where Are We Going?*, 1897–98, oil on canvas, 139 x 375 cm, Museum of Fine Arts, Boston.
Image courtesy of the Museum of Fine Arts, Boston.
D'où venons-nous? Que sommes-nous? Où allons-nous? and License Agreement **{{PD-US}}**

energy deposited in space before that. Does our family tree extend forever backward? Or do its roots terminate? Is the cosmos as impermanent as we are? In all these questions our historic narratives have provided an answer which has gone underground into our subconscious – absolute and unquestioned neutrality.

What would happen if we gave up the historic supremacy of the reductionist ghost of mechanistic neutrality for an exteriority in which the culmination of neutrality, quantum physics arising from classical science, has brought us to a chasm where the meta of scientific objectivity no longer suffices for common sense assumptions. 'Modern times' in which we go around and around in the cogs of neutral mechanism has shown itself in quantum physics as abstraction which can longer suffice. Yet, we go round and round transactionally deluding ourselves into thinking the one with the most toys wins. The gaping void of the 'it-ness' of Being has worn thinner and thinner where desperation and despotism scream that something has gone terribly awry with reactionaries crying against 'wokeness' as if the cultic trance has become necessity. Relativity started us toward the slippery void and co-arising with it, quantum spookiness has exposed the exhaustion of the myths which set the stage for reality and Being. Levinas refers to neutrality as the elemental 'there is' in which sunlight, wind, and rain become 'natural processes' of causality.

> Neutral existence or the there-is (il y a) denoted indeterminate nocturnal being, which gives way to the diurnal being called "the elemental" (sunlight, winds, rain). As noted, being in Levinas thus entails both dynamic forces and a conception of natural processes and causality. *

The sheer 'isness' of *phusis* as inanimate, as 'matter' no longer retains its title and rule as origin, as divine, as absolute in which consciousness bowing to Concept has become immanent neutrality resolving all into totality which has rid itself of self-consciousness.

> Indeterminate, nocturnal being remains 'something'. As the "there is", being fills in all spatial and temporal intervals, whereas consciousness arises out of it thanks to its self-originating awakening and focus. This is Levinas' first comprehensive sketch of being as a totality, in which the self-ego dyad appears as the limited transcendence of neutral being. Over the course of his analyses, this self-ego will hearken to a call. However, the call comes not from being but from an alterity that Levinas compares with death itself. "Just as with death, I am not concerned with an existent, but with the event of alterity" †

Radical exteriority is *phusis* and contemporary physics disappears as the limit of chaos in the quantum wave function which physicists colloquially refer to as *universal*. But as mentioned the notion of universal, many-worlds, etc. may be a useful pedagogical tool at times but it really imposes a kind of ontology and epistemology on the quantum wave function which creates more impediments to understanding than it appears to solve. However, in the esteemed tradition of classicism and particu-

---

* Emmanuel Levinas, *Existence and Existents*, translated by Alphonso Lingis (The Hague: Martinus Nijhoff, 1978), 58–60. See Levinas (Stanford)

† Tamara Albertini, "The Seductiveness of Being," in *The Quantum Self: A New Philosophy of Quantum Mechanics*, edited by Ignazio Licata (Cham: Springer, 2024), 162.
See Levinas (Stanford)

larly Hegel, chaos shows itself as abstraction which must find concretization. From the entire history of classicism chaos can only be the 'not' of order. However, this 'not' has no excess. This 'not' remains a binary duality best described as disorder. Chaos as the 'not' of order or the 'not' of knowledge departs from Hesiod's yawning gap. Isn't this 'chaos' yet another form of violence and mere deference from origin? In this case, disorder is a condition of order and vice versa which really amounts to degrees of the same. Chaos in this case has been reduced to a deficient mode of order. Historically, reductive logic has not had a great track record of maintaining its veracity over time. It seems subsequently, more times than not, there was some excess to the reduction which doomed it.

Any reductionary scheme will exhibit symptoms reminiscent of Hegel's dialectical 'not-ing' process which we are told proceeds from more abstraction to more concrete unities which hold apparent abstractions as a higher unity. Yet, the higher unity finds its 'not' upon which it shows itself as more abstract until yet again a higher more concrete unity evolves from and in these apparent abstract reductions. In this way the intermediate 'not' of reductions evolve towards higher unities which organically and immanently proceed from abstract appearances of oppositional negations to higher concrete unities. As we saw with Hegel's derivation of space and time, the neutralities of space and time increasing eventually lead to self-consciousness as a more concrete manifestation of abstractions of space and time. Hegel's process then gives an apparent solution to the problem of reductionism as a hierarchical movement towards immanent truths which, in a sense, transcend the problem of reductive binary dualities. Sceptics at the time of Hegel and even into critiques of 20th century postmodernism have argued that all these shenanigans are really simply playing with language, language games. When transitions like the neutrality of space and time pass over into self-consciousness our modernist reductions to the sense thinking machine start sounding alarms which reflect on the inappropriate use of language. The reason for this is legitimate to some degree. This is because language itself grew out of ancient intuitions, religions, superstitions, and pretenses to truth which could not stand the test of time. But even modernist truths reflecting a distrust of language and its looseness with empirical observation have also failed to account for discoveries emanating from the very bedrock of empirical positivism in the 20th century as we have seen with quantum physics. But there are other discoveries in almost all 20th century sciences including evolutionary science which have eroded confidence and certainties in empirical observation. One such example of this directly has to do with language and its incongruity with biology.

> Avram Noam Chomsky[a] (born December 7, 1928) is an American professor and public intellectual known for his work in linguistics, political activism, and social criticism. Sometimes called "the father of modern linguistics",[b] Chomsky is also a major figure in analytic philosophy and one of the founders of the field of cognitive science. He is a laureate professor of linguistics at the University of Arizona and an institute professor emeritus at the Massachusetts Institute of Technology (MIT). Among the most cited living authors, Chomsky has written more than 150 books on topics such as linguistics, war, and politics. Ideologically, he aligns with anarcho-syndicalism and libertarian socialism. *

* "Noam Chomsky," *Wikipedia*, last modified March 3, 2026, wikipedia.org.
See Noam Chomsky

Noam Chomsky believed deep and surface language structures were derivation trees of a context-free language. These trees are then transformed by a sequence of tree rewriting operations ("transformations") into surface structures. Chomsky proposed the concept of Universal Grammar, which posits that humans are born with innate biological knowledge about many features of language. This includes structures and rules that are common across all languages. According to this theory, there exists a deep structure underlying language. Outward-facing surface structures relate to phonetic rules (how sounds are produced), while inward-facing deep structures relate to words and conceptual meaning. At the heart of Chomsky's theory lies the innateness hypothesis. He suggests that we come into the world primed to learn language. This hypothesis is anchored by the concept of Universal Grammar, a built-in set of grammatical structures inherent to all human tongues. Chomsky proposed that the first stage of language development occurs immediately upon birth. Infants, though preverbal, possess an innate language acquisition device. The language acquisition device sets them up for language development, allowing them to acquire language effortlessly and consistently across cultures. Two striking features of language are:

- Diversity: We do not all speak the same language.
- Constraints: Most languages cluster in a few tiny bands, showing more similarity than random variation would predict. *

Chomsky's theory aims to explain these features by positing a deep and underlying structure shared universally across languages. Evidence supporting this idea comes from studies showing that children understand syntactic categories even before they express them in their own speech. Chomsky's linguistic theory is rooted in biolinguistics. Biolinguistics asserts that the principles underlying language structure are genetically inherited and preset in the human mind. Just as our organs develop as we mature, language also emerges as a species-wide trait. The innate language faculty knows the rules, allowing children to acquire language, while vocabulary and specific language details are learned. †

In complementarity, just as quantum particles exhibit both wave-like and particle-like behavior, the wavicle, language structures can be viewed from different perspectives (deep vs. surface). Quantum entanglement involves non-local correlations between particles. Similarly, linguistic elements (such as

---

* Noam Chomsky, *The Architecture of Language*, edited by Nirmalangshu Mukherji, Bibudhendra Narayan Patnaik, and Rama Kant Agnihotri (New Delhi: Oxford University Press, 2000), 7–8.

† See Deep structure and surface structure - Wikipedia
And Universal grammar (Wikipedia)
And Linguistics of Noam Chomsky (Wikipedia)
And Chomsky Philosophy
And Chomsky's Theory
And Chomsky Theory
Hartshorne, Joshua K. September 17, 2013. "How to Understand the Deep Structures of Language". Scientific American (online)
And Chomsky language development unlocking the theory of linguistics
And Chomsky's stages language development
And Deep structure and surface structure (Wikipedia)
And Linguistics of Noam Chomsky (Wikipedia)
And The Application of Chomsky's Syntactic Theory in Translation Study (.pdf)

words or phrases) can be entangled in complex ways. Quantum superposition allows particles to exist in multiple states simultaneously. Analogously, linguistic elements can have multiple interpretations or meanings. Studies in the area of linguistics and quantum mechanics have some interesting possibilities. Research is beginning to increase rapidly in many areas involving macro-oriented phenomenon and quantum physics. Much of this research is also spawned on by developments in artificial intelligence and quantum computing known as quantum intelligence. *

Chomsky's theory has intriguing parallels with the quantum wave function. His theory, as for the wave function, blurs the inherited line we get from modernist histories which cleanly divide biology and linguistics. The result is that linguistics tends to get swallowed up into a reduction to the mechanics of biology. The wave function does not allow for such linguistic distinctions in reductions to mechanism. Chomsky's theory also blurs lines between individual and collective as does the branching of the wave function into classicism implicating the brain, consciousness, language, chemistry, biology, evolutionary science and many other disciplines. As with quantum biology, many sciences are getting more and more integrated with quantum physics; we can expect this kind of trend to continue. The premature summation and reduction of the self, the 'my' of self-consciousness, has exasperated itself in modernism.

Exformation Again

Resistance to change is the tendency set up by our uniquely human capacity to already be historic, to already orient ourselves phenomenologically from an all-encompassing narrative. This narrative is shortened in such a word as 'reality' and such actions as practicality. Philosophers call these reductions originating from the historic tradition ontology. Yet now we have physics which has told us in no uncertain terms that at the most fundamental level - reality has no certain state, no 'Being' or even existence. And to add insult to injury there is a spooky relation to observation and complexities which pass into chaos. This is not some wild-eyed Democrats conjuring up the breakdown of social norms. This is physics telling us that 'reality' is necessarily uncertain and has an uncertain dependency on what we are looking for observationally. Furthermore, we shall not call these extremely different observations contradictory, we shall call them euphemistically complementarity. Call it what we may, it wildly exceeds any historic ontological understanding of Being or essence.

What we see is not enduring when we are not observing. This does not mean 'reality' disappears. That would simply be the negative case of the conventional notion of reality which appears in our history. In that case, we simply reaffirm our historic narrative in its denial. Doing this exercise simply shows itself and 'all things' as 'objective reality' OR 'absurd', 'ridiculous', 'nutty', etc. In making these assertions we simply reaffirm our unwavering confidences in historic narrative. Exformation is very different from our historic narratives. Exformation has no origin, no cause. Exformation has no alle-

* See Essay on mesoscopic and quantum brain
And Quantum Physics in Consciousness Studies (.pdf)
And Quantum Linguistic Patterning
And The computation power of living systems is maintained by decoherence-free internal quantum states
And ENTROPY, QUANTUM MECHANICS, AND INFORMATION IN COMPLEX SYSTEMS A PLEA FOR ONTOLOGICAL PLURALISM
And Chapter Nine - Panexperiential materialism: A physical exploration of qualitativeness in the brain

giance to our historic narrative of some 'objective reality' which exists apart from 'my subjectivity'. And neither does it fall into the contrary category of senseless, meaningless, pure psychic hallucinations of absolute subjectivity. These simplistic binary dualities which must of necessity feed upon themselves are reductionary and ultimately shut down thinking as a kind of hermetically sealed environment.

Regarding the absolutes of objectivity and subjectivity, exformation even takes an extraordinary step away from our lived historicizing phenomenology. Exformation breaks down our classic notions embodied as 'absolute'. From this perspective, the absolute then becomes an artifact of a certain anthropology of the West. Information now necessarily includes exformation which can find no place in our 'common sense' culture. The wave function's branching into classicism necessarily means a loss of information called decoherence which cannot be recovered. We think of this simply as the 'existence' of particles.

> First, according to quantum theory, at some point the naive notion of reductionism, in the sense that emergent properties of higher layers of description can be at least in principle explained by the properties of isolated constituents, breaks down, as the mere existence of these constituents itself turns out to be a consequence of isolating the corresponding degrees of freedom from their environment. In other words, the existence of particles is a consequence of a specific perspective: Whenever a particle is observed, the quantum field gets coupled to the measurement device and its unknown environment. It is this interaction and the information loss that implies that the state of the quantum field is observed as a particle. Particles thus are a consequence of decoherence. Particles are, as Zeh argues consistently, emergent themselves (in the sense that a description in terms of constituents neglects information about the entangled total system). [*]

To add insult to injury, this loss of information in branching is directly related to observation. Classicism has conceived 'reality' based on an assumption that: 1) what is real is fully accessible in some type of information whether it be empirical, philosophical, religious, etc. and 2) observation is independent of reality. Heinrich Päs is a professor of theoretical physics at TU Dortmund University in Germany. In his paper entitled "Physics Beyond the Multiverse: Naturalness and the Quest for a Fundamental Theory" [†] he makes some excellent points about quantum physics and what we might

---

* Claus Kiefer, "Quantum Theory, Decoherence and the Problem of Emergence," in *The Quantum Self: A New Philosophy of Quantum Mechanics*, edited by Ignazio Licata (Cham: Springer, 2024), 68.

See Physics Beyond the Multiverse (pdf)

arXiv:1809.06326 **[physics.hist-ph]**

(or arXiv:1809.06326v2 **[physics.hist-ph]** for this version)

https://doi.org/10.48550/arXiv.1809.06326

@article{P_s_2019, "Physics Beyond the Multiverse: Naturalness and the Quest for a Fundamental Theory", volume={49}, ISSN={1572-9516}, url={http://dx.doi.org/10.1007/s10701-019-00247-1}, DOI={10.1007/s10701-019-00247-1}, number={9}, journal={Foundations of Physics}, publisher={Springer Science and Business Media LLC}, author={Päs, Heinrich}, year={2019}, month=mar, pages={1051–1065} }

Heinrich Päs is a professor of theoretical physics at TU Dortmund University in Germany. He has held positions at Vanderbilt University, the University of Alabama, and the University of Hawai'i and has conducted research visits at CERN and Fermilab. He lives in Bremen, Germany.

Also, Why More Physicists Are Starting to Think Space and Time Are 'Illusions'

† See Physics Beyond the Multiverse (pdf)

arXiv:1809.06326 **[physics.hist-ph]**

mean by *fundamental*. I highly recommend reading this especially, pages 8 through 12 of the .pdf. Typically, and from a classical perspective, by fundamental concepts or realities we mean observer independent. But in quantum physics there is no *independence* of the observer. Each observer has a unique and *unanimous* perspective. And what does 'classical' even mean when a definite result applies to one Everett branch? Päs remarks,

> As has been argued by Werner Heisenberg in 1932 [38], the protons and neutrons constituting the atomic nucleus behave so similar that they can be understood as two states of a single particle – in analogy to a quantum spin which can point up and down. Heisenberg thus described proton and neutron together as the two possible states of an isospin doublet. Curiously, this way the apparent material difference of proton and neutron is reduced to the information about the state of the isospin Q-Bit. *

Päs gives some very good examples of how the determination of fundamentals runs into roadblocks in quantum physics. Nevertheless, there are elements in quantum theory of a materialistic component. The spooky notion of locality makes any such classical orientation to 'fundamental' difficult. Päs concludes that,

> Thus the fundamental state of the Universe can not be a constituent, it has to be the total entangled system of observer, measured system, and environment, also known as the quantum Universe itself. †

He also remarks that decoherence, everything we supposed from classicism, is a *lens* not a reality. And this lens is adjusted and custom-fits the eye of the historic beholder.

This quandary leads us inevitably back to consciousness and self-consciousness as somehow indispensable to our notion of reality and also somehow implicated in the spooky quantum riddle of the wave function.

What is Consciousness?

Here is a question that the history of science, philosophy, and religion has asked and answered in many ways but remains, what is consciousness? Is consciousness purely contingent? Is self-consciousness

---

(or arXiv:1809.06326v2 **[physics.hist-ph]** for this version)
https://doi.org/10.48550/arXiv.1809.06326
@article{P_s_2019, "Physics Beyond the Multiverse: Naturalness and the Quest for a Fundamental Theory", volume={49}, ISSN={1572-9516}, url={http://dx.doi.org/10.1007/s10701-019-00247-1}, DOI={10.1007/s10701-019-00247-1}, number={9}, journal={Foundations of Physics}, publisher={Springer Science and Business Media LLC}, author={Päs, Heinrich}, year={2019}, month=mar, pages={1051–1065} }

Heinrich Päs is a professor of theoretical physics at TU Dortmund University in Germany. He has held positions at Vanderbilt University, the University of Alabama, and the University of Hawai'i and has conducted research visits at CERN and Fermilab. He lives in Bremen, Germany.

Also, Why More Physicists Are Starting to Think Space and Time Are 'Illusions'

* Werner Heisenberg, "Über den Bau der Atomkerne. I," *Zeitschrift für Physik* 77, no. 1-2 (July 1932): 1–11.

† Claus Kiefer, "Quantum Theory, Decoherence and the Problem of Emergence," in *The Quantum Self: A New Philosophy of Quantum Mechanics*, edited by Ignazio Licata (Cham: Springer, 2024), 69.

arising from the human history of reflection uniquely or is it common to some or all mammalian and/or biological life? Is consciousness simply an adaptive, evolutionary trait of Homo sapiens? Perhaps consciousness has degrees of self-awareness for other species. Is self-consciousness absolutely unique and non-contingent? Is consciousness, the soul, a gift from God? Is Being onto-theological or ontological? For Heidegger, Being would be ontological, a unique kind of questioning aroused by and arousing the 'there of being', *Dasein*, which must co-arise as the emergent and emerging of beings. This is what Heidegger refers to in his later writings as the 'sway of being', the event of appropriation? Is the fascination of self-consciousness much ado about nothing? Are we simply cogs in a "Modern Times" machine of Chalie Chaplin?*

Certainly, modernity has exacerbated this ludicrous scenario. In the universe of absolute space and time where cause and effect rules absolutely, all of these questions dance upon the field of historic consciousness. Currently, there is a prevalent resignation to finding 'joy' in simply reducing consciousness to mechanism. This paints life and the universe from the canvas of modernity's metaphysical, sense-driven, cause and effect, reduction originating from classical physics. These modernist truth-values have been eroding from at least the beginning of the 20th century. Quantum physics has ushered in a new *phusis* which places historic physics, along with all histories of sciences, philosophies, and religions built on modernism, as a very limited case of a universe which never existed as it was imagined – as merely boundary and membrane taken in the narrative of absolute.

Consciousness is self-awareness. Self-awareness is not based on a subject separated from objects. The self is an entangled quantum system. Entanglement is not solitary, finite, and bound in a universe of cause and effect in spacetime. Entangled systems are not boundaries in the binary way we think of them. Boundaries can be artificially drawn in the process we call renormalization providing us with practical applications like qubits. However, unless one is the metaphysical Laplace demon, infinites are what make the quantum world so complicated.

> What makes it [quantum field theory] so complicated for mathematicians? In a word, infinity.
>
> When you measure a quantum field at a point, the result isn't a few numbers like coordinates and temperature. Instead, it's a matrix, which is an array of numbers. And not just any matrix — a big one, called an operator, with infinitely many columns and rows. This reflects how a quantum field envelops all the possibilities of a particle emerging from the field.
>
> "There are infinitely many positions that a particle can have, and this leads to the fact that the matrix that describes the measurement of position, of momentum, also has to be infinite-dimensional," said Kasia Rejzner of the University of York.
>
> And when theories produce infinities, it calls their physical relevance into question, because infinity exists as a concept, not as anything experiments can ever measure. It also makes the theories hard to work with mathematically.
>
> "We don't like having a framework that spells out infinity. That's why you start realizing you need a better mathematical understanding of what's going on," said Alejandra Castro, a physicist at the University of Amsterdam.
>
> The problems with infinity get worse when physicists start thinking about how two

* See Modern Times (film)

> quantum fields interact, as they might, for instance, when particle collisions are modeled at the Large Hadron Collider outside Geneva.*

In quantum physics boundaries are windows on infinities. Self-awareness could be thought of as a term to describe highly entangled quantum systems with boundaries from the observe effect. But these boundaries entangled with us are entangled with other systems near and far (in the classical way we think of near and far) in spooky ways as wavicles with chaotic and complex field infinites. Self-awareness, awareness, self-consciousness, consciousness, are all 'renormalized' systems which history defines as bodies, minds, etc. The observer effect has some spooky *phusis* effect with informational branching which makes the classical idea of cause-and-effect, things going bump in the night, an outdated nursery tale we tell ourselves. Wave function collapse by observation or spontaneous collapse certainly underlie anything we think of as awareness of awareness. Aristotle's awareness of awareness to be discussed a little later in the section entitled "The Unmoved Mover" is not some metaphysical Latin soul but something much stranger. Whatever we think of as the boundaries, membrane of 'me', my body, etc. are extremely dynamic and unpredictable windows upon unimaginable infinites. Quantum physics is a call to consciousness, to spooky observation, to wake up and view the precipice which stares back at us.

Sensibility, from the senses, was the motif of modernity. Sensibility is still what guides much of 20th century philosophy. Alternately, sensuality is not merely erotic. Sensuality is kinesthetic. Sensuality in the broader sense is affective. Sensuality comports us toward 'how we feel' and orients us in our lived existence. Sensuality ranges from the caresses of our loved ones to the horrid and tragic deaths of those we love in our lives. It ranges from the driving determinations of security to the hollowing out and emptying out of our insecurities. Sensuality does not exclusively harken back to the existentialist determination of the angst of existence, of the entrapment of sensuality in sensibility, in existence. Rather, from the exhaustion of modernism based on sensibility, the abstractions of the thinking machine which senses its way through existence, to having arrived before and inherent in the phenomenologically sensual way we concretize our 'lifeworld'. How do we 'feel' ourselves in spacetime? What is the dynamic of feeling our way through life and the interaction with how we think or narrate ourselves, our history, our being in the world? Where is the sensual breakdown thrust upon us starting with the reevaluations of relativity and moving towards chaotic non-causal exformation? How can 'the good' have anything to do with chaos? What would it mean to try to think of externality which cannot be recuperated into our common sense, our history of language and culture which determines the narrative of common sense, of practicality? Husserl was one of the first philosophers to seriously undertake the seismic consequences of the early 20th century. Husserl introduced the term lifeworld in 1936 with his publication of "The Crisis of European Sciences and Transcendental Phenomenology".†

> In whatever way we may be conscious of the world as universal horizon, as coherent universe of existing objects, we, each "I-the-man" and all of us together, belong to the world as living with one another in the world; and the world is our world, valid for our consciousness as existing precisely through this 'living together.' We, as living in wakeful world-consciousness, are

* Charlie Wood, "The Mystery at the Heart of Physics That Only Math Can Solve," *Quanta Magazine*, June 10, 2021, https://www.quantamagazine.org/the-mystery-at-the-heart-of-physics-that-only-math-can-solve-20210610/.
See The Mystery at the Heart of Physics That Only Math Can Solve

† See The Crisis of European Sciences and Transcendental Phenomenology

> constantly active on the basis of our passive having of the world... Obviously this is true not only for me, the individual ego; rather we, in living together, have the world pre-given in this together, belong, the world as world for all, pre-given with this ontic meaning... The we-subjectivity... [is] constantly functioning.*

For Levinas, consciousness reduced into the boundaries of 'thinking' must always turn back on itself. From my perspective, thinking turned back on itself can never find certainty only continual deferment as the basis for transactionalism. Thinking turns back on itself because the Other (Autrui), the she or he which faces us, is,

> Immemorial, unrepresentable, invisible, the past that bypasses the present, the pluperfect past, falls into a past that is a gratuitous lapse. It can not be recuperated by reminiscence not because of its remoteness, but because of its incommensurability with the present. The present is essence that begins and ends, beginning and end assembled in a thematizable conjunction... †

Perhaps later we will see that love is the trace of externality. Love overtakes us, interrupts us, draws us towards the Other (Autrui) without totalizing the Other (Autrui). The play of my freedom cannot *a priori* take account of exteriority. Memory cannot call love to presence. My subjectivity is the call to account for the Other (Autrui) which I cannot account for. The null site of subjectivity is 'me' which must retreat from exteriority. 'Me' as without recourse is hiding in the cleft of the rock. I cannot face the Other (Autrui) in his or her externality as if from some mystic, ecstatic, experience. I can only exist as the aftermath, the residue as subjectivity, as never knowing and yet held to account as the haunting which, like waves on the shore, beats against my subjectivity. Here again we arrive at "*epekeina tês ousias*" or 'the good beyond being'. All of this will be discussed later.

Phusis or epoch begins with chaos, indetermination, uncertainty and through the centuries acquires certainty in truncation as practical common sense. Now at the end of our epoch we once again for the first time face uncertainty that is stripped bare of any language to reaffirm and found our epoch or perhaps any epoch. We are speechless before the abyss 'to be able to be able'. Yet the Other (Autrui) still speaks to us in a still small voice from an unbreachable abyss. What is it that all this time we have not heard? Even in Husserl's epoché, the bracketing, where consciousness thinks back and forward from its living past and future finding the living present, the internal [lived] time consciousness is sensation congealing in spontaneous regeneration of time as retention (what is dynamically retained as the living past) and protension (what is projected towards the living future).

Husserl calls this time consciousness lived protension and retention. Phenomenology recognized the breakdown of modernism and historic abstractions. It sought to return meaning-bestowing back to life as we live it not as the empty abstractions of history. According to Husserl, phenomenology aims to clarify the constitution of transcendence. In other words, it seeks to understand how consciousness constitutes and relates to objects beyond itself. The term transcendental in Husserl's context refers to several interconnected aspects:

---

* Edmund Husserl, *The Crisis of European Sciences and Transcendental Phenomenology: An Introduction to Phenomenological Philosophy*, translated by David Carr (Evanston, IL: Northwestern University Press, 1970), 108–109.

† Emmanuel Levinas, *Otherwise than Being: Or Beyond Essence*, translated by Alphonso Lingis (Pittsburgh: Duquesne University Press, 1998), 11.

- It signifies the exceptional status of rationality—the philosophical attitude that allows us to explore the relationship between the knower (subject) and the known (object).
  - It encompasses the methodological approach and language used by philosophers engaged in this exploration.
  - Importantly, Husserl's use of traditional terms was far from conventional. His conception of the transcendental differed significantly from the traditional understanding. *

One key aspect of Husserl's approach is the transcendental-phenomenological reduction. This reduction uncovers the ego, revealing the foundational level where everything gains meaning and existence. Simultaneously, it transforms the world into mere phenomena, emphasizing the subjective experience and intentional acts of consciousness. Husserl's transcendental philosophy is lived subjectivity and consciousness and how it constitutes reality. †

In Husserl, time is in nowise associated with the time of classical physics. Husserl's relativity finds some commonality with Einstein's continuous and relative frames of spacetime but also very different, not as the domain of physics but as lived space and time. In a way, sensualism in Husserl is compatible with the epistemological doctrine of empiricism which holds that all pure rational cognitions are derived from the senses. However, Husserl finds no 'senses' of cause-and-effect relation to sensual intentionality but to the surprise and spontaneity of lived internal time consciousness. Husserl finds no such possibility extreme externality can enter into internal time consciousness. In this rather difficult passage for folks that are not steeped in Husserl, Levinas writes,

> In Husserl internal time consciousness, and consciousness as such, are described in the temporality of sensation: "sensing is what we hold to be the originary consciousness of time," and "without the impression consciousness is nothing." Time, the sensorial impression and consciousness are put together. Even at this primordial level which is that of lived experience, in which the flow, reduced to pure immanence, should exclude even any suspicion of objectification, consciousness remains an intentionality, an "intentionality of a specific kind," to be sure, but unthinkable without an apprehended correlate. This specific intentionality is time itself. There is consciousness insofar as the sensible impression differs from itself without differing; it differs without differing, is other within identity. The impression is illuminated by "opening up," as though it plugged itself up; it undoes that coincidence of self with self in which the "same" is smothered under itself, as under a candle extinguisher. It is not in phase with itself; just past, about to come. But to differ within identity, to maintain the moment that is being altered, is "protaining" and "retaining"! Differing within identity, modifying itself without changing, consciousness glows in an impression inasmuch as it diverges from itself, to

* Edmund Husserl, *The Crisis of European Sciences and Transcendental Phenomenology: An Introduction to Phenomenological Philosophy*, translated by David Carr (Evanston, IL: Northwestern University Press, 1970), 97–100.

† See 10 Husserl and the Transcendental
And Husserl's Transcendental Turn
And Husserl's Transcendental-Phenomenological Reduction
And Some differences between Kant's and Husserl's conceptions of transcendental philosophy
And Edmund Husserl (1859—1938)

still be expecting itself, or already recuperating itself. Still, already — are time, time in which nothing is lost. The past itself is modified without changing its identity, diverges from itself without letting go of itself, "becomes older," sinking into a deeper past: it remains identical with itself through retention of retention, and so on. Then memory recuperates in images what retention was not able to preserve, and historiography reconstructs that whose image is lost. To speak of consciousness is to speak of time. *

It is in any case to speak of a time that can be recuperated. At a level which for Husserl is originary, temporality involves a consciousness that is not even intentional in the "specific" sense of retention. Despite the complete overlapping of the perceived and the perception in the Ur-impression, the originary or primal impression, which ought to no longer let the light pass through, despite their strict contemporaneousness which is the presence of the present, despite the non-modification of this "absolute beginning of this generation — the primal source, that from which all others are continually generated," this today without a yesterday or a tomorrow — the primal impression is nonetheless not impressed without consciousness. Does not the originary impression "non-modified," self-identical, but without retention — precede every protention, and thus precede its own possibility? Husserl seems to say this, in calling the "primal impression" the "absolute beginning" of every modification that is produced as time, a primal source which "is not generated" and is born "**through spontaneous generation**." "It does not grow up (it has no seed); it is primal creation." Would not the "real" that precedes and surprises the possible be the very definition of the present, which in this description is indifferent to protention ("generation has no seed!), but would nonetheless be conscious of it? That is certainly the most remarkable point of this philosophy in which intentionality "constitutes" the universe, the prototype of theoretical objectification commands all the modes of intentional positing, be they axiological or practical, and in which in every case a rigorous parallelism between the doxic, axiological and practical theses is constantly affirmed. Husserl will then have liberated the psyche from the primacy of the theoretical neither in the order of know-how with equipment nor in that of axiological emotion, nor in the thought of Being, different from the metaphysics of entities. Rather, objectifying consciousness, the hegemony of re-presentation, is paradoxically surmounted in the consciousness of the present. And this makes understandable indeed the underlying or commanding importance of the still so little explored manuscripts concerning the "living present" in the whole corpus of his research. Though it be a rehabilitation of the "sensorial given" of empiricist sensualism, the primal impression finds again in the context of intentionality (which for Husserl remains all-encompassing) its power to surprise. Through the notion of the living present, the notion of origin and of creation, a spontaneity in which activity and passivity are completely one, tend to become intelligible. When it turns out that this consciousness in the living present, originally non-objectifying and not objectified, is thematizable and thematizing in retention, without thereby losing the "temporal place" which gives "individuation," then we see the non-intentionality of the primal impression fitted back in the normal order, not leading to the hither side of the same or of the origin. Nothing enters incognito into the same, to interrupt the flow of time and interrupt the consciousness that is

* Emmanuel Levinas, *Otherwise than Being: Or Beyond Essence*, translated by Alphonso Lingis (Pittsburgh: Duquesne University Press, 1998), 33.

> produced in the form of this flow. A putting the self-identity of the living present out of phase, a putting of the phases themselves out of phase, in the intentionality of retentions and protentions, the flow looks like a multiplication of modification dispersing from the living present. In Husserl the time structure of sensibility is a time of what can be recuperated. The thesis that the non-intentionality of the primal retention is not a loss of consciousness, that nothing can be produced in a clandestine way, what nothing can break the thread of consciousness, excludes from time the irreducible diachrony whose meaning the present study aims to bring to light, behind the exhibiting of being. *

Rather than being a metaphor taken from the movement of waters in a river, would not flowing be the very temporality of time and the "science" of which "consciousness" (conscience) is made? To speak of time in terms of flowing is to speak of time in terms of time and not in terms of temporal events. The temporalization of time — the openness by which sensation manifests itself, is felt, modifies itself without altering its identity, doubling itself up by a sort of diastasis of the punctual, putting itself out of phase with itself — is neither an attribute nor a predicate expressing a causality "sensed" as a sensation. The temporal modification is not an event, nor an action, nor the effect of a cause. It is the verb to be. (Levinas, Otherwise Than Being or Beyond Essence, 1981, p. 100)

Note that this passage in Husserl has nothing to do with some modernist notion of subjectivity. I hope the reader can at least get a flavor from this rather long passage about how Husserl has much more to do with the early 20th century than what came before. Nevertheless, in any case, Levinas still believes internal time consciousness rests comfortably in the nascent certainty of 'to be' even without the necessity of causality. The divergence of causality from consciousness in Husserl shares a contemporaneous 20th century history with the founders of quantum mechanics.

In quantum mechanics, cause has been lost to exformation. However, exformation is not merely a physics of disembodied objects. Even in quantum physics the self, 'me', is an entangled quantum system. Certainly, there is quite an abyss between quantum entanglement and Levinas' radical alterity of the Other (Autrui). However, one important idea they share is that 'me' is not produced in the totality of 'to be'. Another unsettling idea they share is that all that is certain only vanishes in the misty haze of spooky uncertainty, something which can find no return to itself as essentiality or even sensuality of certainty. In the quantum wave function the observer does not die as history would have it. Information is updated. Could it be that lifeworld branches in some spooky, indeterminate way?

Death is a historic myth we tell ourselves to anesthetize the pain of life, the suffering of old age and tragic loss of our precious loves. We do not die in glory, in Being, as Nietzsche's heroic figure. We die in pain and suffering without any recourse. We die in weakness, absolutely unable to be able. We do not actively and heroically conquer death. Death conquers us. Death lays us waste in weakness and suffering, tragedy, and insufferable loss. We are not challenged by authenticity to be resolute in [towards] death. As Buddhism teaches, life is suffering not glory. Glory is a fiction thrust upon us in abstraction. We yearn for glory but in life we endure tragedy, remorse, pity, and guilt. Virtual reality lives in the possibility of glory and dies in the reality of existence. So, we need to look at death in terms of how our abstractions of it pacify our sensibilities but open a gaping void in our sensuality.

* Emmanuel Levinas, *Otherwise than Being: Or Beyond Essence*, translated by Alphonso Lingis (Pittsburgh: Duquesne University Press, 1998), 34.

Certainly, we will die but to believe that a mechanistic, reductionary, absolute finality is more than a historic myth we tell ourselves is to make assumptions which harken back to the non-spooky, the quickly dimensioning historic avenues which desperately desire to transform the spooky into the dying same. I want to briefly skip ahead at this point to introduce some thoughts of Levinas in this regard. One important relevant point to make here is that Levinas is not the stereotypical type of philosopher who is all brains and no feeling. To the contrary, Levinas cannot be understood without knowing his approach to philosophy is arriving to us from our sensual relation to the other and the world. What lurks behind the scenes when we measure sense data over lived embodiment and the way we sensually interact with the world and others?

❧ 2 ❧

HISTORIES OF ORIGIN

First, let me state that Levinas and Hegel are by far the most difficult philosophers I have ever read. While they are very different from each other, their words are incredibly compacted, rich, and borrow heavily from the history of philosophy. I have already elucidated very simple, probably too simply, some of my issues with Hegel. However, Hegel is not the direction in this volume except to point out some histories of philosophy and some problematic differences in 20th century philosophy with Hegel. With Levinas, I am going to try my best to unpack his ideas, but I will also be working through it yet once again with you while I write. Nietzsche once said that he could write in more in an aphorism than most other philosophers would write in an entire book. While Levinas is also very different than Nietzsche, he does seem to share that quick and concise wit.

I have gone much of the way through this volume with a bit of angst and a lot more reading of Levinas knowing that there is an important link to the previous discussion of physics but not specifically knowing how I am going to convey this in my writing. The beginning here will not be easy for the reader but anyone that might have ventured thus far must already be acclimated to such a task. This beginning will be difficult and risks becoming murky. The only solace here is that after the beginning I will try to build more slowly from the history of philosophy already explored which finds its apex and final stand in the 20th century. For now, let's try to make a probably too hasty beginning.

I want to start this major section with a difficult piece by Levinas. Levinas is not an easy philosopher to read. His thoughts are densely packed with a rare breed of scholarly knowledge of the history of philosophy. Additionally, Levinas was not content to be merely academic. Perhaps part of this could be attributed to his imprisonment during the holocaust. However, I find his orientation to also recognize the postmodernist dilemma which has brought us face to face with the exasperation of classicism. But unlike many postmodernists and beyond, Levinas brings us face to face with the Other (Autrui), the people we face every day, in a highly insightful reckoning with the void we cannot turn from. In this abyss Levinas, in a similar way to the spookiness of quantum physics, brings us to a difficult thought, an awkward inability to be able, to find an easy historic and commonsense answer. Levinas wants to bring us face to face with the Other (Autrui) as radical alterity not merely the 'thing' as other which can still find echoes in the paradigm of even quantum physics. However, unlike physics, radical alterity is not 'spooky' but rather is responsibility to what cannot be brought under my auspices, my narcissistic echoes which simulate the Other (Autrui), dilute the Other (Autrui) as simulacrum, as image and representation in which responsibility is optional, based on my freedom to mold the Other (Autrui) into my own image.

The long quote I will get to near the end of this section will be cryptic and difficult to understand without a more comprehensive background in classical, modern, and postmodern philosophy. But I do not do this to intentionally aggravate the reader. My hope is to offset the awkwardness of this quote with simply a sense of a song in Levinas' style. I will get into the details of this quote later, but I think that in reading with a feeling of chaos, the inability to be able to situate meaning readily may also bring a bit of intrigue in the way that music provokes without comprehension. If the reader can first get a sense of richness of possibilities in Levinas' thought along with a feeling of some excess, some externality, which elicits a sense of what was always intended but never quite grasped – beauty, wonder, and awe in already being indebted before the 'I' can be *said*, then the abrupt departure into this difficult beginning will be worth it. But before the long quote I want to make a few preparatory remarks.

We not only 'live' the world, but we also feel the world. Levinas writes of love, fecundity, eros, desire, the caress, paternity, suffering, death, insomnia, and other lived feelings. Part of this is because of his background in phenomenology and lived experience but beyond that Levinas wants us to understand that we encounter the Other (Autrui) sensually. Philosophers tend to write descriptively even when they reflect on sensuality. However, that is not how we experience ourselves. I encounter others with feelings whether it is retreating such as anger, fear, insecurity, guilt, remorse or attraction as love, concern, intimacy, touch, etc. Even indifference to others is an emotion of solitude, withdrawal, or retreat. Hate, dislike, bigotry, and intolerance are emotions which cling to a concept of the despicable or evil other. These emotions take no time for factuality or critical thought. They retreat into a fictional narrative reductive of the other. They isolate and withdraw either in solitude or in the company of those who validate and devote themselves to the project of fictionalizing the other. There is no other in hate. There is only me and mine. Love and ownership of the Other (Autrui) cannot exist in the same lived space.

Love is a vulnerability to the other. It is an intentional opening of myself to the otherness of the other. It resists and refuses consolidation of the Other (Autrui) into a persistent thought of the Other (Autrui). From risk, love allows and welcomes the Other (Autrui) to break through the plastic castes of their face which Concept, Being, totality, and idea want to universalize. The fictions we cast upon the other are intently allowed to shatter responsibility for my loves, my wife, my child, my friends, strangers, and even my enemies. Love suffers long and believes steadfastly not for the sake of love itself but for the other, the one I love from indebtedness and responsibility. Love does not affirm 'me' as Narcissus' echo. Sensual intimacy which languishes in self-satisfaction is masturbation or auto-affection. Auto-affection cannot survive the otherness of love, the vulnerability of love. Writing of an important early work of Levinas called "Totality and Infinity" Todd McGowan writing of love states,

> Levinas speaks of the importance of avoiding 'fusion', 'possession' and 'knowledge' in love. In Totality and Infinity, he writes: 'Love is not reducible to a knowledge mixed with affective elements which would open to it an unforeseen plane of being. It grasps nothing, issues in no concept, does not issue, has neither the subject-object structure nor the I-thou structure. Eros is not accomplished as a subject that fixes an object nor as a pro-jection, towards a possible. Its movement consists in going beyond the possible'.*

For the most part, Levinas makes a distinction between sensibility and sensuality which I will also try to maintain in this text. Sensibility is given from the modernist senses. I have already spent much time in this series of volumes bringing to the fore how classicism and modernism reductively has given us the persistent narrative of 'practicality' which validates and sanctifies 'common sense'. Let's return

* Todd McGowan, "Existentialism After Finitude: The Transcendence of the Unconscious," in *Death and Love: Psychoanalytic and Philosophical Perspectives*, edited by Todd McGowan and Julie Reshe (Cham: Springer, 2023), 13.

See Re-viewing the Sexual Relation: Levinas and Film

to the historic narrative of death which is shaken to the core by our sensual forebodings and uncertainties as radical externality. Can we begin to think about exformation and where that might lead us?

As I will try to explicate, Levinas wants to turn away from almost everything we have already covered in the history of philosophy. He wants us to sensually make apparent to ourselves how our philosophical inheritance tends to practically determine contexts of our sensual lived experiences. In particular, how do Being, 'reality' and existence already efface the Other (Autrui)? He wants to ask us, to ask ourselves, a really important question. What if the inexplicable gap we now face has a face, the face of the Other (Autrui), which the history of thought cannot account for? What could that possibly mean? What if the self is not its own? What if the self does not arise in spacetime from sensibility and fundamentally even lacks its assumed presence upon reflection but rather, sensually towards radical exteriority, the Other (Autrui) who escapes our knowledge? What if there is an unaccounted-for excess in sensuality which somehow makes it unaccountable to sensibility? Rather, could it be that sensibility has always already been accountable to sensuality?

The rupture of the Other (Autrui) in Levinas is not a smooth transition from phenomenology. Phenomenology wants to impartially observe, to bracket, all presumptions from science, philosophy, and religion. For Husserl any science or philosophy which would not bracket its assumptions whether it be assumptions of objectivity, subjectivity, time, space, rationality, etc. could not be a complete science. Therefore, he thought philosophy should be the science that must account for how we first, individually experience phenomena without presuppositions and second, how these phenomenological observations make any such thing as a lifeworld possible. In this way a phenomenologist is more like a clinical psychologist without theoretical psychological presumptions. Phenomenology would incorporate all the ways we encounter phenomena by holding at bay presumptions and getting back to what shows itself. Levinas cannot maintain the 'showing' of the Other (Autrui), the he or the she, because the Other (Autrui) cannot come to presence or absence. It must exceed logical oppositions and hierarchies which unify them into consciousness and unconsciousness, presence and absence. In Levinas, the externality of the Other (Autrui) cannot 'give account', be brought under auspices of senses into commonsense. But neither can science find a rational basis for the spooky behavior quantum entanglement. Why not use this conundrum in science as a hint for what philosophy must face?

The realistic problem here is that Levinas gets taken as yet another metaphysics or validation of religion in some way. Levinas is fully aware of this danger. His writing is so dense partly because he wants to demonstrate his acute awareness of this susceptibility to his philosophy. At the same time, he wants to show that the history of metaphysics as a history of Error was subducted away from a kernel of the Other (Autrui) which it could not account for. The history of war is the history of religion, metaphysics, philosophy, and rationalism. These traditions lacked an excess to totality. They lacked the alterity of the Other (Autrui). Instead of excess, alterity has been translated in popular culture as alien, as evil other, as monstrous to be abhorred. This has been done in spite of the admonition of Judaism for the sojourner, the immigrant, the stranger. Even in Jesus' compassion and care for the sojourner, we see in current politics that the immigrant has been transformed into the evil drug peddling 'illegal immigrants'. A clear example of this is popular culture's reaction to encouraging diversity. Diversity is seen in the mantra of 'wokeness'. Why is this? Is this because logic, origin, and totality has lost its faith? Or because logic, origin, totality congealing into commonsense has won the day.

* Logic - The other is evil by virtue of being other to the same.
* Origin - Underneath the covers of this logical binary truism, the origin of the 'other' is my idea of the other.
* Totality - My idea and origin of the other as evil, as unacceptably 'different' has everything to do with me without any possibility for an other exterior to the totality of my echo.

Even the 'difference' in the sanctified same and the profane other is preserved under an absolute pretention to 'logic'. If the same is good, the other is evil. But both come under the absolute assumption and reduction to the other as idea, as my idea. In this case, there is no excess to my idea, my logic, my adherence to popular politics. This incessant insistence has become a totality which actually preserves its inaccessible agenda to mindlessly exclude anything other than itself, its logic, its 'idea'. This obsessive insistence hermetically seals itself off from even the possibility of an excess to its sameness in the idea of the 'other'. Feigned religious altruism as morality has become a strategy to divert accusation, persecution, and self-doubt so as, in popular culture, to have nothing to do with guilt and remorse for leveling off and totalizing the other into my idea of the evil other.

Washing my hands of the suffering of the other depends on a self that is hermetically sealed in itself. If I am solitary, absolutely identical to my thoughts of myself, there is feigned equality between me and the other in which I emerge as the hero putting to waste the foe. All this is sanctified by the vernacular of transactionalism. This is the self as solitary, as absolute identity masquerading under the banner of 'me' and my notion of the other. This equality might be thought of as the basis of my branching from supposition in my hermeneutic, my interpretation, of myself as absolute freedom. My duty and obligation to the Other (Autrui) is conceived as equality, defined in terms of binary dualisms of inequality with the bad faith other. Even more, I get to be the judge of what 'equality' means. If the other is lazy and ignorant, 'me' as equality of the 'I' with the other, has judged, in the sanctity of its freedom, how *my* truths of equality and its necessary counterpart – inequalities, are drawn. But what if the Other (Autrui) is not in the self's field of freedom? What if the Other (Autrui), the he or the she, cannot come into presence or even absence in my freedom, my truths which have nothing to do with radical externality but everything to do with Narcissus? What if the other is not equal to my field of freedom but prior to it. By "prior to it" I do not even mean in phenomenologically 'my' lived time, 'my' lived space, 'my' freedom, but as not commensurate in my memory or my common senses or even my sensuality which rest heavily upon the totalities of history, culture, and language. Levinas refers to this as notion as diachrony. Levinas describes diachrony as,

> Immemorial, unrepresentable, invisible, the past that bypasses the present, the pluperfect past, falls into a past that is a gratuitous lapse. It can not be recuperated by reminiscence not because of its remoteness, but because of its incommensurability with the present. The present is essence that begins and ends, beginning and end assembled in a thematizable conjunction; it is the finite in correlation with a freedom. Diachrony is the refusal of conjunction, the non-totalizable, and in this sense, infinite. But in the responsibility for the Other, for another freedom, the negativity of this anarchy, this refusal of the present, of appearing, of the immemorial, commands me and ordains me to the other, to the first one on the scene, and makes me approach him, makes me his neighbor. It thus diverges from nothingness as well as from being. It provokes this responsibility against my will, that is, by substituting me for the other as a

hostage. All my inwardness is invested in the form of a despite-me, for-another. Despite me, for-another, is signification par excellence. And it is the sense of the "oneself," that accusative that derives from no nominative;9 it is the very fact of finding oneself while losing oneself. *

Diachrony, the inability to synchronize temporalities, cannot come into my field of freedom where the other is a cog in a machine of 'my-selves'. This machine synchronizes all its parts into an assembly, a cog in. "Modern Times", a local realism ordered by me. The effect of this is to make all the parts, the 'selves', into pieces which have their place, their time, and accountability to my judgements of how all this works. the Other (Autrui) as "*Immemorial, unrepresentable, invisible, the past that bypasses the present*" does not include me as the arbiter and assembler of all the parts, as Dr. Frankenstein. The other has a time that is not or ever was my time, a space which any space of mine could ever inhabit. There is a gap which cannot be spanned nor recalled by my memory. How can the self, 'my-self', be undone as the arbiter of freedom which is optional and co-equal to responsibility? Am I the sole judge of my responsibility to the other as if I am the origin of the other? Ultimately, Levinas tracks this down as force, as violence to the other. For Levinas, one hint of this force is in a sensual mode which is absolutely incommensurate, irreconcilable, with the identity of the self as pure freedom. This violence is rooted in the sensuality of obsession especially exhibited in the state of insomnia.

Obsession with insomnia finds no end in itself. It cannot take leave of itself. It can never rest. It must ceaselessly feedback upon itself with absolutely no hint of resolve. It makes 'reality' into cardboard where nothing can be but must be. It is inexhaustible thinking which has no end or no origin but only thinking about thinking itself which is unable to lose itself, to sink into oblivious sleep. It can only mock rest as yet another thought. Obsession undermines any such sense or sensibility of 'my-self', as absolute origin. No sense can be made of obsession. The self must historically rest on sense and sensibility. However, such a history as self, myself, finds no place, no rest in obsession. From this analogy, Levinas wants to show how even in the chaos of obsession without the hope of resolve there is a pure signifyingness which is endemic to both signifiers and signified which can find neither. This is not nothingness. Even in this abyssal chaos of pure signification without any object there is a saying which cannot be said, a passivity more passive than all passivity, an inability to be able. Why? Because the radical alterity of the Other (Autrui) can find no origin, no *place* (or time) to lay his or her head, no essence of Being of classicism but only responsibility for what cannot come to the fore remains. Additionally, when Levinas writes "passivity more passive than all passivity" instead of thinking of this as an encrypted statement think analogously of the quantum wave function. The ego is not the self. As we will see, the ego is the retreat from self. The ego is a refinement of a history which is the identity of the 'I'. Identity is a topic we discussed earlier. It is the basis of tautology. It is the basis of logic. It is what necessarily must be true. The 'I' of identity is defined as origin in the history of Being, of ontology. The 'I' arises from the particular into the heights of a universal field of play as Being. Obsession destroys the centrality of the self to 'be' itself. It is a nether world which must always be awake, conscious, and find no rest. In obsession all things flee into a dead neutrality which cannot yet die but must be condemned merely to exist in the *il y a*, *isness*, the sheer 'it is'. The *il y a* has no temporality. There is no interruption of the other. It is the self's freedom which the 'it' of 'is' has no beginning or end, finds nothing as

* Emmanuel Levinas, *Otherwise than Being, or Beyond Essence*, trans. Alphonso Lingis (Pittsburgh: Duquesne University Press, 1998), 11–12.

everything, can only find termination in interminable nothingness, condemned to endlessly endure nothingness.

Levinas takes this as a hint of why the self cannot be what it pretends. Also, the 'inability to be able' as I have mentioned, the haze of birth and infancy, the passive endurance of old age, weakness, suffering and the finality of death from which the self has no power over render us helpless to be the eternal, immortal self. The historically fantasized resurrection of the self into Concept cannot include the sensual self, the self which desires to find immortality, to conquer death. Levinas tells us that the self is a fiction in self-fictionalizing fantasies which never dies upon the altar of its obsessions. Levinas tells us the only way to think of the self is as existential retreat from the radical alterity of the Other (Autrui). The diachrony of my time, my space, my incommensurability to create, sustain, immortalize me is not possible because I am not my own creator, and neither am I the 'not' of my own creator. Neither is the 'not' of my own creator some historical face of my equivalence with the history of Being which takes the name God. The self is the result of a passivity before all [my ideas] of passivity. The self is accused and persecuted because it owes debt it cannot repay. The self does not originate itself. The 'I' cannot congeal the ego into Being. The radical externality of the Other (Autrui) which I cannot assume in my freedom is not 'the not-of my origin of me'. 'The not-of my origin of me' in its assumed binary dualisms does not imply determinism wherein I had no choice, no say so, no freedom to become myself by myself. There is an absolute vulnerability that spawns me which cannot be deemed the 'No Exit' of Sartre or the great nausea of Nietzsche's eternal return of the same. In his or her face I do not 'find me', my 'self' as known-other, the negation of the same as the other in angst. Me, myself, 'I' finds itself situated from an other in which I have no memory, no co-existed space or time. Ipseity, personal identity, individuality, selfhood, is not becoming, the self-inventing itself, from Being and nothingness. The NIH (National Institute of Health) tells us mental illnesses like schizophrenia all find deficiencies in these areas,

> The assessment of psychopathology in most contemporary research is based squarely on signs and symptoms of disorder, often measured in fairly crude checklist-type fashion. This approach has tended to indicate significant overlap in psychotic and other symptoms across disorders, eg, between schizophrenia and bipolar disorder1 and between psychotic disorders and borderline personality disorder. This may partly be the result of the assessment tools and conceptual frameworks being used. By contrast, insights from phenomenological psychiatry and philosophy, focused on disturbed subjectivity, indicate that disturbed self-experience or selfhood may underlie and generate many "surface-level" psychotic symptoms, particularly in schizophrenia spectrum disorders.
>
> There are many different meanings and controversies surrounding the notion of the "self." These controversies mainly concern its ontology or ultimate reality status, eg, as a kind of "substance," object, or process. The experiential, subjective notion of the self (the sense of self) is, however, widely acknowledged, both in the analytic philosophy of mind and in phenomenology. Two levels of the experiential self are typically proposed:
>
> 1. "Minimal" self, also referred to as "basic" or "core" self or as "ipseity." This is a pre reflective, tacit level of selfhood. It refers to the implicit first-person quality of consciousness, ie, the implicit awareness that all experience articulates itself in first person perspective as "my" experience. In other words, all conscious acts are intrinsically self-conscious, a feature sometimes

designated as "self-affection." "Minimal" or "core" self constitutes the foundational level of selfhood on which other levels of selfhood are built.

2. "Narrative" or social self. This refers to characteristics such as social identity, personality, habits, style, personal history, etc. Psychological concepts such as "self-esteem" or "self-image" refer to this level of selfhood. This level is widely understood to presuppose the sense of existing as a subject of experience ("minimal self") and often involves reflective, metacognitive processes, in which one's self is largely an object of awareness. *

In Levinas' late and most well-known work entitled "Otherwise than Being or Beyond Essence" he asks us how can we be sure or *faithful enough to the anarchy of passivity* in the face of the Other (Autrui) and not just caught up unawares in yet another scheme of the history of Being (ontology)? What Levinas is asking is how do we 'know' if we are not weaving yet another tale of origin in which we play the role of the protagonist or the antagonist? Even 'knowing' itself can be active as "I know the answer" or passive as "The answer is known by me". So, what does Levinas mean by "*faithful enough to the anarchy of passivity*"? Levinas writes near the end of the quote, "*Responsibility for another is not an accident that happens to a subject, but precedes essence in it, has not awaited freedom, in which a commitment to another would have been made.*" Levinas has a phrase which refers to this "*passivity more passive than all passivity*". In other words, responsibility to the Other (Autrui) is not a doing or not-doing which I must choose or even not-choose. Rather, responsibility is a recognition of an excess which makes any such thing as a 'me' or a 'not-me' possible. the Other (Autrui) is a diachrony to me, a time not my time. Time is something that is always beyond the individual's control and is constantly slipping away. In this sense, Levinas speaks of a "time not my time" to emphasize the idea that time is not something that can be owned or possessed by the individual. The excess of other *branches* anarchically, not traceable to a self which originates time. The phrase "*passivity more passive than all passivity*" cannot be an active cause or passive effect coopted by a self which has the freedom to choose narratives. Levinas asks our question another way, "*How can all these senses of the facticity and irrevocable passing of time — the passive synthesis of time not entirely convertible into an active synthesis — be intelligibly subordinated to the time-structure of the relationship with alterity?*" (Levinas, Otherwise Than Being or Beyond Essence, 1981, p. 23) This will be a goal of the major section to be addressed. But for now, without being reductionary, in the context of the first two volumes on physics, perhaps we might be able to glimpse some kind of sense of Levinas' phrase "*passivity more passive than all passivity*" could be at least given an allegorical direction in the radical externality of the quantum wave function, its incomprehensible otherness from spacetime, histories, and classicisms. But let's not forget the spooky netherland of the wave function is not merely epistemological or ontological but yet, in its fall into classicism, gives us/me a sense of lived sensuality, feeling of embodiment in the world, which gets taken up as histories of the narrative as 'me here now'.

For now, I also want to note that in his work "Otherwise Than Being or Beyond Essence", Levinas makes a huge distinction with the 'saying' of the other where, for only an instant (dare we say the moment exformation branches into the classical 'me'), the radical alterity of the face of the Other

* Barnaby Nelson, Leonie Pihlens, and Josef Parnas, "The Soul of Psychopathology: The Sense of Self," *Schizophrenia Bulletin* 48, no. 5 (September 2022): 941–942.

See Disturbance of Minimal Self (Ipseity) in Schizophrenia: Clarification and Current Status.

(Autrui) breaks through only to be immediately forgotten, erased, lost, taken up into the form of consciousness as the 'said', the plasticity of 'me' as fallen into classicism. The 'said' is narrative, the history of Being or what I have referred to as the history, language, and culture which '*is*' what we passively take as practicality in common sense. This is the long quote I referred to at the beginning of this section,

> In this exposition of the in itself of the persecuted subjectivity, have we been faithful enough to the anarchy of passivity? In speaking of the recurrence of the ego to the self, have we been sufficiently free from the postulates of ontological thought, where the eternal presence to oneself subtends even its absences in the form of a quest, where eternal being, whose possibles are also powers, always takes up what it undergoes, and whatever be its submission, always arises anew as the principle of what happens to it? It is perhaps here, in this reference to a depth of anarchical passivity, that the thought that names creation differs from ontological thought. It is not here a question of justifying the theological context of ontological thought, for the word creation designates a signification older than the context woven about this name. In this context, this said, is already effaced the absolute diachrony of creation, refractory to assembling into a present and a representation. But in creation, what is called to being answers to a call that could not have reached it since, brought out of nothingness, it obeyed before hearing the order. Thus in the concept of creation ex nihilo, if it is not a pure nonsense, there is the concept of a passivity that does not revert into an assumption. The self as a creature is conceived in a passivity more passive still than the passivity of matter [could he mean the totality of all things which adhere to cause and effect?], that is, prior to the virtual coinciding of a term with itself. The oneself has to be conceived outside of all substantial coinciding of self with self. Contrary to Western thought which unites subjectivity and substantiality, here coinciding is not the norm that already commands all non-coinciding, in the quest it provokes. Then the recurrence to oneself cannot stop at oneself, but goes to the hither side of oneself; in the recurrence to oneself there is a going to the hither side of oneself. 'A' does not, as in identity, return to 'A', but retreats to the hither side of its point of departure. Is not the signification of responsibility for another, which cannot be assumed by any freedom, stated in this trope? Far from being recognized in the freedom of consciousness, which loses itself and finds itself again, which, as a freedom, relaxes the order of being so as to reintegrate it in a free responsibility, the responsibility for the Other (Autrui), the responsibility in obsession, suggests an absolute passivity of a self that has never been able to diverge from itself, to then enter into its limits, and identify itself by recognizing itself in its past. Its recurrence is the contracting of an ego, going to the hither side of identity, gnawing away at this very identity — identity gnawing away at itself — in a remorse. Responsibility for another is not an accident that happens to a subject, but precedes essence in it, has not awaited freedom, in which a commitment to another would have been made. I have not done anything and I have always been under accusation — persecuted. The ipseity, in the passivity without arche characteristic of identity, is a hostage. The word I means 'here I am', answering for everything and for everyone. Responsibility for the others has not been a return to oneself, but an exasperated contracting, which the limits of identity cannot retain. Recurrence becomes identity in breaking up the limits of identity, breaking up the principle of being in me, the intolerable rest in itself characteristic of defini-

> tion. The self is on the hither side of rest; it is the impossibility to come back from all things and concern oneself only with oneself. It is to hold on to oneself while gnawing away at oneself. Responsibility in obsession is a responsibility of the ego for what the ego has not wished, that is, for the others. This anarchy in the recurrence to oneself is beyond the normal play of action and passion in which the identity of a being is maintained, in which it is. It is on the hither side of the limits of identity. This passivity undergone in proximity by the force of an alterity in me is the passivity of a recurrence to oneself which is not the alienation of an identity betrayed. What can it be but a substitution of me for the others? It is, however not an alienation, because the other in the same is my substitution for the other through responsibility, for which, I am summoned as someone irreplaceable. I exist through the other and for the other, but without this being alienation: I am inspired. This inspiration is the psyche. The psyche can signify this alterity in the same without alienation in the form of incarnation, as being-in-one's-skin, having-the-other-in-one's-skin. *

For quantum physics, even as a universe or universes emerge from entanglement, how could the self, myself, rise from what is not my own? Could radical externality arise in the face of the Other (Autrui) which throws us back *as* self, *as* myself? Surely, this could not 'be' but as we have seen 'to be' is implicated in narrative, a narrative of history which has become common sense and practicality. We can no longer afford to be enticed, entranced, by a fiction, an abstraction, which looms as Being, as presence and its *logos* that has congealed and completed itself in absence, in logic. From the previous discussions in these volumes, I hope we know that origin becomes infinite regression that is perpetually condemned to lose itself and re-find itself in the tireless play of cause and effect. We never discover 'origin'. We always only defer it (transactionally?) to yet another replication of rationality. This infinite regression is not from the 'way that it is'. What our history has reductively guaranteed is that the radical externality of the other must go underground, it must be effaced as the absolute 'not' of totality. Becoming must overtake the other as Concept or perhaps as lived practicality which solidifies and covers over externality. It cannot stand the scrutiny of contingent Being and can only persist as universal Concept. Levinas contests the overcoming of Being into Concept with the idea of anarchy. However, this comes with a major caveat.

Anarchy linguistically comes forward from history as the negation of *archy* (*ἀρχή*, origin). The prefix *an* negates *archy*. This is a simple and logical reduction to origin and the 'not' of origin which has no exteriority. As we have found with exformation, there is a hazy middle with *logos*, which cannot be summed up and concluded by logic. Thus, even origin as the history for *phusis* for later Greeks like Aristotle and Plato came with a hazy, chaotic backdrop from their own ancestors. This backdrop has long been lost to modernity. I can almost hear Kierkegaard sarcastically protest that the (Hegelian) System in the "Science of Logic" is not quite complete. However, in the current day the Logic has become completed merely in the feigned *logic* of politics, practicality, and common sense. Ironically, instead of science positively and finally finishing the System, it appears to be collapsing into an *ad hoc* finality which could damn well extinguish us, Homo sapiens.

The abyss we face makes no promises for species survival, planetary survival. The reactionary

---

* 1 Emmanuel Levinas, *Otherwise than Being, or Beyond Essence*, trans. Alphonso Lingis (Pittsburgh: Duquesne University Press, 1998), 113–115.

promise of logic completed has run its course through time. The genealogy of cause and effect beginning in origin, whether it be the gods, God, or empirical objectivity has exhausted itself in the face of its own science of quantum *phusis*. Quantum *phusis* undermining the futile promises of the logic of cause and effect, spacetime, the mythic completion of the System has arrived at a gaping abyss. The indeterminate, spooky haze of the observer effect which cannot mechanically shed itself of Victor Frankenstein. Being taken up into an eschatology of Concept has exhausted itself in the Logical *telos* of absolute certainty. Only the repetition of authoritarian force and violence dialectically guarantees extinction. But ancient and vague intuitions of *phusis* as chaos which cannot be taken up into binary dualities of classicism leave us before a yawning gap which cannot be summed up cleanly by promises of being and not-being, absolute Concept which must forget itself to *be* absolute, order and disorder, origin and anarchy where anarchy is the 'not' of origin. *Phusis* in its ancient hazy intuitions gave way to strife to usher in classicism's logic and totalities. However, at the end of the epoch of strife the void we face is the between and exterior to Being and Nothing (as the 'not' of Being and *vice versa*). From the yawning gap of *phusis* and the exhaustion of strife quantum physics brings exteriority before us. Chaos is excess and branching into classicism's logic as subjective and objective where subjective resists the force of modernist objectification in reactionary fanaticisms and extremisms. The logic of dialectical oppositions of order and disorder, origin and the 'not' of origin must always concede to lifting itself up into higher and higher unities until Concept remains absolute even when and because it must forget itself. Any excess must finally and teleologically be vindicated in the absolute. Exceeding itself must be absolute in which even the 'not' of absolute must recursively define Identity - one without another. Even excess must be discounted by the history of metaphysics which must bow before immanentism. However, the path of metaphysics' failure and quantum physics leads us in an 'other' direction. All the 'nots' of binary dualisms cannot be brought under domination and force by their thesis. However, the 'nots' of binary dualisms need not be subjugated to their masters. They are not allowed to exceed their masters as an unrecoverable excess, they cannot be lost in the nether lands and underworlds of anarchy but must persist in sheer obsession in which even extinction becomes yet another obsession. Quantum physics gives way and allows classism's narratives but as a limited case of the observer. An exteriority to the 'not' of anarchy, of origin, a "*passivity more passive than all passivity*" behind the violence of origin and Narcissus' echo.

Levinas speaks of anarchy as not simply origin and no-origin but as what origin and no-origin cannot conceive. This is what Levinas calls the hither side of identity, the tautological 'I' which totalizes itself in the identical universalism of logic. Despite the pretense, the System and the 'I', can never be completed or finalized. The pretense for Being to rise into Concept cannot take absolute account but only exhausts itself in its violent attempts. Sure, it can play in the polarities of logic and its histories but similarly to classical physics, it can never take full account. It can only be a limiting case with historic assumptions that abstract us away from any such thing as exformation. Levinas died on December 25, 1995. I am sure he knew something about quantum physics, but I know of nowhere that he writes about it. However, the failure of history and Being to finally take full account in Concept and its frenetic progeny is because the Other (Autrui) is not Concept. The Other (Autrui) is not taken up into Concept. Anarchy is oblique in Levinas. In terms of this volume, I have used Levinas' sense of anarchy in the term 'chaos'. Anarchy cannot rest in negation of origin. It always has a residue which origin and not-origin cannot totalize. Science is at the least the historic surety of Western truth which aspires to finalizing logic as identity from the completion of a theory of everything which has no excess to be discovered, which lays all truth bare to facts at least in theory but only exists as obsession in *reality*. Now theory in science, truth in science, has led science back towards a beginning in philosophy which once again faces an unaccounted-for excess.

The conventional wisdom in physics is for a positivism that will meld perfectly with empiricism. This would be the scientific *telos* (final completion or goal) of Hegel's Concept. This direction conserves the philosophical underpinnings of science whose 'legitimacy' is preserved by guarding against religion and metaphysics. Simultaneously, the virtue of science is its perseverance on observation of the other, the physicality of what shows itself in phenomena. It has historically resisted losing the otherness of *phusis* at any cost...the disproving of antiquated theories, the strewn and discredited careers of scientists throughout history. In keeping with Thomas Kuhn's "The Structure of Scientific Revolutions" science has a messy process of adopting new paradigms which involve a lot more than just the naive beliefs of truer truths (e.g., capital acquisition, political pressures, stable career pathways led by consensus, etc.). In spite of this messy process, it has to a notable degree kept its end goal on phenomenon, the other, which does not collapse in on merely establishing self-serving power interests. Capitalism, religion, and philosophy have historically demonstrated a susceptibility to vested interest, interests which by any means will usurp themselves over and against the other. To the degree that science has not relinquished its focus on the other as phenomena is why it now faces a ghostly phantasm which philosophically undergirds much of its history - rationalism. With the advent of quantum physics, we find an emergent spooky alliance in science with its other half, philosophy. This inevitable reunion feels like a chasm, a chaos, a yawning gap. This is the *phusis*, the abyss, which stared back into early Greek civilization that we lost along the way. I don't want to be presumptuous because I have not studied other civilizations enough, but I suspect that this chasm has, in one way or another, haunted all early civilizations. Similarly, Levinas writes of an anarchy which cannot be subsumed as the 'not' of *archy*, of origin. Levinas writes of a "*passivity more passive than all passivity*". All of this is excess to the absoluteness of identity in logic, the perpetual deferment of origin in its negation, the tautology of presence and absence, the *not* of presence. But absence leaves a leftover residue which comes in the morning, yet again, with the she or he we face.

Histories of Origin

I know this beginning has been difficult, more like a sensual plunge than a contextual approach which moves more slowly. To do this more contextually, I need to back up and start anew. However, I hope this beginning can give the reader a sense of where this is going and make the journey more amenable (if that is possible with my limited abilities). The next section will philosophically explore the path which has led us here with an emphasis on the 20th century. This will be a bit of a nerdier philosophical step back into histories of origin. However, I will throw out some ideas which will tie back into quantum physics as simply possible stepping-off points for speculation. I will not adhere strictly to temporal ordering of these topics as the way they filter down to our everyday lives and commonsense have little relevance to that kind temporal ordering. This will eventually wind its way back towards implications for us, for me, and the 'I' with proximity to the Other (Autrui), the he or she which faces us. What was the hope born of Enlightenment and modernity and borne by classicism? How did these histories as well-trod world-scape lead us to the abyss?

On Nature as Love and Strife

Earlier in the section entitled, "The Failure of Intuition and Phusis" I brought up Empedocles. I want to go into more detail and bring a quote up I mentioned before. But first, here is a bit of background on Greek mythology.

- Zeus, the chief deity of the ancient Greek pantheon, reigns as the ruler of the gods from his celestial abode atop Mount Olympus. His influence extends over the sky, weather, and natural forces. Zeus holds the title of father to both divine beings and mortal humans. His role as a protector and ruler encompasses all realms of existence. According to Greek poet Homer, Zeus's heavenly domain is the summit of Mount Olympus, the highest peak in Greece. There, he governs the affairs of gods and mortals, rewarding virtue and punishing wrongdoing. Zeus is often depicted as an older man with a beard, and his symbols include the lightning bolt and the eagle. In a Cretan myth, Zeus's mother, Rhea, saved him from being swallowed by his father, Cronus, who feared one of his children would dethrone him. Rhea replaced Zeus with a stone and hid him in a cave on Crete. After growing to manhood, Zeus led a revolt against the Titans and dethroned Cronus, sharing dominion over the world with his brothers, Hades and Poseidon. Zeus's influence extends beyond the celestial; he is also the protector of cities, homes, property, strangers, guests, and supplicants. His complex character combines authority, justice, and a touch of divine mischief. Sky and Thunder God: Zeus is the embodiment of the sky and weather. He wields thunderbolts, sends lightning, rain, and winds. His traditional weapon is the mighty thunderbolt. [*]
- Hera, known as Hērā (Ἥρα) in Greek, holds a prominent place in ancient Greek mythology. She embodies several significant roles. Hera is the queen of the gods and the patroness of marriage. She oversees the sacred union between spouses and the bonds of matrimony. Her role extends beyond mere partnership; she symbolizes the harmony and commitment within marriage. As the goddess of women, Hera safeguards their interests and well-being. She watches over women during childbirth, offering protection and guidance during this crucial phase of life. Hera's influence extends to family life. She is associated with motherhood, fertility, and the welfare of children. Her maternal aspect emphasizes her nurturing and protective qualities. Hera is one of the twelve Olympian deities who reside on Mount Olympus. She holds a position of authority and power among the divine assembly. Her regal demeanor reflects her status as the queen of this celestial pantheon. While Hera embodies nobility and authority, she is also known for her jealousy and vengeful nature. Her husband, Zeus, often tested her patience with his numerous affairs. Hera ensured that each of Zeus's consorts faced challenges and trials.

---

[*] *Encyclopaedia Britannica*, s.v. "Zeus," accessed March 7, 2026; "Zeus," Wikipedia, last modified January 25, 2024; *Encyclopaedia Britannica*, s.v. "Hera"; "Hera," Wikipedia, last modified February 12, 2024; "Aidoneus," *Greek Myth Index*, accessed March 7, 2026.

See Zeus (Britannica)
And Who is Zeus (Britannica)
And Zeus (Wikipedia)

Hera is depicted as both a majestic figure and a complex goddess with human emotions. Her stories intertwine with those of other gods and heroes, making her an integral part of Greek mythology. [*]

- Aidoneus, also known as Hades, is a significant figure in ancient Greek mythology. Aidoneus was a mythical king of the Molossians in Epirus. He is often depicted as the husband of Persephone. The story involving Aidoneus is closely tied to the abduction of Persephone. After Theseus and Pirithous concealed Helen at Aphidnae, they sought to reward Pirithous by obtaining Kore (another name for Persephone) as his wife. When Aidoneus discovered their intention to take his wife, he had Pirithous killed by Cerberus, the three-headed dog guarding the entrance to the underworld. Aidoneus then captured Theseus, who was later released upon the request of Heracles. The legend of Aidoneus essentially mirrors the myth of Hades kidnapping Persephone, with the mythical king serving as a substitute for the god Hades himself. The name "Aidoneus" is a lengthened form of the Greek word for Hades (ᾍδης), which infrequently appears in original sources, including Homer's works. As the ruler of the subterranean realm inhabited by the shades of the dead, Aidoneus plays a crucial role in the Greek pantheon. [†]
- Nêstis was the daughter of Zeus and Demeter. Her uncle Hades abducted her, and she became queen of the underworld and goddess of death. Hades would later marry her. Every spring she could return to the surface which resulted in the return of spring. Nêstis is found in much earlier myths pre-Greek all over the world with different names going all the way back to Proto-Indo-European. Nêstis, in ancient Greek mythology, is associated with water. She represents various forms of water, including mist, rain, moisture, and steam. However, her most significant role is as the queen of the underworld. Due to the reverence and caution surrounding her name, she is often referred to by other euphemistic titles such as Cora or Kore. Interestingly, Nêstis is also linked to moistening mortal springs with tears, as mentioned in a text attributed to the philosopher Empedocles. Her dual nature as both a water goddess and a ruler of the underworld adds depth to her character in Greek mythology.

Empedocles (born early fifth century BC) [‡] a student of Parmenides of Elea [§] in the late sixth century of early fifth century BC wrote a work referred to as "*On Nature*" where he states,

---

[*] *Encyclopaedia Britannica*, s.v. "Zeus," accessed March 7, 2026; "Zeus," Wikipedia, last modified January 25, 2024; *Encyclopaedia Britannica*, s.v. "Hera"; "Hera," Wikipedia, last modified February 12, 2024; "Aidoneus," *Greek Myth Index*, accessed March 7, 2026.

See Hera (Wikipedia)
And Hera (Britannica)
And Greek Goddesses Hera

[†] *Encyclopaedia Britannica*, s.v. "Zeus," accessed March 7, 2026; "Zeus," Wikipedia, last modified January 25, 2024; *Encyclopaedia Britannica*, s.v. "Hera"; "Hera," Wikipedia, last modified February 12, 2024; "Aidoneus," *Greek Myth Index*, accessed March 7, 2026.

See Aidoneus
And Hades

[‡] See Empedocles

[§] See Parmenides (Wikipedia)

> Hear first of all the four roots of all things:
> Zeus the gleaming, Hera who gives life, Aidoneus,
> And Nêstis, who moistens with her tears the mortal fountain.*

The 'roots' are the four deities Zeus, Hera, Aidoneus, and Nêstis (Persephone). Zeus is the son of Cronus. Chronus is related to Cronus. In Greek mythology, Cronus (also spelled Kronos) was the Titan god of time and the leader of the Titans. He was known for devouring his own children, fearing that they would overthrow him. However, his son Zeus managed to escape and eventually defeated Cronus, becoming the ruler of the gods. Chronus (sometimes spelled Khronos) is a concept in Greek mythology representing time itself. Unlike Cronus, who was a specific deity, Chronus embodies the abstract notion of the passage of time. In art and literature, Chronus is often depicted as an old man with a flowing white beard, symbolizing the inexorable march of time like Zeus. So, while Cronus is a specific character from mythology, Chronus represents the broader concept of time, which transcends individual gods and mortals. Zeus has to defeat father time in order to be. Hera is the goddess of union, she watches over women and childbirth, offspring, fertility, maternal and nurturing, and growth. She is the guardian of well-being. She also knows her husband Zeus can be a pain in the ass, obstinate, overly assertive, thinking too highly of himself, yet still understanding of his trials and tribulations. Aidoneus as the king of Hades, death, who abducts his niece Nêstis and forcibly betroths her but even though she is horribly abused she is still the origin of spring and even empathizes with her abuser not because of him and who he is but because of her and who she is. She cannot be subsumed under even the threat of hell. She cannot lose herself. But she cannot be named. Interestingly,

> Of the four deities of Empedocles' elements, it is the name of Persephone alone that is taboo – Nêstis is a euphemistic cult title[e] – for she was also the terrible Queen of the Dead, whose name was not safe to speak aloud, who was euphemistically named simply as Kore or "the Maiden", a vestige of her archaic role as the deity ruling the underworld. Nestis means "the Fasting One" in ancient Greek. †

In my opinion our present-day cult of history still cannot speak ***her*** name which should make all of us weep. Even in the dry thirst of eternal Hell she "*moistens with her tears the mortal fountain*". If this does not bring to the fore passion, love and strife, all wrapped in the garb of existence, the necessity of Hades awaits us until perhaps eternal echoes usher in yet another spring which may yet become.

In fragment 17 of Diels-Kranz' sixth edition entitled "*Die Fragmente der Vorsokratiker*" (The Fragments of the Pre-Socratics), apparently speaking of the physical world as a whole, Empedocles states his fundamental thesis about the relation of roots: ‡

---

* Empedocles, *On Nature*, quoted in M. R. Wright, *Empedocles: The Extant Fragments* (London: Bristol Classical Press, 1995), 165.
(B6 = D 57), Diels, Hermann and Walther Kranz, 1951–2, Die Fragmente der Vorsokratiker: griechish und deutsch, 6th edition, 3 volumes, Berlin: WeidmannscheVerlagsbuchhandlung.
See Empedocles
And Parmenides (Wikipedia)

† "Persephone," Wikipedia, last modified March 3, 2024, [Insert Page Number].
See Persephone (Wikipedia)

‡ See Hermann Alexander Diels (Wikipedia)

Twofold is what I shall say: for at one time they [i.e., the elements] grew to be only one
Out of many, at another time again they separate to be many out of one.
And double is the birth of mortal things, double their death.
For the one [i.e., birth] is both born and destroyed by the coming together of all things,
While the other inversely, when they are separated, is nourished and flies apart (?).
And these [scil. the elements] incessantly exchange their places continually,
Sometimes by Love all coming together into one,
Sometimes again each one carried off by the hatred of Strife.
<Thus insofar as they have learned to grow as one out of many,>
And inversely, the one separating again, they end up being many,
To that extent they become, and they do not have a steadfast lifetime;
But insofar as they incessantly exchange their places continually,
To that extent they always are, immobile in a circle.*

Much of this is thought to have come from Empedocles teacher Parmenides. I will discuss the ancient idea of time in this passage soon but for now time is not how we think of time for the ancient mind. Time was more like what we think of as a quality. Therefore, in the first sentence of the quote above think of time as a quality, as growth, as 'grew to be one'. Even as early as the late sixth century BC, Parmenides and his student Empedocles are speculating on growth and its unique capacity to be 'many' and 'one'. This odd observation of phenomenon causes 'doubling'. Doubling might be referred to as binary dualisms much later in history. Life and death have doubles. For each is 'born' and 'destroyed' at the same *time,* or more precisely as the *telos* of growth. 'Coming together' and 'falling apart' have a simultaneity. They both 'depend on each other' and 'exchange places continually'. Love is 'coming together'. Strife is 'falling apart'. 'Becoming' does not have a 'steadfast lifetime'. Becoming interchanges 'continually like a circle'.

All of this is based on the claim that everything is composed of four roots which are moved by two opposing forces, Love and Strife. Since the roots are identified by the names of deities (Zeus, Hera, Aidoneus, Nêstis) —and not by the traditional names for the elements fire, earth, air, and water—there are rival interpretations of which deity is to be identified with which root. Nevertheless, there is general agreement that the passage refers to fire, earth, air (= aether, the upper, atmospheric air, rather than the air that we breathe here on earth) and water (cf. B 109 = D 207). Aristotle credits Empedocles with being the first to distinguish clearly these four elements (*Metaphysics*. 985a31–3). However, the fact that the roots have divinities' names indicates that each has an active nature and is not just inert matter (Rowett 2016). These roots and forces are eternal and equally balanced, although the influence of Love and of Strife waxes and wanes.†

Immediately one is struck by the comprehensive symmetry of this scheme. It seems to address

* Empedocles, "Fragment 17," in *Die Fragmente der Vorsokratiker,* 6th ed., ed. Hermann Diels and Walther Kranz (Berlin: Weidmann, 1951).

See Empedocles, (B 17.1–13 = D 73) , Diels, Hermann and Walther Kranz, 1951–2, Die Fragmente der Vorsokratiker: griechish und deutsch, 6th edition, 3 volumes, Berlin: WeidmannscheVerlagsbuchhandlung

† (B 6 and B 17.14–20 = D 57 and D 73.245–51) , Diels, Hermann and Walther Kranz, 1951–2, Die Fragmente der Vorsokratiker: griechish und deutsch, 6th edition, 3 volumes, Berlin: WeidmannscheVerlagsbuchhandlung

See Empedocles

coming-to-be and passing-away, birth and death, and it does so with an elegant balance. The four roots come together and blend, under the agency of Love, and they are driven apart by Strife. Empedocles also describes a time when Strife had separated the roots. This separation is total and is the opposite pole from the *circle*, which is a total mixture under the influence of Love.

> When Strife has reached the deepest depth
> Of the vortex, and Love has come to be in the center of the whirl,
> Under her dominion all these [i.e., the elements] come together to be only one,
> Each one coming from a different place, not brusquely, but willingly *

First of all, this somewhat mysterious description suggests that the means by which Strife separates the roots from the beginning is a vortex. Heavier elements like earth settle in the middle and lighter ones like fire are pushed to the periphery. This reference to the vortex also implies that dominance by Strife is characterized by the whirling motion of the cosmos as we know it. In addition, this fragment suggests the end of the rule of Strife and the beginning of the rule of Love, as this principle begins to insinuate itself into the elements. The latter part of this passage describes the unifying effect of Love.† Additionally, somehow strife and love seem to still have relevance for daily life.

These very ancient intuitions have their own meanings to those that lived them which makes it difficult if not impossible at times to decipher. However, I want to make some way out of bounds comments which I have no doubt some would be abhorred by. Could there be a kind of allegorical intuition in all this with the quantum wave function? Perhaps, we could venture some creative speculation that there is a type of doubling which cannot be reduced or resolved into one or the other and yet have some odd characteristics,

- the 'elements' as 'one' or 'many' as the quality of 'growth' perhaps reminiscent of the wave function and its many branchings where 'growth/becoming' is the observer effect falling into classicisms (e.g., spacetime/quanta)
- the branchings of the observer effect have a kind of 'doubling' in simultaneity with the wave function and narratives of classicisms
- these doublings form doubling dualisms of 'birth/death', 'come together/falling apart', 'love/strife', steadfast as lifetimes and histories/exchanging places continuously as overlapping/simultaneous many-worlds of possibilities, being as mobile in becoming/immobile as a circle which ever goes around but stays the same - all symbiotic on each other
- all of these doublings bring to the fore love and strife, the many as the one/the one as the many, the coming together and falling apart but somehow all bringing us to the possibility

* Empedocles, "Fragment 35," in *Die Fragmente der Vorsokratiker*, 6th ed., ed. Hermann Diels and Walther Kranz (Berlin: Weidmann, 1951), 1:326–327.

(B 35.20–23 = D 75.3–6) , Diels, Hermann and Walther Kranz, 1951–2, Die Fragmente der Vorsokratiker: griechish und deutsch, 6th edition, 3 volumes, Berlin: WeidmannscheVerlagsbuchhandlung

See Empedocles

† See Empedocles

of growth, of a sense of radical exteriority which has something to do with responsibility and not whether we respond but how we respond.
- can we think - rhizomes and the tree of life?

Ok, a stretch to say the least but the oddness of how Empedocles phrases this reminds me (at least) of the oddness of the quantum wave function which cannot find easy categories and patterns of thinking. And the later quote where strife has reached the deepest depth – can we think the violent vortex as Big Bang singularity, wormhole, where exformation is chaos and love, branching, the part of spacetime we inhabit which gets centered in life and love from the spooky observer effect and the one and many which calls itself 'me'. Empedocles calls this the domain of "her", Nêstis which finds no name, the stranger, the sojourner, where each one becomes in a 'different place' not from rudeness, bluntness, or curtness but tactfully and willingly.

Obviously, I am not implying the Empedocles had the quantum wave function in mind here! But what I am doing is showing how to speculate, to start to engage actively in making connections no matter who thinks it is right or wrong, true or false, etc. This is how we start to engage in growth, learning, becoming one with a process which is a journey and not a destination. There are woefully too many folks out there that find joy in raining on parades. But instead of resorting to a kind of defensive gamesmanship with them why not learn the practice of looking for the value in what the other brings to us regardless of what we imagine their motivations may be? What matters is whether to die in transactional gamesmanship or live in an open quest for an alterity of the other which finds no contentment in hearing one's own echo. Let's see if we can hear another voice to our echo of time in an ancient and other temporality.

Time in Aristotle's aion

We all have an idea of time as eternal. However, time as eternal has its roots in Latin Christianity, the monotheistic God who was one and never changed. The Greek word *aion* was translated to mean 'eternity' in Latin. In Heleen M. Keizer's paper "'ETERNITY' REVISITED: A Study of the Greek Word *aion*" she states,

> The history of aiôn starts with aiôn being a word for 'life'; indeed, in Homer aiôn is far from being a word for 'time' but rather has the connotation of 'force of life'. A word with which aiôn in Homer is combined, is psuchê. The Greek language has yet two other words to designate life: there is zôê, indicating the state of being alive (not yet in Homer), and bios, indicating the ways and means
>
> of maintaining that state. Thus while zôê refers to the life 'bred' and bios to the life 'led', aiôn can well be characterized as designating the life 'had'. My investigation of aiôn in Greek literature from Homer to Hellenism has shown me a coherent complex of meaning of the word, which turned out to be built up from the following three notions: 'life', 'time', and what I call 'completeness', 'wholeness', or 'entirety'. Aiôn refers to 'life' as a 'whole' of 'time' — hence, besides 'lifetime' it can also designate life's 'lot'...
>
> Gathering my conclusions about the role allotted to aiôn successively in the philosophy of Empedocles, Plato, Aristotle, the Stoa and Epicurus, I observe that this role is a cosmological

one. In Empedocles (fragment B16 Diels-Kranz) we seem to meet aiôn for the first time on a cosmic scale as the 'life' of the cosmos, coinciding with the whole of time. For Plato (discussed in more detail in §2a), aiôn — usually translated 'eternity' — is the unitary whole of 'life(time)' on the intelligible level, which chronos (time) displays or 'counts out' on the sensible level. In Plato for the first time we find aiônios as the adjective form of aiôn ('aiônic'). Aristotle (more about him in §2b) defines aiôn as that which encompasses the infinite time of the cosmos, in analogy of aiôn as

that which encompasses the time of an individual person's life; the particular term Aristotle uses here is telos ('the telos encompassing the time...'). In Plato's system, aiôn is reserved for the intelligible, transcendent world; in Aristotle, aiôn (called divine, and as telos touching upon the first, transcendent, principle) is the comprising sum of the immanent, sensible world of time itself...

Aristotle's first definition of aiôn, "the telos which encompasses the time of everyone's life", fits in perfectly with what I have indicated above as characterizing aiôn in earlier writers ('the ancients'), viz. that the word refers to a person's life as a complete(d) whole with the inherent aspect of time. Aristotle subsequently points out that aiôn, thus defined, applies to the entire cosmos as well — in doing so he implicitly follows Plato...

In the first book of De caelo, chapter 9, Aristotle offers an almost lexicographic description of what the word aiôn in his view conveys. De caelo, in Greek Peri ouranou, is a work on physics, dealing with cosmology: Aristotle uses the word ouranos (Lat. caelum) not only to designate the heavens but also the cosmos, or universe. The universe, Aristotle argues in I 9, is made of the totality of matter, and so 'outside the universe' there is neither matter nor what is associated with matter, viz. space and time. When Aristotle subsequently talks about the things or beings 'over there', De caelo as a work on physics touches upon the divine: it appears to refer to the prime, unmoved Mover, also elaborated as a plurality of unmoved Movers, which is the divine transcendent principle discussed by Aristotle in book XII, chapters 6-8, of his Metaphysics. I now quote De caelo I 9

279a18-30:

Therefore, those-over-there are not such as to be in place, nor does time cause them to age, nor does change work in any way upon any of those that are arrayed beyond the outermost motion: unalterable and impassive, in having the best and most self-sufficient life (zôê) they continue (throughout) the whole aiôn.

a22 Indeed, that name [sc. aiôn] has been divinely uttered by the ancients. For the completion (telos) which encompasses the time of everyone's life (zôê), which cannot in nature be exceeded, has been named everyone's aiôn. Along the same line of thought also the completion of the whole universe, the completion which encompasses time as a whole and infinity, is aiôn, having taken the name from aei einai [to be always], being immortal and divine.

From there depends for all other things, for some more directly, for others more obscurely, being and life (zên)... At the start of the second book of De caelo, Aristotle summarizes what he has

> demonstrated in the first book. I quote De caelo II 1 283b26-30:
> the universe as a whole neither has come into being nor admits of destruction, as some assert that it does, but is one and everlasting (aïdios) with no beginning or end (teleutê) of the (/its) whole aiôn, but containing and encompassing in itself the infinite time (chronos). *

For Aristotle the idea of 'eternity' was not endless time as we might imagine it now. Eternity as *aiôn* expressed a quality first not a temporality. The quality expressed was teleological. The Greek notion of *telos* was thought as the ultimate completion or fulfillment of something whether it was the building of a house which found its *telos* in the end result - the actual house, or the result of a life well lived. This also gives us a hint of how temporality showed itself to the ancient Greek mind. We tend to think anachronistically meaning we think everyone at all times experienced times in the linguistic ways that we do. The ancient Greeks had no way to think about 'eternity' the way we do. They had no classical science to draw from. They thought in terms of qualities in which ages or epochs were *telos*. For our purpose, *telos* means the end of some action was the quality of completion and fulfillment of that action. Even more *telos* was used with an individual meaning as not just the quality of one's lifetime but also as a quality of goal or completion for a civilization. I find this to be evident as well in another Greek notion of time as *Kairos*. Permeating ancient Greece from the earliest to the latter citizens and thinkers and very important to Aristotle, was the notion of *kairos* meaning 'the opportune moment'.† This was the moment of passion and fulfillment. It was purely qualitative. It was more like what we think of as a profound moment when something momentous happens. For Aristotle, the uncaused, the "immovable", was the *telos* of cause and effect, motion, which was uncaused and motionless. It was responsible for the cosmos. Therefore, it was the 'good', the fulfillment of its end. It was also the 'good' for something because all existence was the result of the immovable.

Also, oddly enough, let's highlight this remark from Aristotle's discussion a little further down, "*it [the good] causes motion as being an object of love*". Plato had another odd notion which is intriguing and has spawned much scholarly debate which translated means, 'the good beyond being'. Aristotle states, "For the final cause is not only the good for something," but also "the good which is the end of some action." In the latter sense it applies to immovable things, although in the former it does not; and it causes motion as being an "object of love", whereas all other things cause motion because they are themselves in motion. Aristotle tells us that everything that moves must have a cause. Therefore, there must be a final cause that will have several characteristics:

1. The "final cause" or origin must be the good as it is "good for something", i.e., the cosmos.
2. The 'good' generates motion or, for our purposes, time and space, cause and effect.
3. The 'good' cannot have a cause. It is the unmoved mover.

What is the unmoved mover? As already discussed, the unmoved mover was not in time or space and not the cause of anything else.

---

* Heleen M. Keizer's study "'Eternity' Revisited: A Study of the Greek Word Aiôn" was published in *Vigiliae Christianae* 53, no. 1, in February 1999, spanning pages 74–77. The paper examines the semantic range of the Greek word *aion* [1].
See 'Eternity' Revisited: A Study of the Greek Word αἰών

† See A Detour of Time

Origin and the Good Beyond Being

Plato tells us something strange in the Republic with the phrase "*epekeina tês ousias*" or 'the good beyond being'. Ever since Plato uttered this phrase scholars have discussed it many times over the centuries especially in the 20th century. What did Plato have to say concerning this seemingly odd idea?

Plato, a mentor of Aristotle, echoes his notion of the good in the Republic (Book 6, Section 509a through Book 7,518e). This is Plato's allegory of the cave. Socrates is having a dialogue with both of his brothers Adiemantus and Glaucon. The dialog happens in the home of an old man named Cephalus. The setting in Athens looks very grim for Socrates as the Athenians had just been defeated in the Peloponnesian War in 404 BC. A pro-Spartan oligarchy called the Thirty Tyrants is about to tragically take over Athens. The "Great Beast" is the society which is about to take over Athens. Socrates wants to show how the Athenians can gain freedom over the Beast.

More generally, the allegory is about Plato's notion of the Forms (Greek *Eidos*, Ideas). The Forms are the realities behind what shows itself to us in phenomena. The Forms cannot be directly perceived. This dialog does not exemplify the later metaphysical and classical notions of a comprehensible origin. Rather, it emphasizes what the ancient Greek philosophers called *steresis* or privation. All humanity lives in a cave where shadows of the Forms are cast onto the walls. Outside the cave are the Forms. However, humanity is condemned to only seeing the shadows. This is privation or incapacity to see what is truly real. Plato was fully aware of an older backdrop to his Forms and privation in the ancient Ionians. *Peiron* in '*apeiron*' is the Ionic Greek for boundary or limit. For Plato, humanity lives in privation as boundary or limit, the cave. The older form of boundary or limit, *peras*, also meant 'beyond' or 'further'. Thus, *apeiron* in Ionic Greek from Anaximander is the alpha privative, the privation of boundary and limit or without boundary or limit. Plato was also aware of the Greek history of origin and chaos from Hesiod, Anaximander, and Heraclitus. Plato, in the chaotic darkness of Athens to come, looks into *apeiron* and glimpses the Forms. I am going to include much of the dialog below because I think it is worth reading instead of summarizing. I will highlight some of Plato's remarks which demonstrate privation.*

Plato's Republic

> Socrates,
>
> **it is right to deem light and vision sunlike, but never to think that they are the sun, so here it is right to consider these two their counterparts, as being like the good or boniform, but to think that either of them is the good is not right**. Still higher honor belongs to the possession and habit of the good."
>
> Glaucon,
>
> "An inconceivable beauty you speak of," he said, "if it is the source of knowledge and truth, and yet itself surpasses them in beauty. For you surely cannot mean that it is pleasure."
>
> Socrates,
>
> "Hush," said I, "but examine "In like manner, then, you are to say that the objects of

* Plato, *Republic*, 508a–517a, in *Plato in Twelve Volumes*, vols. 5 & 6, trans. Paul Shorey (Cambridge, MA: Harvard University Press, 1969).

knowledge not only receive from the presence of the good their being known, but their very existence and essence is derived to them from it, though **the good itself is not essence** but still transcends essence in dignity and surpassing power."

Glaucon,

And Glaucon very ludicrously said, "Heaven save us, hyperbole can no further go."

Socrates,

"The fault is yours," I said, "for compelling me to utter my thoughts about it."

Glaucon,

"And don't desist," he said, "but at least expound the similitude of the sun, if there is anything that you are omitting."

Socrates,

"Why, certainly," I said, "I am omitting a great deal."

Glaucon,

"Well, don't omit the least bit," he said.

Socrates,

"I fancy," I said, "that I shall have to pass over much, but nevertheless so far as it is at present practicable I shall not willingly leave anything out."

Glaucon,

"Do not," he said.

Socrates,

"Conceive then," said I, "as we were saying, that there are these two entities, and that one of them is sovereign over the intelligible order and region and the other over the world of the eye-ball, not to say the sky-ball, but let that pass. You surely apprehend the two types, the visible and the intelligible."

Glaucon,

"I do."

Socrates,

"Represent them then, as it were, by a line divided into two unequal sections and cut each section again in the same ratio (the section, that is, of the visible and that of the intelligible order), and then as an expression of the ratio of their comparative clearness and obscurity you will have, as one of the sections of the visible world, images. By images I mean, first, shadows, and then reflections in water and on surfaces of dense, smooth and bright texture, and everything of that kind, if you apprehend."

Glaucon,

"I do."

Socrates,

"As the second section assume that of which this is a likeness or an image, that is, the animals about us and all plants and the whole class of objects made by man."

Glaucon,

"I so assume it," he said.

Socrates,

"Would you be willing to say," said I, "that the division in respect of reality and truth or the opposite is expressed by the proportion: as is the opinionable to the `knowable so is the likeness to that of which it is a likeness?"

Glaucon,

"I certainly would."

Socrates,

"Consider then again the way in which we are to make the division of the intelligible section."

Glaucon,

"In what way?"

Socrates,

"By the distinction that there is one section of it which the soul is compelled to investigate by treating as images the things imitated in the former division, and by means of assumptions from which it proceeds not up to a first principle but down to a conclusion, while there is another section in which it advances from its assumption to a beginning or principle that transcends assumption, and in which it makes no use of the images employed by the other section, relying on ideas only and progressing systematically through ideas."

Glaucon,

"I don't fully understand what you mean by this," he said.

Socrates,

"Well, I will try again," said I," for you will better understand after this preamble. For I think you are aware that students of geometry and reckoning and such subjects first postulate the odd and the even and the various figures and three kinds of angles and other things akin to these in each branch of science, regard them as known, and, treating them as absolute assumptions, do not deign to render any further account of them to themselves or others, taking it for granted that they are obvious to everybody. They take their start from these, and pursuing the inquiry from this point on consistently, conclude with that for the investigation of which they set out."

Glaucon,

"Certainly," he said, "I know that."

Socrates,

"And do you not also know that they further make use of the visible forms and talk about them, though they are not thinking of them but of those things of which they are a likeness, pursuing their inquiry for the sake of the square as such and the diagonal as such, and not for the sake of the image of it which they draw? And so in all cases. The very things which they mould and draw, which have shadows and images of themselves in water, these things they treat in their turn as only images, but what they really seek is to get sight of those realities which can be seen only by the mind."

Glaucon,

"True," he said.

Socrates,

This then is the class that I described as intelligible, it is true, but with the reservation first that the soul is compelled to employ assumptions in the investigation of it, not proceeding to a first principle because of its inability to extricate itself from and rise above its assumptions, and second, that it uses as images or likenesses the very objects that are themselves copied and adumbrated by the class below them, and that in comparison with these latter are esteemed as clear and held in honor."

Glaucon,

"I understand," said he, "that you are speaking of what falls under geometry and the kindred arts."

Socrates,

"Understand then," said I, "that by the other section of the intelligible I mean that which the reason itself lays hold of by the power of dialectics, treating its assumptions not as absolute beginnings but literally as hypotheses, underpinnings, footings, and springboards so to speak, to enable it to rise to that which requires no assumption and is the starting-point of all, and after attaining to that again taking hold of the first dependencies from it, so to proceed downward to the conclusion, making no use whatever of any object of sense but only of pure ideas moving on through ideas to ideas and ending with ideas."

Glaucon,

"I understand," he said; "not fully, for it is no slight task that you appear to have in mind, but I do understand that you mean to distinguish the aspect of reality and the intelligible, which is contemplated by the power of dialectic, as something truer and more exact than the object of the so-called arts and sciences whose assumptions are arbitrary starting-points. **And though it is true that those who contemplate them are compelled to use their understanding and not their senses, yet because they do not go back to the beginning [arche] in the study of them but start from assumptions you do not think they possess true intelligence about them although the things themselves are intelligibles when apprehended in conjunction with a first principle.** And I think you call the mental habit of geometers and their like mind or understanding and not reason because you regard understanding as something intermediate between opinion and reason."

Socrates,

"Your interpretation is quite sufficient," I said; "and now, answering to these four sections, assume these four affections occurring in the soul: intellection or reason for the highest understanding for the second; assign belief to the third, and to the last picture-thinking or conjecture, and arrange them in a proportion, considering that they participate in clearness and precision in the same degree as their objects partake of truth and reality."

Glaucon,

"I understand," he said; "I concur and arrange them as you bid.

Socrates,

Next," said I, "compare our nature in respect of education and its lack to such an experience as this. **Picture men dwelling in a sort of subterranean cavern with a long entrance open to the light on its entire width. Conceive them as having their legs and necks fettered from childhood, so that they remain in the same spot, able to look forward only, and prevented by the fetters from turning their heads. Picture further the light from a fire burning higher up and at a distance behind them, and between the fire and the prisoners and above them a road along which a low wall has been built, as the**

exhibitors of puppet-shows have partitions before the men themselves, above which they show the puppets."

Glaucon,

"All that I see," he said.

Socrates,

"See also, then, men carrying past the wall implements of all kinds that rise above the wall, and human images and shapes of animals as well, wrought in stone and wood and every material, some of these bearers presumably speaking and others silent."

Glaucon,

"A strange image you speak of," he said, "and strange prisoners."

Socrates,

"Like to us," I said; "for, to begin with, tell me do you think that these men would have seen anything of themselves or of one another except the shadows cast from the fire on the wall of the cave that fronted them?"

Glaucon,

"How could they," he said, "if they were compelled to hold their heads unmoved through life?"

Socrates,

"And again, would not the same be true of the objects carried past them?"

Glaucon,

"Surely."

Socrates,

"If then they were able to talk to one another, do you not think that they would suppose that in naming the things that they saw they were naming the passing objects?"

Glaucon,

"Necessarily."

Socrates,

"And if their prison had an echo from the wall opposite them, when one of the passersby uttered a sound, do you think that they would suppose anything else than the passing shadow to be the speaker?"

Glaucon,

"By Zeus, I do not," said he.

Socrates,

"**Then in every way such prisoners would deem reality to be nothing else than the shadows of the artificial objects.**"

Glaucon,

"Quite inevitably," he said.

Socrates,

"**Consider, then, what would be the manner of the release and healing from these bonds and this folly if in the course of nature something of this sort should happen to them: When one was freed from his fetters and compelled to stand up suddenly and turn his head around and walk and to lift up his eyes to the light, and in doing all this felt pain and, because of the dazzle and glitter of the light, was unable to discern the**

objects whose shadows he formerly saw, what do you suppose would be his answer if someone told him that what he had seen before was all a cheat and an illusion, but that now, being nearer to reality and turned toward more real things, he saw more truly? And if also one should point out to him each of the passing objects and constrain him by questions to say what it is, do you not think that he would be at a loss and that he would regard what he formerly saw as more real than the things now pointed out to him?"

Glaucon,

"Far more real," he said.

Socrates,

"**And if he were compelled to look at the light itself, would not that pain his eyes, and would he not turn away and flee to those things which he is able to discern and regard them as in very deed more clear and exact than the objects pointed out?**"

Glaucon,

"It is so," he said.

Socrates,

"And if," said I, "someone should drag him thence by force up the ascent which is rough and steep, and not let him go before he had drawn him out into the light of the sun, do you not think that he would find it painful to be so haled along, and would chafe at it, and when he came out into the light, that his eyes would be filled with its beams so that he would not be able to see even one of the things that we call real?"

Glaucon,

"Why, no, not immediately," he said.

Socrates,

"Then there would be need of habituation, I take it, to enable him to see the things higher up. And at first he would most easily discern the shadows and, after that, the likenesses or reflections in water of men and other things, and later, the things themselves, and from these he would go on to contemplate the appearances in the heavens and heaven itself, more easily by night, looking at the light of the stars and the moon, than by day the sun and the sun's light."

Glaucon,

"Of course."

Socrates,

"And so, finally, I suppose, he would be able to look upon the sun itself and see its true nature, not by reflections in water or phantasms of it in an alien setting, but in and by itself in its own place."

Glaucon,

"Necessarily," he said.

Socrates,

"And at this point he would infer and conclude that this it is that provides the seasons and the courses of the year and presides over all things in the visible region, and is in some sort the cause of all these things that they had seen."

Glaucon,

"Obviously," he said, "that would be the next step."

Socrates,

"Well then, if he recalled to mind his first habitation and what passed for wisdom there, and his fellow-bondsmen, do you not think that he would count himself happy in the change and pity them?"

Glaucon,

"He would indeed."

Socrates,

"And if there had been honors and commendations among them which they bestowed on one another and prizes for the man who is quickest to make out the shadows as they pass and best able to remember their customary precedences, sequences and co-existences, and so most successful in guessing at what was to come, do you think he would be very keen about such rewards, and that he would envy and emulate those who were honored by these prisoners and lorded it among them, or that he would feel with Homer and "'greatly prefer while living on earth to be serf of another, a landless man,'" [Homer Odysseus 11.489] and endure anything rather than opine with them and live that life?"

Glaucon,

"Yes," he said, "I think that he would choose to endure anything rather than such a life."

Socrates,

"And consider this also," said I, "if such a one should go down again and take his old place would he not get his eyes full of darkness, thus suddenly coming out of the sunlight?"

Glaucon,

"He would indeed."

Socrates,

"Now if he should be required to contend with these perpetual prisoners in 'evaluating' these shadows while his vision was still dim and before his eyes were accustomed to the dark—and this time required for habituation would not be very short—would he not provoke laughter, and would it not be said of him that he had returned from his journey aloft with his eyes ruined and that it was not worth while even to attempt the ascent? And if it were possible to lay hands on and to kill the man who tried to release them and lead them up, would they not kill him?"

Glaucon,

"They certainly would," he said.

Socrates,

"This image then, dear Glaucon, we must apply as a whole to all that has been said, likening the region revealed through sight to the habitation of the prison, and the light of the fire in it to the power of the sun. And if you assume that the ascent and the contemplation of the things above is the soul's ascension to the intelligible region, you will not miss my surmise, since that is what you desire to hear. But God knows whether it is true. **But, at any rate, my dream as it appears to me is that in the region of the known the last thing to be seen and hardly seen is the idea of good**, and that when seen it must needs point us to the conclusion that this is indeed the cause for all things of all that is right and beautiful, giving birth in the visible world to light, and the author of light and itself in the intelligible world being the authentic source of truth and reason, and that anyone who is to act wisely in private or public must have caught sight of this."

Glaucon,

"I concur," he said, "so far as I am able."

Socrates,

"Come then," I said, "and join me in this further thought, and do not be surprised that those who have attained to this height are not willing to occupy themselves with the affairs of men, but their souls ever feel the upward urge and the yearning for that sojourn above. For this, I take it, is likely if in this point too the likeness of our image holds"

Glaucon,

"Yes, it is likely."

Socrates,

"And again, do you think it at all strange," said I, "if a man returning from divine contemplations to the petty miseries of men cuts a sorry figure and appears most ridiculous, if, while still blinking through the gloom, and before he has become sufficiently accustomed to the environing darkness, he is compelled in courtrooms or elsewhere to contend about the shadows of justice or the images that cast the shadows and to wrangle in debate about the notions of these things in the minds of those who have never seen justice itself?"

Glaucon,

"It would be by no means strange," he said.

Socrates,

"But a sensible man," I said, "would remember that there are two distinct disturbances of the eyes arising from two causes, according as the shift is from light to darkness or from darkness to light, and, believing that the same thing happens to the soul too, whenever he saw a soul perturbed and unable to discern something, he would not laugh unthinkingly, but would observe whether coming from a brighter life its vision was obscured by the unfamiliar darkness, or whether the passage from the deeper dark of ignorance into a more luminous world and the greater brightness had dazzled its vision. And so he would deem the one happy in its experience and way of life and pity the other, and if it pleased him to laugh at it, his laughter would be less laughable than that at the expense of the soul that had come down from the light above."

Glaucon,

"That is a very fair statement," he said.

Socrates,

"Then, if this is true, our view of these matters must be this, that education is not in reality what some people proclaim it to be in their professions.

What they aver is that they can put true knowledge into a soul that does not possess it, as if they were inserting vision into blind eyes."

Glaucon,

"They do indeed," he said.

Socrates,

> **"But our present argument indicates," said I, "that the true analogy for this indwelling power in the soul and the instrument whereby each of us apprehends is that of an eye that could not be converted to the light from the darkness except by turning the whole body. Even so this organ of knowledge must be turned around from the world of becoming together with the entire soul, like the scene-shifting periaktos act in the theater, until the soul is able to endure the contemplation of essence and the brightest region of being.**
>
> And this, we say, is the good, do we not?"
>
> Glaucon,
>
> "Yes."
>
> Socrates,
>
> **"Of this very thing, then," I said, "there might be an art, an art of the speediest and most effective shifting or conversion of the soul, not an art of producing vision in it, but on the assumption that it possesses vision but does not rightly direct it and does not look where it should, an art of bringing this about."**
>
> Glaucon,
>
> "Yes, that seems likely," he said. "Then the other so-called virtues of the soul do seem akin to those of the body. For it is true that where they do not pre-exist, they are afterwards created by habit and practice. But the excellence of thought, it seems, is certainly of a more divine quality, a thing that never loses its potency, but, according to the direction of its conversion, becomes useful and beneficent,"*

For Plato, the good beyond being would be the Forms but 'being' is in the state of privation. Also, throughout this discussion Plato has an important undercurrent unfolding in the imminent extinction of Athenian democracy. We see this in Plato's reluctant consignment to the almost unescapable fate of those, humanity, fettered and watching shadows on the cave walls thinking that the shadows are reality.

> Picture men dwelling in a sort of subterranean cavern with a long entrance open to the light on its entire width. Conceive them as having their legs and necks fettered from childhood, so that they remain in the same spot, able to look forward only, and prevented by the fetters from turning their heads. Picture further the light from a fire burning higher up and at a distance behind them, and between the fire and the prisoners and above them a road along which a low wall has been built, as the exhibitors of puppet-shows have partitions before the men themselves, above which they show the puppets.†

Humanity is condemned to shadows and images. These shadows taken as reality portray humanity's origin as chaotic, as fundamental privation, which cannot ever rise to overcome itself. On the other

* Plato, *Republic*, 508a–517a, in *Plato in Twelve Volumes*, vols. 5 & 6, trans. Paul Shorey (Cambridge, MA: Harvard University Press, 1969).

See Plato, Republic 509a through Plato, Republic 518e

† Plato, *Republic*, 514a–514b, in *Plato in Twelve Volumes*, vol. 7, trans. Paul Shorey (Cambridge, MA: Harvard University Press, 1969).

See Plato, Republic 509a through Plato, Republic 518e

hand, in my view, Plato first espouses the possibility of intelligibility, rationality as, in some circuitous fashion, having the possibility of understanding or acquiring a glimmer of the Forms. This glimmer of hope might have been filled out more by the later histories from his student Aristotle, Christianity, and rationalism. Aristotle disagreed with Plato's notion of the Forms as they were static. Here is a thought, can we find any similarities between privation and the observer effect, branching into classicism from the quantum wave function? ...just saying...

The Unmoved Mover

Aristotle took as a founding premise that everything moved. He also had the notion of the unmoved mover. On the surface, we may want to think of this as everything moves = actuality and unmoved mover = potentiality but there is more to it than that.

> In Chapter 6, Book L, of the Metaphysics, Aristotle begins a discussion about "substances." One of the substances he describes is that of an "unmoved mover" which, he argues, exists by necessity and is eternal. For something to be eternal, it is neither created nor destroyed, but always has and always will exist. For something to be a substance, it exists by virtue of itself ("kath'auton") in the sense that its existence is not dependent on anything else--it just is. In contrast, Aristotle describes things that have "accidental" existence ("kata symbebekos") whose existence depends and adheres to an underlying subject.*

The previous quote illustrates a problem with many Greek translations. The word translated as "substance" actually comes from Latin (*substantia*) not Greek. So, the phrase tên ousian prôton ti estin Aristotle asks is translated as what is it to be a substance. However, the word ousian comes from the ancient Greek word *ousia* was also incorrectly translated in Latin as essence (*essentia*) or substance (*substantia*). It became a Christian concept in the Latin world as the divine essence of God. However, this transformation obscured the more ancient Greek notion of *ousia (οὐσία)*.

> The term οὐσία is an Ancient Greek noun, formed on the feminine present participle of the verb εἰμί, eimí, meaning "to be, I am", so similar grammatically to the English noun "being". There was no equivalent grammatical formation in Latin, and it was translated as essentia or substantia. Cicero coined essentia and the philosopher Seneca and rhetorician Quintilian used it as equivalent for οὐσία, while Apuleius rendered οὐσία both as essentia or substantia. In order to designate οὐσία, early Christian theologian Tertullian favored the use of substantia over essentia, while Augustine of Hippo and Boethius took the opposite stance, preferring the use of essentia as designation for οὐσία. Some of the most prominent Latin authors, like Hilary of Poitiers, noted that those variants were often being used with different meanings. Some modern authors also suggest that the Ancient Greek term οὐσία is properly translated as essentia (essence), while substantia has a wider spectrum of meanings.†

* Aristotle, *Metaphysics*, 1071b–1072a, trans. Hugh Tredennick (Cambridge, MA: Harvard University Press, 1933). See The Unmoved Mover in Aristotle's Metaphysics

† "Ousia," Wikipedia, last modified January 15, 2024 See Ousia

In Aristotle we see that the ultimate, final good may apply to immovable objects. The final good is for something but also the final good at the end of some action (telos). Aristotle writes in what he entitled "Lambda" but Latin scholars entitled, "Metaphysics 12" tells us,

> *That the final cause may apply to immovable things is shown by the distinction of its meanings. For the final cause is not only "the good for something," but also "the good which is the end of some action." In the latter sense it applies to immovable things, although in the former it does not; and it causes motion as being an object of love, whereas all other things cause motion because they are themselves in motion. Now if a thing is moved, it can be otherwise than it is. Therefore if the actuality of "the heaven" is primary locomotion, then in so far as "the heaven" is moved, in this respect at least it is possible for it to be otherwise; i.e. in respect of place, even if not of substantiality. But since there is something—X—which moves while being itself unmoved, existing actually, X cannot be otherwise in any respect. For the primary kind of change is locomotion, and of locomotion circular locomotion; and this is the motion which X induces. Thus X is necessarily existent; and qua necessary it is good, and is in this sense a first principle. For the necessary has all these meanings: that which is by constraint because it is contrary to impulse; and that without which excellence is impossible; and that which cannot be otherwise, but is absolutely necessary.* [*]

Continuing further actuality, which is Aristotle's unmoved mover is inclusive of thinking but not as concept as one pleasure among many. 'Actuality' here is not ultimate Concept but retains a sense of sensuality, enjoyment, pleasures, and is also likened to *theos* which is translated as 'God'. First of all, *theos* was not capitalized. Capitalization was a Latin and later translation implying the Christian God. *Theos* was not monotheistic until the time of Constantine in Rome. *Theos* is singular for god. *Theoi* is plural for gods. Either form could be masculine or feminine.[†] Aristotle continues in Metaphysics 12,

> Such, then, is the first principle upon which depend the sensible universe and the world of nature. And its life is like the best which we temporarily enjoy. It must be in that state always (which for us is impossible), since its actuality is also pleasure. (And for this reason waking, sensation and thinking are most pleasant, and hopes and memories are pleasant because of them.) Now thinking in itself is concerned with that which is in itself best, and thinking in the highest sense with that which is in the highest sense best.] And thought thinks itself through participation in the object of thought; for it becomes an object of thought by the act of apprehension and thinking, so that thought and the object of thought are the same, because that

---

* Aristotle, *Metaphysics*, 12.1072b, trans. Hugh Tredennick (Cambridge, MA: Harvard University Press, 1933).
See Aristotle, Metaphysics, Book 12
Also ARISTOTLES_CONCEPT_OF_GOD

† See theo-.
From Latin theo- ("god"), combining form of theos ("god"); from Ancient Greek θεό- (*theó-*, "god"), combining form of θεός (*theós*, "god").
Also, θεός
From Proto-Hellenic *tʰehós (whence also Mycenaean Greek (*te-o*)), a thematicization of Proto-Indo-European *dʰéh₁s, from *dʰeh₁- ("to do, to put, to place") + *-s. Cognate with Phrygian δεως (*deōs*, "to the gods"), Old Armenian դիք (*dikʿ*, "pagan gods") and Latin fēriae ("festival days"), fānum ("temple") and fēstus ("festive").
Despite its similarity in form and meaning, the word is not related to Latin deus; the two come from different roots. A true cognate of deus is Ζεύς (*Zeús*).

> which is receptive of the object of thought, i.e. essence, is thought. And it actually functions when it possesses this object. Hence it is actuality rather than potentiality that is held to be the divine possession of rational thought, and its active contemplation is that which is most pleasant and best.If, then, the happiness which God always enjoys is as great as that which we enjoy sometimes, it is marvellous; and if it is greater, this is still more marvellous. Nevertheless it is so. Moreover, life belongs to God. For the actuality of thought is life, and God is that actuality; and the essential actuality of God is life most good and eternal. We hold, then, that God is a living being, eternal, most good; and therefore life and a continuous eternal existence belong to God; for that is what God is. *

Notice the amplification of marvelous in the quote above. Also, note the strangeness of the Greek mind to the translation onto Latin history. The unmoved mover is pure actuality not something potential. Sensibility and sensation are not the five senses, it is pleasant. The good is not theology it is sensuality, pleasure. All this speaks to embodiment not some doxology. Eternal is *aiṓnios* from the Greek word *aiṓn* meaning literally age or eon meaning long-lasting, a fixed time period, lasting for life and by extension age of ages, ages of ages, and the like has been translated eternal but that is probably more from the Latin influence. However, the literal interpretation is primarily meant as previously discussed as a quality of *telos*. What does Aristotle mean by the unmoved mover?

Erwin Sonderegger (1942- present) states that the Latin history of Christianity consistently mistranslated *ἀκίνητον κινοῦν* meaning the first or unmoved mover in Greek to God's knowledge both of himself and the world) or Unmoved Mover as masculine. The Latin preference was given because the Christian God had to be male and first 'substance'.

> It is taken for granted that οὐσία [ousia, being] is both correctly translated with substance and adequately reproduced in its content. But, 'substance' translates ὑπόστασις, not οὐσία, and, it is ὑποκείμενον, subject, that has the function of being the foundation, of being and saying. The translation of Aristotle's οὐσία as 'substance' is not only a wrong translation of the word, but moreover substance does not correspond to what οὐσία means in terms of content either.
>
> The term 'substance' is, as probably all terms, also the first ones, not absolutely given but has a genesis, a history which can be reconstructed. However, the origin of this concept does not lie with Aristotle, even if many think so. I can only give some hints on important stations of the origin of the concept of substance, to go into the details would need much more space. The period of its origin reaches very far, from Hippocrates to the Church Fathers, and influences of quite different kind and origin can be found in it. †

More importantly, the unmoved mover is ***noesis*** *noeseos* or 'awareness of awareness'. In other words, being aware of awareness. From this '**noesis** noeseos **noesis**' would then be 'awareness that is an awareness of awareness'. Often this gets mistranslated as 'thinking that is a thinking of thinking'.

---

* Aristotle, *Metaphysics*, 12.1072b, trans. Hugh Tredennick (Cambridge, MA: Harvard University Press, 1933).
See Aristotle, Metaphysics, Book 12
Also ARISTOTLES_CONCEPT_OF_GOD

† Erwin Sonderegger, "Two Dogmas That Many Readers of Aristotle's Metaphysics Share" (Manuscript, PhilArchive, 2020).
See https://www.academia.edu/44759750/Two_dogmas_that_many_readers_of_Aristotle_s_Metaphysics_share

Could 'awareness that is an awareness of awareness' also mean consciousness? Sonderegger writes in his article, "Aristotle, Metaphysics Λ: Introduction, Translation, Commentary - A Speculative Sketch devoid God" (note: this is a bit of a hard read),

> A second firm conviction claims that Aristotle advocates a theology whose core consists in a *πρῶτον* or *ἀκίνητον κινοῦν*, the "First" or "Unmoved Mover" (with preference in capitals), as the expression is consistently mistranslated masculine. According to the standard interpretation, Aristotle develops a concept of God in Metaphysics Λ with the essential provisions of *νόησις νοήσεως* and actus purus [pure act, the absolute perfection of God], and that the proof of his existence is given. Aristotle asks, one says, in this text for God as the first substance and he wants to give an existence proof of this primary substance.
>
> In the statement
>
> ...*καὶ ἔστιν νόησις νοήσεως νόησις*,
>
> ...and awareness1 is awareness2 of awareness1
>
> [awareness as the structure of the world (awareness1) in which each individual instance of being aware (awareness3) realizes (awareness2) a certain structure in the world or node in a net]
>
> awareness1 means the structure of a world in which each individual instance of being aware (awareness3) realizes (awareness2) a certain node in a net; the result of this process is that both the perceiving and the perceived are. To be in this way means to enter in a noetic framework (**noesis**1 as awareness1), and this provides the speculative answer to the question of what founds becoming.
>
> If ousia [being] is the heading for the question about being then it can not be translated with substance, it must be understood in the many ways Aristotle registers. At the core of Aristotle's reflection is being, Sein, just as it has been the theme beginning with Parmenides and mediated by Plato's Sophist. In the Metaphysics this question does, of course, take many ramifications. That question includes one additional moment not addressed by Aristotle, one not even made explicit by Plato. Instead, Plato dramatized that moment in the character of the guest from Elea. The question about being is a question which is questioning the questioner himself...
>
> My aim is to call into question the suppositions that Met. Λ has anything to do with theology and that god has an important role to play in this text. Since there are some scholars who agree so far, I would also like to deny that this theology can be replaced by a metaphysics of the substance. Rather, I propose a reading of Met. Λ as a speculative sketch about the meaning of being (ousia, Sein).*
>
> For all those interested in this text, the **noesis** noeseos [awareness of awareness] stands at the center of all considerations. Indeed this is a strong common assessment, which I too share. In almost an equal number, on the other hand, are those who concur in understanding this term as denoting 'God's thinking' although this is nowhere in fact stated in the text. This claim is supported only if the explicit theme of chapter 9, the nous [mind], is implicitly identified with God. That identification would entail the problem that the nous is moved by what it thinks, so it is on the wrong side of the systoichia [a spatial gathering of elements]. That should

* See Aristotle, Metaphysics (pdf)

not be the case for God. Almost all interpreters seem to take the **noesis** in the famous sentence, once as intentio prima [primary intention] then as intentio secunda [secondary intention]. In this case, then, it seems to represent a reflective attitude. In fact the term is employed three times. As far as I can tell, this fact has perturbed no one up to now. The whole phrase reads as follows:

Therefore it <: the nous> thinks itself, if it is the strongest <being>, and <its> thinking is the thinking of thinking.

or, in the translation of D. W. Ross:

Therefore it must be of itself that the divine thought thinks (since it is the most excellent of things), and its thinking is a thinking on thinking.

The standard interpretation takes two occurrences as identical (so one of them can be omitted), and the 'is' as defining. So the phrase means: "The 'thinking of the nous' is to be understood as 'thinking of its own thinking' by definition." Neo-scholastic authors and authors close to them without hesitation give to subvene in the term a personal God (see e. g. the translation by D. W. Ross). In that way nous becomes a conscious subject explicitly. Many authors who do not mention these presuppositions share them too. This conviction is widespread not only in neo-scholastic literature but representatives of philological and historical research hold to it too. All these schools or circles effectively prolong the medieval line of thought...

It is worth noting that the proposals for understanding **noesis** noeseos [thought of thought] offered by the interpreters never transcend the line of aporias [an expression of deliberation with oneself regarding uncertainty or doubt as to how to proceed] formulated by Aristotle himself in Met. Λ 9: What does the nous think? Does it think something or nothing? Does it always think only itself or other things? If other things, then always the same things or different things in succession? Does it think of accidental things or of beautiful things? Isn't it better not to think of some things at all? All 'solutions' remain within the frame established by these aporias. Met. Λ has been landed in a dense and almost inextricable net of interpretations, which were not seldom expressed through a questionable vocabulary and with an incorrect translation of to proton kinoun. The task now is to trace and understand the thematic references in Met. Λ to other Aristotelian and Platonic texts and to do so without the dogmatic determinations resulting from the history of reception...

This view of the result to be achieved later on should give the background for that which now could seem arbitrary. In the commentary we will find the reasons for this. We normally think of **noesis** [pure intellectual apprehension, for Plato the highest kind of knowledge or knowledge of the eternal forms or ideas] as a thinking of a subject, of a consciousness, of a person. But we have now to think of a **noesis** without a personal subject endowed with consciousness. How are we to speak about **noesis** without asking "Whose **noesis** is this?" We look at the world and see many things. Why is it that we are able to see something? – We see something because we have already seen what there is to be seen, not in detail but in principle. A particular being, which we have seen, activates a network of associations or an order which has been seen before. By this 'before' is meant the basic distinctions, the basic concepts and values of prevailing opinions which go to forming a world (Grundmeinungen einer Meinungswelt). What is seen before as the world, in this sense: this is the **noesis**1 [awareness]. This

noesis makes possible another **noesis**, which consists in our particular understanding, seeing, recognizing. We see in actuality what has in fact already been seen as a structure, this is the meaning of **noesis** noeseos [thought of thought]. Finally we are brought to consider that it is not **noesis** that is the final point in Met. Λ, but the sense of to be as {ousia← **noesis** →energeia / Doxa}...

The old gods were already toppled by 384 C. E. in the battle over the altar of Victoria, fought between Ambrosius and Symmachus. Efforts in the 20th century to revive the old gods in a modern manner have failed too. Leaving aside the negative stance adopted against K. Kerényi many scholars were only bemused by W. F. Otto's Die Götter Griechenlands (1929), who tried quite rightly to take seriously the Wirklichkeit der Götter. The history of religion is permitted a rather unconcerned positivism as exemplified by M. P. Nilsson in his work...

Despite the dawn of new research in the 19th century the current understanding of Aristotle remains an understanding determined by presuppositions arising during the Middle Ages, its pillars being substance and realism. In its time this was a good appropriate understanding. It helped people to lead a good life, but it became rigid and stereotypical, without connection to our contemporary world and its exigencies, even leaving aside the fact that the historical correctness of that earlier understanding is open to doubtful reservations. If, interpreting Aristotle, we sustain premises deriving from the Middle Ages, then we maintain an obvious anachronism...

We may well ask why there obtains a permanent dispute over a metaphysics of substance the uselessness to our context of which is demonstrable? M. Heidegger alone took the effort to articulate and clarify his own hermeneutical situation in relation to Aristotle in his "Phänomenologische Interpretationen zu Aristoteles."4 There is no comparable effort in the historical or philosophical research on Aristotle. Scholars simply assume the validity of their unreflective and immediate understanding and the critique derived from that. Unsuccessfully did P. Natorp argue on the basis of the composition of the Met. Λ that there is no theology to be discerned there. It seems that his thesis finds an opportunity for revival today. R. Bodéüs, has pointed to some dubious basic issues in the standard interpretation, and H. Lang too resists that interpretation. B. Botter reaffirms this doubtfulness about the theological standard interpretation in her investigations of the use of the words theos and theios...

This opening phrase of Met. Λ is far from being the claim "That is so and so," quite otherwise it is the summary of a long-term discourse, in which have participated Presocratics, Sophists, Plato, the Old Academy. The phrase also denotes the method to be followed in that inquiry: theoria. Sometimes theoria means the same as methodos but in Met. Λ it refers to a special method. The word appears only three times in this text, here at the beginning, Λ 1, 1069a18; then at the speculative high point, Λ 7.19, 1072b24; and again at Λ 8.6, 1073b6. Two references are highly exposed, that at the beginning and the occurrence at Λ 7.19, where, as later in the Nicomachean Ethics K 8, 1178b29–32, it is the name for the highest possible realization of us human beings. The second phrase names the respect in which the theme will be treated. The ousiai, the beings, shall be investigated with respect to ousia, Sein, which is their principle and origin. That corresponds exactly with the aim formulated by Theophrastus, in Met. § 1. Further Theophrastus points out that the theme ousia has become urgent in the context of the question about becoming; and finally he names the same key factor in the

> present issue as Aristotle does, namely, metabole, change.7 Aristotle says that we have to examine the reasons of becoming (Met. Λ 2.2). It has resulted from the previous investigations that being came to be seen as the first cause of becoming. Thus the next question to be asked is: How can being (Sein) be the first cause of movement? In Met. Λ Aristotle gives an answer in three steps. Being (Sein) is primary in becoming and it is the cause of becoming
>
> 1. as actuality (energeia, Met. Λ 6.8), then
> 2. as a moving cause like a goal which is not moved itself (Met. Λ 7), and finally
> 3. as **noesis**, awareness, (Met. Λ 7 und 9).
>
> These answers must not be understood as if they had to be added together. What I called 'three steps' is, in fact, to be understood as a climax which makes it possible to understand by speculation what ousia is. Energeia, proton kinoun akineton, **noesis**, "actuality, first unmoved moving, awareness" are three manners of speaking of the same, namely ousia (being, Sein), under three different aspects. [*]

The Greek phrase "*καὶ ἔστιν νόησις νοήσεως νόησις*" in the previous quote can translate to "and there is an understanding of understanding, an understanding." This intriguing repetition emphasizes the concept of understanding within understanding itself. [†] But I like the way Sonderegger breaks this down. I will come back to this later in this volume as well.

I have been using the term 'unmoved mover' because that is a term handed down from centuries after Aristotle that most readers will be familiar with. However, 'unmoved mover' as God has more to do with Latin Christianity and not with Aristotle's meaning of πρῶτον ("First") or ἀκίνητον κινοῦν ("Unmoved Mover"). What does Aristotle tell us about the unmoved mover? Sonderegger asks the question how can ousia or being get associated with the unmoved mover? Becoming is that actuality of being.

> Aristotle says that we have to examine the reasons of becoming (Met. Λ 2.2). It has resulted from the previous investigations that being came to be seen as the first cause of becoming. Thus the next question to be asked is: How can being (Sein) be the first cause of movement? In Met. Λ Aristotle gives an answer in three steps. Being (Sein) is primary in becoming and it is the cause of becoming
>
> 1. as actuality (energeia, Met. Λ 6.8), then
> 2. as a moving cause like a goal which is not moved itself (Met. Λ 7), and finally
> 3. as **noesis**, awareness, (Met. Λ 7 und 9).
>
> These answers must not be understood as if they had to be added together. What I called 'three steps' is, in fact, to be understood as a climax which makes it possible to understand by speculation what ousia [being, Sein] is. Energeia, proton kinoun akineton [primal unmoved moving], **noesis** [awareness], "actuality, first unmoved moving, awareness" are three manners of speaking of the same, namely ousia (being, Sein), under three different aspects.[‡]

---

* 1 Erwin Sonderegger, "Aristotle, Metaphysics Λ: Introduction, Translation, Commentary—A Speculative Sketch Devoid of God," *PhilArchive*, December 15, 2020, [Insert Page Number].

See Aristotle, Metaphysics Λ Introduction, Translation, Commentary A Speculative Sketch devoid God (pdf)

† See Interpretation Problems in Aristotle's Metaphysics Λ. The Case of the Sentence: καὶ ἔστιν ἡ νόησις νοήσεως νόησις

‡ Erwin Sonderegger, "Aristotle, Metaphysics Λ: Introduction, Translation, Commentary—A Speculative Sketch Devoid of

The unmoved mover was pure actuality. Pure actuality is prior to potentiality (Met. Λ 6.8) for the unmoved mover. The unmoved mover "is totally unmoved and without any potentiality, i.e., pure actuality (Met. Λ 6.8)."* Sonderegger continues,

> Aristotle is – by a long and strong tradition – prisoner of a sort of metaphysics which he never dreamed of. This is the result of an interpretation of some of his texts not much later in the Peripatos, but mostly by the fathers and in Middle Ages. In that period such an interpretation was a good one because it stood in connection with everyday life. But now, in the 20th and 21st century we must be aware of the incongruity and the ideologic distortion of this interpretation. If we read the text in Greek, not in English, German or otherwise, the fact will be obvious. Most of these translations, as good as they may be, are profoundly imbued with the tradition, the mentality, the conceptions of the Middle Ages.†

Just a musing interlude but could wave function collapse and superposition have some similarities with the unmoved mover? Could it be that actuality as a moving cause which is not moved has some similarity to the simultaneity of all possibilities in the quantum wave function which branches with the observer effect into classical awareness? Could there be similarities with what was meant by *ousia* to both individual and historic branching of the wave function's narratives, as ontic, individual narratives of my being, and ontological, historic narratives of Being. Even more so, the observer effect actualizing all simultaneous potentialities into many-world confluences is spookily contrasted with the oddity of the quantum wave function which 'sees' no branching but simply the simultaneous potentials which already included the observer effect. Maybe more simply, the potentiality of the quantum wave function collapses, branches as actuality. In its collapse it moves but in superposition it does not move. Could the wave function then be thought of as a kind of unmoved mover? If so, for Aristotle this then is the good. Could the observer effect be similar to Aristotle's *ousai*, awareness. There is also a long history of debate about whether or not the quantum wave function is ontological or epistemological. In other words, is it real or a vehicle of knowledge? It seems to have both in a spooky sort of way. Could it be that reflective knowing in experience effectively as the quantum wave function, is a kind of awareness of awareness, self-awareness? Remember from the previous quote,

> ...καὶ ἔστιν νόησις νοήσεως νόησις,
>
> ...and awareness1 is awareness2 of awareness1
>
> [awareness as the structure of the world (awareness1) in which each individual instance of being aware (awareness3) realizes (awareness2) a certain structure in the world or node in a net]
>
> awareness1 means the structure of a world in which each individual instance of being aware (awareness3) realizes (awareness2) a certain node in a net; the result of this process is that both the perceiving and the perceived are. To be in this way means to enter in a noetic frame-

God," *PhilArchive*, December 15, 2020.

See Aristotle, Metaphysics (pdf)

* See Aristotle, Metaphysics (pdf), page 28

† Erwin Sonderegger, "Aristotle, Metaphysics Λ: Introduction, Translation, Commentary—A Speculative Sketch Devoid of God," *PhilArchive*, December 15, 2020.

See Erwin Sonderegger

work (**noesis**1 as awareness1), and this provides the speculative answer to the question of what founds becoming. *

Could it be that,

- awareness as the structure of the world (awareness1) ≅ Quantum wave function
- in which each individual instance of being aware (awareness3) ≅ observer effect, branching or collapse without specifying how the observer effect comes about, e.g., a person, spontaneous, etc.
- realizes (awareness2) ≅ consciousness whether it be human, animal, plant, alien, etc.

Edmund Husserl (1859-1938) is considered the founder of modern phenomenology. I will write more about him later but for now, notice that two very important concepts of Husserl are ***noesis*** and ***noema***.

> Husserl introduced two Greek words to capture his version of the Bolzanoan distinction: **noesis** and **noema**, from the Greek verb noéō (νοέω), meaning to perceive, think, intend, whence the noun nous or mind. The intentional process of consciousness is called **noesis**, while its ideal content is called **noema**. The **noema** of an act of consciousness Husserl characterized both as an ideal meaning and as "the object as intended". Thus the phenomenon, or object-as-it-appears, becomes the **noema**, or object-as-it-is-intended. The interpretations of Husserl's theory of **noema** have been several and amount to different developments of Husserl's basic theory of intentionality. (Is the **noema** an aspect of the object intended, or rather a medium of intention?) †

Husserl is not implying that the content of thought, **noema**, is 'objective' and the 'intentionalizing' act, **noesis,** is not mere subjectivity. He is telling us that these kinds of boundary conditions are abstract and artificial. Think of it analogously more like the wavicle which from the observer effect is a particle, the content of thought, in a field of quantum waves which is more like a smorgasbord of infinite possibilities as in all possibilities in superposition. Aristotle's unmoved mover is more akin to this than the Latin transformation of an omnipresent, omniscient, omnipotent God who stands behind the scenes intervening in actualities in human or cosmic affairs. Later scholastic Latin transformations of Aristotle would associate Aristotle's unmoved mover with the monotheistic God which I think was a massive transformation of Aristotle. It seems to me that Husserl, in keeping with the early 20th century findings of Einstein's relativity found a philosophical way to rethink Aristotle's notion of 'pure actuality' in a more authentic and less abstract fashion. Actuality is not the abstract notion of simple movement. Lived actuality is not sheer 'movement'. Lived actuality is always as an intention towards

* Erwin Sonderegger, "Aristotle, Metaphysics Λ: Introduction, Translation, Commentary—A Speculative Sketch Devoid of God," *PhilArchive*, December 15, 2020.

† David Woodruff Smith, "Phenomenology," in *The Stanford Encyclopedia of Philosophy*, ed. Edward N. Zalta (Fall 2021 Edition), sec. 1.

See Phenomenology

<fill in the blank>. Lived actuality intends towards some object, some goal, some perception. Even as sheer hallucinations, Husserl would still find this dynamic at work.

In Aristotle's use of the word *energeia* there is the idea of propulsion, some actional intention. Additionally, the intention 'towards'... has some goal, some object, some idea of where it is directed – can we harken back to the discussion on Aristotle's *telos*? In the lived action of 'intention towards' the goal is not moved. Sure, it can change to another goal but then that goal becomes the unmoved mover. The 'intention towards' is ***noesis***. The goal is ***noema***. Sonderegger after Heidegger tells us the goal or object of intention is direct towards Being, towards what 'is'. But Husserl later will tell us that sometimes 'intention towards' can be mistaken in its goal. If we are walking on a path and see a snake, we jump away, only to find that it was a stick meaning we had an intention to not get bitten by a snake but then the object turned out to be something different than we intended. Intention towards <fill in the blank> also intends a certain lived temporality and spatiality, a certain quality of existence or Being that is teleological. But the lived practicality of the intention does not make it so, make it absolute, make it true. So, for Husserl phenomenology was about bracketing our beliefs from acquired habitual intentions and simply opening up the possibility that the objects of our intentions may not 'be' the truths we assume them to 'be'.

To recap,

Socrates tells us,

> it is right to deem light and vision sunlike, but never to think that they are the sun, so here it is right to consider these two their counterparts, as being like the good or boniform, but to think that either of them is the good is not right. Still higher honor belongs to the possession and habit of the good.
>
> Glaucon responds,
>
> "An inconceivable beauty you speak of," he said, "if it is the source of knowledge and truth, and yet itself surpasses them in beauty. For you surely cannot mean that it is pleasure."
>
> Socrates replies,
>
> "Hush," said I, "but examine "In like manner, then, you are to say that the objects of knowledge not only receive from the presence of the good their being known, but their very existence and essence is derived to them from it, though **the good itself is not essence** but still transcends essence in dignity and surpassing power."
>
> Glaucon responds,
>
> And Glaucon very ludicrously said, "Heaven save us, hyperbole can no further go."
>
> Socrates replies,
>
> "The fault is yours," I said, "for compelling me to utter my thoughts about it."
>
> This is Socratic humor.
>
> Finally, Socrates replies,
>
> "This image then, dear Glaucon, we must apply as a whole to all that has been said, likening the region revealed through sight to the habitation of the prison, and the light of the fire in it to the power of the sun. And if you assume that the ascent and the contemplation of the things above is the soul's ascension to the intelligible region, you will not miss my surmise, since that is what you desire to hear. But God knows whether it is true. But, at any rate, my dream as it appears to me is that in the region of the known the last thing to be seen and hardly

> seen is the idea of good, and that when seen it must needs point us to the conclusion that this is indeed the cause for all things of all that is right and beautiful, giving birth in the visible world to light, and the author of light and itself in the intelligible world being the authentic source of truth and reason, and that anyone who is to act wisely in private or public must have caught sight of this." *

Here 'the good' is that 'seeing' must point us to the conclusion that there must be a cause for all things, a cause of all that is right and beautiful, a giving birth in the visible world to light which is the author of light and the author for the intelligible world and being the authentic source of truth and reason, and anyone who is to act wisely in private or public must have caught sight of this. The good is what gives birth to light. The good transcends existence and essence as objects of knowledge not only receive from the presence of the good, their being known, but objects' very existence and essence are derived from the good, though the good itself is not essence but still transcends essence in dignity and surpassing power. For Socrates the "cause of all things" which gives birth to light and the visible world is 'the good' beyond Being and what Plato calls the Forms.

Note here that light and presence has not yet overtaken the good as privation. This means that the good is not only light and presence but also the absence of light and what does not come into presence. Privation still reflects the unknowing of the earlier Ionian's unlimited (*aperion*), flux, and chaos. Also, note that soon after with Aristotle *aperion* begins to disassociate affirmative older ideas of Greek flux and chaos in associations of more negative connotations of logical negation and contradiction.

> What sort of operation is negation? In the Categories and De Interpretatione, Aristotle partitions indicative-mood declarative sentences into affirmation and negation/denial (apophasis from apophanein "deny, say no"), which respectively affirm or deny something about something. As a mode of predication, the predicate denial of Aristotelian term logic, while resulting in wide-scope negation opposed in truth value to the corresponding affirmative, is syntactically distinct from the unary "it is not the case that" connective of Stoic and Fregean logic.†

In a much earlier period of the Greeks, in Hesiod, we have Uranus (father sky) and Gaia (mother earth). Sky suspends, stands off, provides perspective. Sky is the son and husband of earth. Therefore, earth is generative. As the first of the gods, Earth is yet to be differentiated, it is undifferentiated. Earth is the origin of sky. Thus, Earth is *arche* or origin. In Hesiod, Earth was the first of the gods. Yet, Hesiod's Muses tells us that first of all was *khaos*, chaos. Chaos means the 'yawning gap', a void. Thus, chaos differentiates and separates (the heaven and the earth). Earth is permeated through and through with chaos, undifferentiated but fertile and generative.‡

But later and increasingly, *peras* or form has been transformed from boundary or limit in the sea of unlimited (*aperion* -> unlimited, without form, void, chaos filling privation) towards conditions of negation and noncontradiction. Almost unnoticed, the 'good beyond being' will eventually be driven

---

* Plato, *The Republic*, trans. Paul Shorey (Cambridge, MA: Harvard University Press, 1937), 509a–517c.

† Laurence R. Horn and Heinrich Wessel, "Negation," in *The Stanford Encyclopedia of Philosophy*, ed. Edward N. Zalta (Spring 2022 Edition), sec. 1.1.

See Negation

‡ See The Work of Days (revisited)

towards an ethics (*ethos*) of logic which becomes the foundation of rationalism. The good has become the greater good determined by rationality, by ratio and proportion. Additionally, Being as emerging from presence equates to existence. Uncertainty moves towards certainty. Note that ethics has been understood in terms of presence as what shows itself from a history which already displaces and without notice assumes place of the other as Being, as already understood, in terms of conditions set by the history of rationalism.

> In Eros we can begin to understand that classic Greek thinking is not merely the transition from mythos to logos. The Greeks understood and puzzled deeply over emotion, simple sensations, love, concern, affection, sexual intimacy. When they thought of privation, withdrawal, infinite deeps of Darkness and deep Abyss with dark Chaos they assumed Love or Desire as a first principle in things. The distance with which much of modern scholarship has approached classic antiquity reflects more on the 'transformation of the interpretation' than the infinite proximity of the texts which remain and in doing so has denied the excess of proximity to origin, my origin and the chaotic gap, the disruption of origin (which in nowise should be leveled off into mere neutrality, *il y a*). *

The inability of Greek mythos to resolve itself sets up a perpetual strife in the early Greek *psuchê* or psyche. The lack of cosmogonic resolve in Hesiod's earliest moment of chaos reflects an unsettled disruption that fuels the inquiry of *phusis*, drama, sculpture, and poetry which was inspired by an unaccounted-for excess not in affirmation but in the richness of lack, privation (*steresis*) unsettled by the twilight of chaos. Privation could not yet find a finality in negation but overflowed as contingent desire *orexis* † and haunted us in what later became the passion play of metaphysics and still haunts us even to this day.

Interesting to note that boundary conditions abound in quantum mechanics not as barriers but as truncations of infinites which highlight, draw a circle on the face of infinites, a particular *telos*, an experimental goal of observation. As quoted previously, Matsuno writes,

> Thus subsequent internal boundary conditions (of which initial conditions are only a special case) are inseparable from the internal quantum dynamics. Boundary conditions, well-known in Newtonian mechanics, are thus an integral part of the internal quantum dynamics. The "observers" are, in this way, internally specified and fused with the dynamics. ‡

Truncations, approximations, limits are not absolutes but contingents of observation. This is the yes/no of Wheeler's 'it from bit' which obliges astute observations but owns no necessity to judge-

* William Desmond, *The Eros and the Ethos: Morphologies of Desire* (Charlottesville: University of Virginia Press, 2024), 16. See Eros

† From Latin orexis ("longing; appetite"), from Ancient Greek ὄρεξις (*órexis*, "desire"), from ὀρέγω (*orégō*, "I reach, stretch"). See orexis

‡ Thomas J. McFarlane, "The Quantum Model of Positive Emptiness," in *Center Voice: The Newsletter of the Center for Sacred Sciences* (Fall-Winter 1995), 1.

See The Internalist Stance: A Linguistic Practice Enclosing Dynamics

ments derived thereof. For Plato Form has no boundary as *apeiron*, the excess of 'not', of shadows on our cave walls.

For Plato, language can only show origin based on privation, its absolute inability to be able as the good, as love. This is not negation as we think in the modern sense. It is the highest virtue and service language provides us. We can know what is not as in the limit of which *apeiron* refers in its 'a' of privation to the Forms. Limit certainly plays a role in *apeiron* but only to make way for what it cannot be, cannot come into existence. 'What the Forms cannot be from *steresis* (privation) is what being as such cannot be. It is what 'Idea' cannot think. It is an excess which cannot be neuter, cannot be extinguished in light, reason and thought. For Plato, the good can only be faced in privation. Privation exceeds the 'not' of interiority. Privation ever flows towards exteriority which finds no footing in interiority and its exclusive or inclusive middle of logic. For Levinas, exteriority cannot maintain this bipolar duality. In attributing the good as love we have supplemented exteriority with the human face; the he or she who faces us in their radical exteriority.*

Additionally, note here that falsity also remains in itself as a deprived mode of truth often erroneously taken as an opposite. However, falsity can only *be* what it *is* as a contrary distinction to truth, as a circle drawn on the face of infinites which connote an inside and an outside. Falsity is essentially dependent on truth to be what it is just as contradiction is necessarily related to identity and truth, non-contradiction – dialectical opposition. This also means that truth can only assert itself as what it isn't in its falsity. Therefore, truth and falsity are both known as such in privation, a lack of what they are when taken as mere opposites, shadows on the cave walls. For Plato, for the forms to be what they are, to remain in themselves as separate and essentially dependent, they must remain as privation, as excess which cannot be gathered into presence and absence. In the historic dialectic which follows the Greeks, light is only known through darkness, good is only known through bad, etc. Only by the essential 'not' and 'knot' of binary oppositions can Truth make its eternal claim. Truth remains in logos as an essential condition for sight, for logic to assert its priority in its isolation or perhaps clearing for later Heidegger. Specifically, the essential condition for sight, for logic, we refer to as being or ontology, as the essential pillar of language which must erase its trace of/as essence. This is found in the copula which founds all possibilities of thinking, writing, and the repetitious 'foundings' of narratives which long ago lost any trace of an anarchic excess to origin. Ontology in this case is what sifts and retains identity from Hesiod's random chaos (χάος), what allows itself as 'must' and 'can' only showing itself in the clearing of light.† The good as other than Being, existence, comprehension, logic and rationality from the early Greeks long ago left excess, uncertainty, the without of deterministic causality. What other possibilities can we think of in Aristotle's non-being as privation? Could we think a "passivity prior to receptivity" thought as non being?‡ Richard Cohen, a Levinasian scholar, writing about origin states:

> How can what has not yet begun be continued? Can there be a beginning prior to the origin? If being has an origin, as philosophy has always maintained (regardless of whether philosophy can or cannot discover that origin), then what is the sense of "is" when one says that "there is a

* See The Problem of Metaphysics

† See The Impossible Possibility of Paradox – Part One

‡ See On Origin

> beginning prior to the origin"? Such a beginning would already be a challenge to the firstness of first philosophy, would already be a challenge to the "is" that attempts to discover its origin, of itself, by itself, courageously, invoking all of history and nature if that is what it takes to be free of outside help.*

The question of origin was a primary concern for the ancient Greeks and confluences from many other ancient civilizations. † This was long before Latin Christianity and 'metaphysics' in Roman Constantinople. Aristotle's works called 'Metaphysics' by early Latin writers is quite misleading to the ancient Greek mind.

> The word 'metaphysics' is notoriously hard to define. Twentieth-century coinages like 'meta-language' and 'metaphilosophy' encourage the impression that metaphysics is a study that somehow "goes beyond" physics, a study devoted to matters that transcend the mundane concerns of Newton and Einstein and Heisenberg. This impression is mistaken. The word 'metaphysics' is derived from a collective title of the fourteen books by Aristotle that we currently think of as making up "Aristotle's Metaphysics." Aristotle himself did not know the word. (He had four names for the branch of philosophy that is the subject-matter of Metaphysics: 'first philosophy', 'first science', 'wisdom', and 'theology'.) At least one hundred years after Aristotle's death, an editor of his works (in all probability, Andronicus of Rhodes) entitled those fourteen books "Ta meta ta phusika"—"the after the physicals" or "the ones after the physical ones"—, the "physical ones" being the books contained in what we now call Aristotle's Physics. The title was probably meant to warn students of Aristotle's philosophy that they should attempt Metaphysics only after they had mastered "the physical ones," the books about nature or the natural world—that is to say, about change, for change is the defining feature of the natural world.‡

Aristotle's investigations into physics were an essential part of his overall study of *phusis,* part of

---

* Richard A. Cohen, *Ethics, Exegesis, and Philosophy: Interpretation After Levinas* (Cambridge: Cambridge University Press, 2001), 1.

Philosophy Today
Volume 32, Issue 2, Summer 1988
Richard A. Cohen
Pages 165-178
DOI: 10.5840/philtoday198832222
"LEVINAS, ROSENZWEIG, AND THE PHENOMENOLOGIES OF HUSSERL AND HEIDEGGER"
Richard A. Cohen

† See Philosophy Series 8 | Musings (mixermuse.com)

‡ Peter van Inwagen, "Metaphysics," in *The Stanford Encyclopedia of Philosophy*, ed. Edward N. Zalta (Winter 2020 Edition), sec. 1.

See Aristotle Metaphysics

The first major work in the history of philosophy to bear the title "Metaphysics" was the treatise by Aristotle that we have come to know by that name. But Aristotle himself did not use that title or even describe his field of study as 'metaphysics'; the name was evidently coined by the first century C.E. editor who assembled the treatise we know as Aristotle's Metaphysics out of various smaller selections of Aristotle's works. The title 'metaphysics'—literally, 'after the Physics'—very likely indicated the place the topics discussed therein were intended to occupy in the philosophical curriculum. They were to be studied after the treatises dealing with nature (ta phusika). In this entry, we discuss the ideas that are developed in Aristotle's treatise.

See Metaphysics

which we now call physics and metaphysics in a reductionary, contrarian fashion. An excess implied in ancient Greek thought to *phusis* will be hinted at in this volume. This excess is also hinted at in Andronicus of Rhodes description of Aristotle's works as *"the after the physicals" or "the ones after the physical ones"*. Here the 'after' is thought in terms of what does not show itself as physical but is the completion or fulfillment of the physical (*telos*). In the Greek mode of thinking the completion or fulfillment was the beginning and end, the final purpose or goal, and what we might think as what makes the physical, actuality, possible - emphasizing the necessity of origin. Even now physics has pursued Aristotle's notion of origin as cosmology.

The Big Bang, Big Bounce, and Big Crunch asks after the origin and Aristotle's concern. Even evolution looks for hints about the origin of life and the conditions which made it possible. The more proper notions of metaphysics in Latin Christianity such as God, Genesis, heaven, and hell were all irrelevant and unknown to Aristotle's inquiry. Much of my "Philosophy Series" has to do with the ancient Greek inquiries into origin.* In this discussion I want to continue towards the notion of origin in terms of recent findings in quantum physics. One important distinction I want to bring out in this volume is that origin necessarily implies cause and effect. For Aristotle movement or motion requires potentiality and actuality or cause translated to effect. Aristotle tells us in Metaphysics,

> It is the principles and causes of the things which are that we are seeking; and clearly of the things which are qua [as] being. There is a cause of health and physical fitness; and mathematics has principles and elements and causes; and in general every intellectual science or science which involves intellect deals with causes and principles, more or less exactly or simply considered. But all these sciences single out some existent thing or class, and concern themselves with that; not with Being unqualified, nor qua Being, nor do they give any account of the essence ['first philosophy', 'first science', 'wisdom', and 'theology']; but starting from it, some making it clear to perception, and others assuming it as a hypothesis, they demonstrate, more or less cogently, the essential attributes of the class with which they are dealing. Hence obviously there is no demonstration of substance [being] or essence [first principle] from this method of approach, but some other means of exhibiting it. And similarly they say nothing as to whether the class of objects with which they are concerned exists or not; because the demonstration of its essence and that of its existence belong to the same intellectual process. And since physical science also happens to deal with a genus of Being (for it deals with the sort of substance which contains in itself the principle of motion and rest), obviously it is neither a practical nor a productive science. For in the case of things produced the principle of motion (either mind or art or some kind of potency) is in the producer, and in the case of things done the will is the agent—for the thing done and the thing willed are the same. Thus if every intellectual activity is either practical or productive or speculative, physics will be a speculative science; but speculative about that kind of Being which can be moved, and about formulated substance for the most part only qua inseparable from matter. But we must not fail to observe how the essence and the formula exist, since without this our inquiry is ineffectual.†

---

* See A Detour of Time, The Origin, An Interlude to Anaximander, On the Way to Anaximander: Language and Proximity, On Origin.

† Aristotle, *Metaphysics*, trans. Hugh Tredennick (Cambridge, MA: Harvard University Press, 1933), 1025b.
See Aristotle, Metaphysics 1025b

First of all Chaos came-to-be

Here, Aristotle is asking the question of essence, *ousia,* which has haunted humans from our earliest beginnings, "What is the origin". But in thinking *ousia* as Aristotle does, we still must think it as being, as boundary, not as later Latin metaphysics of essence. Unfortunately, origin has been subducted to also mean 'by authority' and 'by rule' as a signet and seal of legitimacy. Through subsequent histories boundary implies binary totalities, absolute certainties, and all the mechanisms from which power originates and maintains itself. However, from the earliest Greek thinkers we find Hesiod in 700 BC telling us in the Theogony,

> Hail, children of Zeus! Grant lovely song and celebrate the holy race of the deathless gods who are for ever, those that were born of Earth and starry Heaven and gloomy Night and them that briny Sea did rear. Tell how at the first gods and earth came to be, and rivers, and the boundless sea with its raging swell, and the gleaming stars, and the wide heaven above, and the gods who were born of them, givers of good things, and how they divided their wealth, and how they shared their honors amongst them, and also how at the first they took many-folded Olympus3. Tell me all of this, you Muses who have your homes on Olympus, from the beginning [archê, ἀρχῆς], tell who first of them (the gods) came-to-be [genet', γένετ'].
>
> First of all Chaos came-to-be [genet', γένετ']; but then afterwards Broad-breasted earth, a secure dwelling place forever for all [the immortals who hold the peaks of snowy Olympus], and misty Tartara in the depths under the wide-wayed grounds and Eros who, handsomest among the deathless gods a looser of limbs, in all the gods and in all human beings overpowers in their breasts their intelligence and careful planning. And from Chaos came-to-be both Erebos [ρεβος, the god of deep darkness, shadow] and dark night, and from night, in turn, came-to-be both Aither [the god of upper air, the mist of bright, glowing light, home of the gods] and day, whom she conceived and bore after joining in love with Erebos. But earth first begat, as an equal to herself, starry sky, so that he might cover her on all sides, in order to be a secure dwelling place forever for all the blessed gods, and she begat the tall mountains, pleasing haunts of the goddess-nymphs who make their homes in the forested hills, and also she bore the barren main with its raging swell, the sea, all without any sweet act of love; but then next, having lain with sky, she bore deep-swirling ocean" *

Chaos, better translated as yawning gap, was before the gods. Origin is not God or gods. It is not any kind of induction of authority or rule. Origin as chaos is gap which cannot be bridged, given footing in history and language. Authenticity and sovereignty find no shelter or retreat to/in/as, origin. In this sense our notions of origin are not original meaning what dominated early Greek thought and other ancient cultures until later, more prevalent notions, in later thinkers perhaps barely beginning in Aristotle. Moreso, origin as authority or rule was not the notion Aristotle was articulating in the books of 'Metaphysics'. It was anarchic with regard to rule or authority. Greek thinkers shortly after Hesiod would call this *apeiron* (ἄπειρον) meaning unlimited, boundless, infinite, indeterminate. *Apeiron* is the negation (ἀ) of *peirar* (πεῖραρ) or end, limit, boundary. *Apeiron* is more like the fertile void of

* Hesiod, *Theogony*, trans. Hugh G. Evelyn-White (Cambridge, MA: Harvard University Press, 1914), lines 104–133.

Buddhism denoted by *Śūnyatā*. While the term origin has been subducted by language and history as order, rule, God. etc., origin in the most ancient Greek thought was permeated with tones of what we think as anarchic, as chaos, yawning gap, apeiron. Many ancient civilizations including the Hebrew account of Genesis which states "the earth was without form and void"* have also echoed this contrarian view from popular religious and secular opinions. Contemporaneously, if origin was the Big Bang, what came before that?

If origin was the Big Bounce where the expansion and subsequent compression of the universe results in another Big Bang, what came before all the bounces? According to the earliest ancient Greeks and many cosmological myths, the cosmos finds its beginning and end, its form, boundary, and limit in indeterminacy. Cause and effect must have a chain of beginnings and ends from which determinacy is achieved. However, the question of cause and effect necessarily implies something of a vastly different kind or order than an endless chain of causal relations. Aristotle recognizes this as the unmoved mover *not* as God. The question of cause and effect necessarily requires a notion of origin which has no external determinacy or dependence, no beginnings before the beginning, no prior cause meaning not subject to time or space. Classical physics is absolutely deterministic with regard to cause and effect.

As Aristotle deduced, any physics which asks the question of origin, cause and effect, and motion must necessarily require something very different with regard to its essence (*ousia*) which implies it cannot be subject to time and space or an endless chain of cause and effect which for Aristotle was the unmoved mover – now we can state, it requires indeterminacy, quantum mechanics. The idea of eternal cause and effect does not seem to suffice for the essential terms required by the question of origin. Origin is a Gordian Knot which cannot be untied, a riddle which cannot be solved. Aristotle continues his inquiry in book 12 of Metaphysics,

> That the final cause may apply to immovable things is shown by the distinction of its meanings. For the final cause is not only "the good for something," but also "the good which is the end of some action." In the latter sense it applies to immovable things, although in the former it does not; and it causes motion as being an object of love, whereas all other things cause motion because they are themselves in motion. Now if a thing is moved, it can be otherwise than it is. Therefore if the actuality of "the heaven" is primary locomotion, then in so far as "the heaven" is moved, in this respect at least it is possible for it to be otherwise; i.e. in respect of place, even if not of substantiality. But since there is something—X—which moves while being itself unmoved, existing actually, X cannot be otherwise in any respect. For the primary kind of change is locomotion, and of locomotion circular locomotion; and this is the motion which X induces. Thus X is necessarily existent; and qua necessary it is good, and is in this sense a first principle. For the necessary has all these meanings: that which is by constraint because it is contrary to impulse; and that without which excellence is impossible; and that which cannot be otherwise, but is absolutely necessary. †

As Aristotle deduced, any physics which asks the question of origin, cause and effect, and motion

* See Genesis 1:2

† Aristotle, *Metaphysics*, trans. Hugh Tredennick (Cambridge, MA: Harvard University Press, 1933), 1072b. See Aristotle, Metaphysics 12.1072b

must necessarily require a different kind of *ousia* meaning indeterminacy with regard to *ousia* as causal. This implies it cannot be subject to time and space or an endless chain of cause and effect. Before I go any further, let's clarify the terms 'essence' and indeterminacy in the previous sentence.

By 'indeterminate' I mean not finite or mutable, a product of change or for Aristotle, motion. Mutability implies change which implies determinacy and cause. However, for Aristotle "*motion cannot be either generated or destroyed, for it always existed; nor can time, because there can be no priority or posteriority if there is no time.*" Aristotle tells us in "Metaphysics" Book 12,

> Thus we have stated what the principles of sensible things are, and how many they are, and in what sense they are the same and in what sense different.
>
> Since we have seen that there are three kinds of substance, two of which are natural and one immutable, we must now discuss the last named and show that there must be some substance which is eternal and immutable. Substances are the primary reality, and if they are all perishable, everything is perishable. But motion cannot be either generated or destroyed, for it always existed; nor can time, because there can be no priority or posteriority if there is no time. Hence as time is continuous, so too is motion; for time is either identical with motion or an affection of it. But there is no continuous motion except that which is spatial, of spatial motion only that which is circular. [*]

And again,

> That the final cause may apply to immovable things is shown by the distinction of its meanings. For the final cause is not only "the good **for something**," but also "the good which is **the end of some action**." In the latter sense it applies to immovable things, although in the former it does not; and it **causes motion as being an object of love**, whereas all other things cause motion because they are themselves in motion. Now if a thing is moved, it can be otherwise than it is. Therefore if the actuality of "the heaven" is primary locomotion, then in so far as "the heaven" is moved, in this respect at least it is possible for it to be otherwise; i.e. in respect of place, even if not of substantiality. But since there is something—X—which moves while being itself unmoved, existing actually, X cannot be otherwise in any respect. For the primary kind of change is locomotion,[1] and of locomotion circular locomotion; and this is the motion which X induces. Thus X is necessarily existent; and qua necessary it is good, and is in this sense a first principle. For the necessary has all these meanings: that which is by constraint because it is contrary to impulse; and that without which excellence is impossible; and that which cannot be otherwise, but is absolutely necessary.[†]

By the previous use of my term 'indeterminacy' I am referring to Aristotle's argument here where he makes the case that cause and effect cannot be a sufficient explanation in itself. In our colloquialism we might think what came first, the chicken or the egg? Aristotle's unmoved mover is the first cause

* Aristotle, *Metaphysics*, trans. Hugh Tredennick (Cambridge, MA: Harvard University Press, 1933), 1071b.

† Aristotle, *Metaphysics*, trans. Hugh Tredennick (Cambridge, MA: Harvard University Press, 1933), 1072b. See Aristotle, Metaphysics 12.1071b through Aristotle, Metaphysics 12.1072b

(genesis as uncaused, as anarchy exceeding the 'not' of causal origins). Aristotle was fully aware that his unmoved mover was more of what we now would call a 'theological' issue.

> In Book E, Aristotle adds another description to the study of the causes and principles of beings qua beings. Whereas natural science studies objects that are material and subject to change, and mathematics studies objects that although not subject to change are nevertheless not separate from (i.e., independent of) matter, there is still room for a science that studies things (if indeed there are any) that are eternal, not subject to change, and independent of matter. Such a science, he says, is theology, and this is the "first" and "highest" science. Aristotle's identification of theology, so conceived, with the study of being qua being has proved challenging to his interpreters.*

Let's recall the ancient Greek history which Aristotle was fully aware of from Hesiod to Anaximander and many others before him concerning chaos and night in Greek mythology. Further in the previous section I quoted he tells us in his Metaphysics,

> Therefore Chaos or Night did not endure for an unlimited time, but the same things have always existed, either passing through a cycle or in accordance with some other principle—that is, if actuality is prior to potentiality. Now if there is a regular cycle, there must be something[3] which remains always active in the same way; but if there is to be generation and destruction, there must be something else[4] which is always active in two different ways. Therefore this must be active in one way independently, and in the other in virtue of something else, i.e. either of some third active principle or of the first. It must, then, be in virtue of the first; for this is in turn the cause both of the third and of the second. Therefore the first is preferable, since it was the cause of perpetual regular motion, and something else was the cause of variety; and obviously both together make up the cause of perpetual variety. Now this is just what actually characterizes motions; therefore why need we seek any further principles?†

It seems Aristotle was content to leave the matter of the unmoved mover as not in time or space and not the cause of anything else. Later history would define the notion of 'not in time and space' and 'uncaused' as eternal. However, the early Greeks did not have the idea of 'eternity' as we do. The word in Greek was *αἰών* which in this context in the earlier Greek period meant an age or a limited span of time. In the later classical and late antique periods [3rd to 7th century AD] it may have taken on the metaphysical idea of eternity or at least ages of ages.‡

Can we associate these notions with the quantum wave function, at least perhaps in some whimsical fashion? The wave function is not static, is not absolute, determinate, is not origin and not even that 'not'. It moves but is not moved upon. It exceeds from all its branchings not from Frankenstein's

* S. Marc Cohen and C. Reeve, "Aristotle's Metaphysics," in *The Stanford Encyclopedia of Philosophy*, ed. Edward N. Zalta (Winter 2021 Edition), sec. 3.
See Aristotle Metaphysics

† Aristotle, *Metaphysics*, trans. Hugh Tredennick (Cambridge, MA: Harvard University Press, 1933), 1072a.
See Aristotle Metaphysics, 12.1072a

‡ See Ancient Greek

graveyards of body parts but from an excess which moves everything but is not moved by anything. The wave function as pure actuality becomes potentiality for the observer which actuates into classical narratives, as *ousia*. *Ousia* as Being is the culmination of space and time, of worlding, not the mechanics of body parts which can be alchemically transformed onto life. Let's focus more on temporality, not our contemporaneous idea of temporality but a much earlier notion of temporality, as eon, as age.

For early Greeks, an age was a specific duration of time which had a beginning and an end. It could also be used for very short durations of time like a day or a lifetime. The Greeks used many ideas of time and ages moving in phrases like 'ages of ages' which could be more like epochs. It could also be used for an indefinite period of time.

> According to scholar David Bentley Hart: "Much depends, naturally, on how content one is to see the Greek adjective *αιωνιον*, *aionios*, rendered simply and flatly as "eternal" or "everlasting." It is, after all, a word whose ambiguity has been noted since the earliest centuries of the church. Certainly the noun αἰών, aion (or aeon), from which it is derived, did come during the classical and late antique periods to refer on occasion to a period of endless or at least indeterminate duration; but that was never its most literal acceptation. Throughout the whole of ancient and late antique Greek literature, an "aeon" was most properly an "age," which is simply to say a "substantial period of time" or an "extended interval."*
>
> Here are other early meanings of *αἰών*,
>
> In early Greek αἰών means 'life' (often in the sense of 'vital force'), 'whole lifetime', 'generation'. It was perhaps through application to the *kosmos*, the lifetime of which is never-ending, that the word [later] acquired the sense of eternity.†
>
> On essence...
>
> Aristotle turns in Z.4 to a consideration of the next candidate for substance: essence. ('Essence' is the standard English translation of Aristotle's curious phrase to ti ên einai, literally "the what it was to be" for a thing. This phrase so boggled his Roman translators that they coined the word essentia to render the entire phrase, and it is from this Latin word that ours derives. Aristotle also sometimes uses the shorter phrase to ti esti, literally "the what it is," for approximately the same idea.) In his logical works, Aristotle links the notion of essence to that of definition (horismos)—"a definition is an account (logos) that signifies an essence" (Topics 102a3)—and he links both of these notions to a certain kind of per se predication (kath' hauto, literally, "in respect of itself," or "intrinsically")—"what belongs to a thing in respect of itself

* David Bentley Hart, *The New Testament: A Translation* (New Haven: Yale University Press, 2017), 539.
See Unto the Ages of Ages

† Liddell, Henry George, and Robert Scott, *A Greek-English Lexicon*, rev. Henry Stuart Jones and Roderick McKenzie (Oxford: Clarendon Press, 1940), 45.
See Oxford Classic Dictionary - Aion

belongs to it in its essence (en tôi ti esti)" for we refer to it "in the account that states the essence" (Posterior Analytics, 73a34–5).*

The phrase *'to ti ên einai' did not just boggle* Roman translators. It has been the subject of many philologists who study these texts in the original Greek but also many philosophers after Aristotle, not the least of which was Martin Heidegger. Suffice it to say that this is a very involved technical area of philosophy. I point this out here just to make an explicit note that the translations into 'essence' and 'substance' is an important limitation of the Latin world which did not apprehend the depth of Greek thought. However, this limitation played a significant role in Western history which followed. All of this will come back in later parts of this volume when Levinas and radical alterity play a very different role than what the end of the Classical Greek period and beyond fashioned as the good.

A Deeper Dive - In the Beginning, the Metaphysics of Eternity was Not

A typical Latin or later Greek anachronism (Christian doxa) which gets imputed to the earlier ancient Greek mind is the idea of eternity. While the idea of 'eternity' may have helped the early Catholic church become 'universal' against competing factions of Christianity (as Elaine Pagels discusses in her book, "The Gnostic Gospels"), it obscured earlier Greek ideas. Even in some English translations of Greek passages quoted in this volume, the anachronistic derivation *'aeon' (also eon, aon, age, epoch)* of the adverb root *'aei'* is translated as eternity or eternal.

In 'ETERNITY' REVISITED A Study of the Greek Word *αἰών* Heleen Keizer concludes,

> Discerning as I do in the (extra-biblical) meaning of aiôn three notions, I have described the first as 'life', the second as 'time', and the third variously as 'whole', 'completeness', 'totality', or 'entirety'. The third notion distinguishes aiôn when used as a word for 'life' from the other words zôê and bios, and when aiôn is used as a word for 'time' this notion adheres to its meaning no less. Aiônis the 'entirety' of time; 'eternity' is too much an 'anachronistic', misleading or unclear rendering.†

In 1880 John Hansen wrote,

> "We submit the following derivation of aion, to which we invite the scrutiny of scholars:—In Greek a noun with ac-cented omega (long o) in the last syllable of the nominative case singular, signifies a container; that is, the on indicates that the preceding syllable is contained in it. Thus, a Greek's loutr-dn is his bath-place ; a dendr-dn is his grove-place ; a rhod-dn is his rose-place, etc. An aion is, therefore, something that contains an aei, or aia—a something containing the

* S. Marc Cohen and C. Reeve, "Aristotle's Metaphysics," in *The Stanford Encyclopedia of Philosophy*, ed. Edward N. Zalta (Winter 2021 Edition), sec. 4.1.
See Aristotle Metaphysics

† Thomas J. J. Altizer, *The Apocalyptic Trinity* (New York: Palgrave Macmillan, 2000), 54.
See Aion and Time in Aristotle

earth, aia', or duration, aei; or breathings, ad; existence, life;—duration, signified by breathings, is perhaps the best etymology of the word.*

In the Wiktionary excerpt below many of the definitions through history for the Greek words *αεί* and *αἰών* are shown:

ἀεί

See also: αεί, αει- and αεί-
Ancient Greek
Alternative forms
αἰεί (aieí) – Epic, Ionic, poetic
αἰέν (aién) – Homeric
ἀέ (aé), ἀές (aés), αἰές (aiés), αἰή (aiḗ) – Doric
αἰέ (aié) – unknown
αἶι (aîi) – Aeolic
ἤι (ēi) – Boeotian
αἰϝεί (aiweí) – Delphic
Etymology

From Proto-Hellenic *aiweí, from Proto-Indo-European *$h_2$eyu- ("lifetime, long time"), ultimately from Proto-Indo-European *$h_2$ey- ("vital energy, life"). Equivalent to locative case of αἰών (aiṓn), with or without the -ν.

Adverb
ἀ̄εί or ἀ̆εί • (āeí or aeí) (Attic)
(at all times): always, ever
Synonym: πάντοτε (pántote)
Νῦν καὶ ἀεί.
Nûn kaì aeí.
Now and always.
(for all time): forever, everlastingly, infinitely
Synonyms: διὰ παντός (dià pantós), διὰπαντός (diàpantós)
(for (an) eternity): eternally

(for an indefinitely long period of time, without specification regarding past present or future): abidingly, at length, indefinitely, in perpetuity, lengthily, perpetually

(for an indefinitely long period of time including the present time): for the time being

(for an indefinitely long period of time including a specific point in time (generally in the past) which is not the present time): at that time, at the time; back then

(at all times from a distant past through the present or some other time): from the get-go, from the start, heretofore, since the beginning, to date, up to now, until this day, "jusqu'aujourd'hui"; theretofore

---

* John Wesley Hanson, *Aiōn-Aiōnios: An Excursus on the Greek Word Rendered Everlasting, Eternal, Etc.* (Chicago: Jansen, McClurg, & Co., 1880), 20.
See Aion-Aionios (pdf)

(at all times from the present or some other time through a distant future): endlessly, evermore, forevermore

Usage notes

Remaining true to its Indo-European root, the adverb ἀεί imparts the sense that a verbal action occurs continuously for "a long time" in a variety of specifications, or indeed for all time. When specifying a temporally bounded period of time, ἀεί may refer to an action occurring: either from the far past through the present, from the present until a distant future, for a relatively long time in either the past or in the future, or from the far past through a distant future. Contrast πώποτε (pṓpote), which (where not negated) indicates that a verbal action might occur at any time over a long period of time, but not necessarily continuously.

Descendants

Greek: αεί (aeí)[*]

αἰών

See also: Αἰών

Ancient Greek

Etymology

From earlier αἰϝών (aiwṓn), from Proto-Indo-European *$h_2$eyu- ("vital force, life, long life, eternity"), whence also ἀεί (aeí, "always"). Cognate with Latin aevum, English aye.

Pronunciation

IPA(key): /ai̯.ɔ̌ːn/ → /ɛˈon/ → /eˈon/

Noun

αἰών • (aiṓn) m (genitive αἰῶνος); third declension

lifetime

generation

a long period of time, eon, epoch, age

the current world

eternity

Inflection

Third declension of ὁ αἰών; τοῦ αἰῶνος (Attic)

Synonyms

καιρός (kairós)

χρόνος (khrónos)

Derived terms

αἰώνιος (aiṓnios)

Descendants

English: aeon, eon

Greek: αιώνας (aiónas)

Latin: aeon[†]

[*] "ἀεί," Wiktionary, The Free Dictionary, last modified March 7, 2026, s.v. "Ancient Greek." See ἀεί

[†] "αἰών," Wiktionary, The Free Dictionary, last modified March 7, 2026, s.v. "Ancient Greek." See αἰών

Alchemy and the Royal Society

From the Latin transformation of *phusis* into substance we are beginning to acquire the modern idea of the mind/body 'thinking/thing' of Descartes blossoming in the history of Enlightenment. * This distance the ancient Greeks had from our modern-day notions of neutrality, the 'it', gave the Greeks from the epic writers to Plato and Aristotle a more unsettled and richer inquiry into *phusis.* The emergence of neutrality showed itself in Newton's writings. However, even then, Newton wrote extensively about God and alchemy - the chemical 'transubstantiation' of base metal into gold which dominated until its demise from England's Royal Society.† As late as the 20th century Maurice Blanchot writes of the insidious abstraction of neutrality which levels off our relations with others. In "The Gaze of Orpheus" Maurice Blanchot's essay "The Narrative Voice" tells us,

> "What Kafka teaches us – even if this expression cannot directly be attributed to him – is that storytelling brings the neuter into play. Narration governed by the neuter is kept in the custody of the "he", the third person that is neither the third person, nor the simple cloak of impersonality. The "he" of narration in which the neuter speaks is not content to take the place usually occupied by the subject, whether the latter is a stated or implied "I" or whether it is the event as it takes place in its impersonal signification. The narrative "he" dismisses all subjects, just as it removes every transitive action or every objective possibility. It does this in two forms: 1) the speech of the tale always let us feel that what is being told is not being told by anyone: it speaks in the neuter; 2) in the neuter space of the tale, the bearers of speech, the subjects of the action- who used to take the place of characters – fall into relationship of non-identification with themselves: something happens to them, something they cannot recapture except by relinquishing their power to say "I" and what happens to them has always happened already: they can only account for it indirectly, as self-forgetfulness, the forgetfulness that introduces them into the present without memory that is the present of narrating speech." (Blanchot, 1981, p. 140)‡

* See George Orwell and Emmanuel Levinas Introspective: Socialism and the Other

† See The last of the alchemists

‡ Maurice Blanchot, *The Infinite Conversation*, trans. Susan Hanson (Minneapolis: University of Minnesota Press, 1993), 385. See Language: Animism and Illusion

3

ORDER AND DESIRE IN KANT AND HEGEL

Neutrality as the object of metaphysical desire can never be established, touched, or known as an end in itself. It must endlessly defer to the desire of transactionalism, the next and future shiny, metal object where value is denominational. In Kantian philosophy, the thing-in-itself (German: *Ding an sich*) refers to the status of objects as they exist independently of representation and observation. The noumenon (or thing-in-itself) stands in contrast to the phenomenon, which represents how things appear to us. While we can study phenomena through our senses and reason, the noumenon remains elusive and inaccessible to direct observation. Immanuel Kant (1724-1804) argued that the empirical world—which encompasses all objects—is a complex of appearances. These appearances exist only within our representations. However, underlying these appearances is the thing-in-itself, which remains beyond our direct knowledge. We can only grasp its manifestations, not its true nature. Kant expressed this idea in his work, stating: "*And we indeed, rightly considering objects of sense as mere appearances, confess thereby that they are based upon a thing in itself, though we know not this thing as it is in itself, but only know its appearances, viz., the way in which our senses are affected by this unknown something*". *

The thing-in-itself represents the true nature of objects, beyond how they appear to us. It exists independently of our sensory experiences and mental constructs. Kant posited that while we can perceive and understand the world through our senses (phenomena), there exists a deeper reality that remains unknowable—the noumenal realm. Essentially, the thing-in-itself is the underlying reality behind the appearances we encounter. The term "noumenon" is closely related to the thing-in-itself. It refers to the unknowable reality that lies beyond our sensory perceptions. Kant argued that human sensibility is receptive, allowing us to know the phenomenal world (appearances). However, the noumenal world remains beyond our grasp. While we recognize the existence of the noumenon, we cannot directly access or conceive it. Our cognitive faculties are limited to understanding phenomena. In his doctrine of transcendental idealism, Kant emphasized that the empirical world—the sum of all objects—is a complex of appearances. These appearances arise from our mental representations. Kant acknowledged that these appearances are grounded in a thing-in-itself, even though we lack direct knowledge of it. We only perceive its effects on our senses. Here's how Kant expressed it:

> "And we indeed, rightly considering objects of sense as mere appearances, confess thereby that they are based upon a thing in itself, though we know not this thing as it is in itself, but only know its appearances, viz., the way in which our senses are affected by this unknown something." — Prolegomena, § 32 1 †

Kant challenged the idea that our minds passively mirror external reality. In his work Critique of

* See Thing-in-itself

† Immanuel Kant, *Prolegomena to Any Future Metaphysics*, trans. Paul Carus (Chicago: Open Court Publishing, 1902), § 32.

Pure Reason (1781), he argued that our perception of the world is not a direct reflection of how it truly exists. Instead, our minds construct a picture of reality based on sensory experiences. This groundbreaking idea implied that we cannot directly experience God through our senses; our knowledge of God is therefore a matter of faith rather than empirical evidence. [*] But sensory experiences are mediated by the will as practical reason: Kant believed that the will is not merely a blind force driven by sensations and desires. Instead, he saw it as practical reason. This means that our will is guided by rational considerations, not just impulsive cravings. Kant's approach combines reason and desire in a unique way, shaping much of his moral philosophy.

Kant believed in individual autonomy and self-legislation. Kant famously emphasized autonomy—the idea that morality originates from the will's self-legislation. Morality's principles and motivating power don't depend on external sources; they emerge from within the will itself. This concept challenges the notion that desires alone dictate our actions. [†] Kant thought of desire as hedonism. Kant's view of desire differs from traditional notions. He considered desire as a natural empirical motive, implying determinism and being essentially beyond rational control. Moreover, desires are subjective, contingent states, making them an inadequate basis for moral motivation and universal rules. Kant also connected desires with pleasure, emphasizing their hedonistic nature. [‡] Kant understands freedom as higher than desires. Kant's libertarian freedom involves being a slave to one's own desires. He didn't classify desires as rational fuel for moral judgments. Instead, he set boundaries for individual morals, emphasizing rationality over mere desire. Many of our founding fathers' assumptions built into our Constitution are based on Kant's idea that individuals would reason and act morally. [§]

F. H. Jacobi (1743-1819) was among the first to critique the concept of a thing-in-itself. He expressed the paradox that he couldn't enter Kant's system with this concept yet couldn't remain in it without it. G. E. Schulze (1761-1833) also challenged Kant's notion, arguing that things-in-themselves cannot cause appearances because causality applies only to objects of experience. If the thing-in-itself represents some mysterious essence behind appearances—a realm beyond our senses and reason, forever veiled in philosophical intrigue, how can it come into our experience, how could we sense it, how could it be a cause for sensation? Kant's transcendental idealism tells us noumena cannot be experienced, sensed, or even known. Only phenomena can be experienced and known but only by perceptions, appearances, and concepts. However, even the word noumena from Greek *noein* means to conceive or apprehend. [#] Essentially, the thing-in-itself is the underlying reality that gives rise to the appearances we perceive.

Kant's thing-in-itself has an interesting relationship to physicist John Wheeler's (1911-2008) famous slogan "it from bit". According to Wheeler, everything in the physical world — whether it's a particle, a field of force, or even the fabric of spacetime itself — has an underlying source. This source is immaterial and deeply fundamental. It arises from the binary choices we make when observing the world: yes/no, true/false, on/off. These binary decisions are akin to the basic units of information,

---

[*] Explainer: the ideas of Kant
[†] See 2 - Reason, desire, and the will
[‡] See Kant on the Nature of Desires (.pdf)
[§] See Kant's take on Freedom
[#] See Thing-in-itself
And thing-in-itself
And noumenon

known as bits. Wheeler suggests that our experience of reality emerges from these binary decisions. When we observe an object, event, or phenomenon, we are essentially making a series of yes-no choices. These choices shape our perception of reality. In other words, reality is participatory — it arises from our interactions with the world. Information is theoretic origin meaning all things physical have their roots in information theory. The very existence and meaning of physical entities derive from the answers to binary questions. Whether it's a particle's behavior, the curvature of spacetime, or the properties of a field, it all traces back to information-based decisions. "It from Bit" symbolizes the idea that our reality is intricately woven from the fabric of information. Every aspect of the physical world ultimately arises from the binary dance of yes and no, making our universe a participatory and information-theoretic realm. Like Kant, there is no such thing as 'reality' only binary judgements which comes from perception. *

In both Kant and Wheeler, we must rely on the already understood ontological notion of 'it' as if that were a given. Even more so, the 'it' refers to and is the basis for perception and judgement and the concepts we form. However, for Kant the 'it' as the thing-in-*it*self could not come into phenomena but for Wheeler the 'it' and phenomena are not sealed off from phenomena but intrinsic to phenomena. But even this, what we think as a very simple notion of 'it', was not always around from the beginning, the origin. The notion of 'it' as neutrality is not universal in history. Many ancient religions we call animistic did not even have a word for a 'thing' which was lifeless. The ancient animistic religions were animated by male gods and female goddesses. What we think of as inanimate objects did not exist in any form in many ancient cultures which saw all nature as energized in some mysterious way. Even Hesiod's idea of chaos [*χάος*] was not neuter in the way we think of it.

> The 'it' comes ready-made by history and language. 'It' is 'common sense'. If we want to think philosophically we must understand that these read-made notions were not available to Hesiod. If we simply take over our contemporaneous idea of the 'it' when we read Hesiod we do what scholars of ancient Greek literature (called philologists) call anachronistic (belonging to a period other than that being portrayed). In order to understand Hesiod we must let χάος [chaos] speak to us in proximity to what we can glean from the texts we have and what scholars tell us of their context. We must relieve ourselves of the assumption that chaos was an 'it', devoid of life or gender. While the myths of the ancient Greeks indicate a kind of animate vitalization of phusis as we see in the pantheon of gods, the notion of 'neuter' in ancient Greek was 'neither one nor the other' as in 'not masculine or feminine'. † The neuter was the 'not' in ancient Greek of gender. The negative does not specify a positive term as modern usage does of the 'it' (i.e., substance, inanimate objects, not alive, thing, etc.). Furthermore, as we will see

* See It from Bit: Pioneering Physicist John Archibald Wheeler on Information, the Nature of Reality, and Why We Live in a Participatory Universe

And Bit from It

And It from bit?

† See neuter – not either, neither of the two

neuter (adj.) – late 14c., of grammatical gender, "neither masculine nor feminine," from Latin neuter "of the neuter gender," literally "neither one nor the other," from ne- "not, no" (from PIE root *ne- "not") + uter "either (of two)" (see whether). Probably a loan-translation of Greek oudeteros "neither, neuter." In 16c., it had the sense of "taking neither side, neutral."

later in this text χάος in Ancient Greek can be used in different ways including the feminine gender.*

Individual autonomy and freedom as rational may be a moral imperative for Kant. However, these days individual autonomy has found another way towards freedom which is not based on rational imperative. Our modern notion of neutrality as simply 'it' has no moral imperative. The universal and absolute of classicism has been transformed into inanimate and anonymous 'matter', stuff, a thing which only has use value for us as Heidegger might tell us – standing reserve. In modernism this has found its way into individual autonomy and freedom. The metaphysical assumption of neutrality has congealed as 'common sense'. This brings with it another meaning to the idea of freedom. Freedom defined by neutrality cannot contest me. The 'it' we face is transformed into 'my freedom' which must always come before any consideration of the other. Freedom has become the monad of individual autonomy. Freedom is the sacrosanct right to unrestrained echoing of oneself back to oneself, what is mine, what I want, what I do not have, and acquisition is ethic-based on transaction. In this sense freedom has found its way to another concept in Kant – the thing-in-itself. Individual autonomy as right and privilege, as monadic echo, which has no restrictions is impenetrable with no reference to externality. The entire history of modernism, the thinking thing, has overtaken Kant's moral imperative not to treat others as a means to an end but as a noumenal and ethical end in themselves, as an other which exceeds the thinking thing that has become recursively internalized, impenetrable to externality.

On Hegel

At the end of the 18th century, we saw a similarity to exformation with Kant's notion of *a priori* in that Kant did not attribute a cause to analytic judgement based on *a priori* categories of understanding. Kant did not attribute causality to analytic judgments. He recognized that *a priori* categories lie outside the concept of something happening, something in motion or change. Therefore, cause and effect cannot be understood as a synthetic judgment; instead, cause and effect itself arises from an *a priori* category of understanding meaning cause and effect is an analytic judgement. In the mid-19th century Freud employed the notion of *a priori* in terms of the unconscious. Freud used the analogy of the common child's toy, the mystic writing pad, to illustrate how when the plastic sheet is lifted up and the drawing disappears, there is still an imprint on the pad which he thought as the unconscious. The unconscious was still there apart from the agent of consciousness and worked behind the scenes of consciousness to influence behavior. The unconscious had a hazy relation with cause and effect in terms of consciousness. The unconscious was not a temporal event but a-temporally held events, unassembled information (exformation) which might be expressed in tensions between the id and the superego. He called these tensions the ego.

In the 19th century, the German idealist Hegel fundamentally changed Aristotle's and Kant's indeterminacy with regard to origin and categories of understanding. Hegel believed he was able to accu-

* Mark Randall Dreher's blog post "On Origin," published October 15, 2024, on MixerMuse, provides the source material for the specified quote. The reference is formatted for inclusion in the book *Quanta, Alterity and Love*. For the full post, visit MixerMuse.com.

See On Origin

rately account for their uncertainties through dialectic. Hegel found an absolute knowing in dialectic in which knowing is immanent and can only be the condition for truth. At the end of Hegel's "*Phenomenology of Spirit*" he concludes that we must overcome the separation of knowing and truth. He reinterpreted and rearticulated classical philosophy through Aristotle and Kant in a way that is still highly researched today. Hegel's methodology is dialectic which is a system of logic which was to give a full and accurate account of all phenomena from human 'being' to time and space. Let's go through an example of what he wrote at the beginning of probably his greatest work, *Science of Logic*. In the beginning movements of his famous work "The Science of Logic", he writes:

> Being is being, and nothing is nothing, only as held distinct from each other; in their truth, however, in their unity, they have vanished as such determinations and are now something else. Being and nothing are the same and, precisely because they are the same, they no longer are being and nothing but possess a different determination; in becoming they were coming-to-be and ceasing-to-be; in existence, which is another determinate unity, they are again moments but differently determined. This unity now remains their base from which they no longer surface in the abstract meaning of being and nothing. *

The genius of Hegel was to convert Aristotelian potentiality from non-existent to what he terms 'nothing' and Aristotle's actuality to 'Being' at the start in his premiere work entitled "The Science of Logic". In so doing, Becoming originates from Being and nothingness. In juxtaposing nothingness and Being, "*held distinct from each other*", he finds an inner unity in that they are "*the same*". They are both moments of the same "*in becoming they were coming-to-be and ceasing-to-be; in existence, which is another determinate unity, they are again moments but differently determined*" into the transformation of becoming. Hegel converts actuality and potentiality, existence and non-existence to a unity as "the same" which then is differentiated in the transformation of becoming. In analogous terms of what has been discussed for quantum mechanics, becoming has now obviated spooky exformation as a-causal, atemporal, a-spatial, etc. and in its place transformed it into sheer becoming with no excess, no 'ex'. Hegelians would say the excess to Being and nothing is subsumed, taken up, entirely into becoming which maintains its predecessors but also unites and exceeds them. In effect, becoming concretizes the oppositional abstraction of Being and nothing.

Much later in the dialectical progression of Hegel's work, logic must find its completion in Concept or the Idea. But in order for Concept to achieve this, Concept must comprehend itself. This is self-comprehension. However, in order to achieve self-comprehension, Concept must also alienate itself. In order for the Idea to achieve self awareness, to realize itself, it must experience alienation—a distancing from its own essence as self-externality. Hegel compares the dialectic of self-consciousness (not yet self-realized) with the famous dialectic of the master and slave relationship. Just as the slave overcomes alienation by recognizing their dependence on the lord, consciousness achieves absolute knowing by transcending self-alienation. Concept's dual requirements of self-awareness and alienation requires that it emerge as a 'here' and 'now', as alone and eternal which can also be abstracted as a point. But as the 'here' and 'now' and solitude to be what it is, it must have a 'before' and after'. It

* Georg Wilhelm Friedrich Hegel, *The Science of Logic*, trans. George di Giovanni (Cambridge: Cambridge University Press, 2010), 82.

must have others, a collectivity, a state. Therefore, the abstract point of solitude must emerge as time, as temporality. In Hegel's "Logic," being itself is temporality. The past is timeless and involved in the actualization of the present. The present is comprehended through *Gewordensein* (becoming), which expresses inherent possibilities of actuality. Hegel's philosophy embraces plasticity, acknowledges the dynamic nature of time, and allows for future transformations. Let's look at how Hegel tells us space-time emerges. *

In Hegel's "Philosophy of Nature", he starts with the most general abstract notion of nature. Hegel writes,

> The first or immediate determination of nature is the abstract generality of its self-externality,- its unmediated indifference, space. It is the wholly ideal juxtaposition, because it is being outside of itself and absolutely continuous, because this being apart from itself is still entirely abstract, and has no specific difference within itself... †

As abstract and general nature comes before us self-externality. It is entirely abstract externality and is undifferentiated, "*no specific difference within itself*".

> Space has, as the concept in general (and more determinate than an indifferent self-externality) its differences within it: (a) in its indifference these are immediately the three dimensions, which are merely diverse and quite indeterminate... ‡

But when we look more concretely at space, we see three dimensions. But these three dimensions are themselves abstract and indeterminate.

> But the difference of space is essentially a determinate, qualitative difference. As such it is (a) first, the negation of space itself because this is immediate and undifferentiated self-externality, the point. (b) The negation as negation, however, is itself spatial, and the relation of the point to space is the line, the first otherness of the point. (c) The truth of the otherness is, however, the negation of the negation. The line, therefore, passes over into the plane, which on the one

---

* See Understanding Hegel's Theory on Time
And Fichte and Hegel On The Definition of Concepts
And Key Concepts of the Philosophy of G W F Hegel
And Concept of Alienation in Hegel's Social Philosophy
And Hegel and the Concepts of Time
And SPACE AND TIME IN HEGEL'S PHILOSOPHY
And Concept of Alienation in Hegel's Social Philosophy
And Hegel on Others and the Self
And Hegel, Alienation, and the Phenomenological Development of Consciousness
And Hegel: Social and Political Thought
And Hegel's Theory of Space-Time (No, Not That Space-Time)
And Hegel's Metaphilosophy of Idealism
And Hegel's Philosophy of Nature

† Georg Wilhelm Friedrich Hegel, *Philosophy of Nature*, trans. A. V. Miller (Oxford: Clarendon Press, 1970), 28. See Hegel's Philosophy of Nature

‡ Georg Wilhelm Friedrich Hegel, *Philosophy of Nature*, trans. A. V. Miller (Oxford: Clarendon Press, 1970), 30. See Hegel's Philosophy of Nature

> hand is a determinacy opposed to line and point, and thus is plane in general, but on the other hand is the suspended negation of space, and thus the re-establishment of spatial totality, which, however, now contains the negative moment within itself an enclosing surface, which splits off an individual, whole space. *

But looking at space more concretely we see that the quality of space is first a point, one-dimensional. But if we negate a single point, we get multiple points as a line, two-dimensional. And if we negate the line, we get multiple lines which pass over into a plane, three-dimensional space. If we negate three-dimensional space itself, we get the totality of space. But negating the totality of space as individual space we see a geometric enclosing like a sphere perhaps.

> That the line does not consist of points, nor the plane of lines, follows from their concepts, for the line is the point existing outside of itself relating itself to space, and suspending itself and the plane is just as much the suspended line existing outside of itself.-Here the point is represented as the first and positive entity, and taken as the starting point. The converse, though, is also true: in as far as space is positive, the plane is the first negation and the line is the second, which, however, is in its truth the negation relating self to self the point. The necessity of the transition is the same.-
>
> The other configurations of space considered by geometry are further qualitative limitations of a spatial abstraction, of the plane, or of a limited spatial whole. Here there occur a few necessary moments, for example, that the triangle is the first rectilinear figure, that all other figures must, to be determined, be reduced to it or to the square, and so on.-The principle of these figures is the identity of the understanding, which determines the figurations as regular, and in this way grounds the relationships and sets them in place, which it now becomes the purpose of science to know...
>
> Negativity, which as point relates itself to space and in space develops its determinations as line and plane, is, however, in the sphere of self-externality equally for itself and appearing indifferent to the motionless coexistence of space. Negativity, thus posited for itself is time... †

In the quote above Hegel goes on to exhibit how all this negating of the point, the line, the plane culminating in the sphere must also be negated as the self-externality of the sphere is static and motionless. So, negating the static and motionless sphere gives rise to temporality in order to make space dynamic and changing.

> Time, as the negative unity of being outside of itself, is just as thoroughly abstract, ideal being: being which, since it is, is not, and since it is not, is. ‡

---

* Georg Wilhelm Friedrich Hegel, *Philosophy of Nature*, trans. A. V. Miller (Oxford: Clarendon Press, 1970), 32. See Hegel's Philosophy of Nature

† Georg Wilhelm Friedrich Hegel, *Philosophy of Nature*, trans. A. V. Miller (Oxford: Clarendon Press, 1970), 33–34. See Hegel's Philosophy of Nature

‡ Georg Wilhelm Friedrich Hegel, *Philosophy of Nature*, trans. A. V. Miller (Oxford: Clarendon Press, 1970), 34. See Hegel's Philosophy of Nature

But if time is negated, we get being which is and is not. This is a binary duality in which being is evoked.

> Time, like space, is a pure form of sensuousness, or intuition; but, as with space, the difference between objectivity and a contrastingly subjective consciousness does not matter to time. If these determinations are applied to space and time, then space is abstract objectivity, whereas time is abstract subjectivity. Time is the same principle as the I = I of pure self-consciousness; but the same principle or the simple concept still in its entire externality, intuited mere becoming, pure being in itself as sheer coming out of itself. Time is just as continuous as space, for it is abstract negativity relating itself to itself and in this abstraction there is as yet no real difference. *

Now Hegel tells us that both time and space negated as being is the pure form of sensuous which he also calls intuition. But objectivity and its negation of subjectivity or consciousness does not matter to time. But there is a relativity at work in that if we apply negations to space and time then space is objective, and time is abstract subjectivity. This is where identity is abstractly established as I = I. This is pure self-consciousness. But for self-consciousness to come out of itself, negate itself, it abstractly intuits becoming.

> **In time, it is said, everything arises and passes away, or rather, there appears precisely the abstraction of arising and falling away. If abstractions are made from everything, namely, from the fullness of time just as much as from the fullness of space, then there remains both empty time and empty space left over; that is, there are then posited these abstractions of exteriority.-But time itself is this becoming, this existing abstraction, the Chronos who gives birth to everything and destroys his offspring.-That which is real, however, is just as identical to as distinct from time.** Everything is transitory that is temporal, that is, exists only in time or, like the concept, is not in itself pure negativity. To be sure, this **negativity is in everything as its immanent, universal essence**, but the temporal is not adequate to this essence, and **therefore relates to this negativity in terms of its power**. Time itself is eternal, for it is neither just any time, nor the moment now, but time as time is its concept. The concept, however, in its identity with itself I= 1, is in and for itself absolute negativity and freedom. Time, is not, therefore, the power of the concept, nor is the concept in time and temporal; on the contrary, the concept is the power of time, which is only this negativity as externality.-The natural is therefore subordinate to time, insofar as it is finite; that which is true, by contrast, the idea, the spirit, is eternal. Thus the concept of eternity must not be grasped as if it were suspended time, or in any case not in the sense that eternity would come after time, for this would turn eternity into the future, in other words into a moment of time. And the concept of eternity must also not be understood in the sense of a

* Georg Wilhelm Friedrich Hegel, *Philosophy of Nature*, trans. A. V. Miller (Oxford: Clarendon Press, 1970), 35.

> negation of time, so that it would be merely an abstraction of time. For time in its concept is, like the concept itself generally, eternal, and therefore also absolute presence. *

But pure abstract self-consciousness is powerless. When self-consciousness is as pure abstraction, it negates as powerless. But abstract powerlessness must give way to power - power as origin I = 1, a start. In turn, when self-consciousness as the power of origin is negated, we get spirit as eternal. But Hegel warns us not to understand the negation of the concept of eternity as a moment of time. That does not proceed from more abstract to more concrete. Therefore, eternity, absolute spirit, is absolute presence. This is the progression of 'logic' for Hegel and is still actively researched and published even today in Hegelian scholarship.

You may have noticed that Hegel leaves out Einstein's relativity and no mention of quantum physics' negations as well. Of course, that is because Hegel's "Philosophy of Nature" was published in 1817. Einstein and quantum mechanics did not come around until the start of the 1900's. It should also be evident that a certain classical interpretation of time and space and self-consciousness has the archaic ring of classical science of Hegel's day. It would be interesting if some of the current Hegelian scholars would take a crack at that, but it seems to me that delving too deeply into such matters might necessitate a postmodern 're-reading' of Hegel. Perhaps we might call this the anti-Hegel or the negation of Hegelianism.

In Hegel's "Philosophy of Nature", he further delves into the essence of motion. According to him, motion embodies the immediate unity of space and time. This unity implies that time now has real existence within a space that is truly differentiated by it. In other words, time and space become actual in motion. This convergence signifies that what time and space are in Concept aligns with what they are in appearance. Hegel views time as introducing diversity (or negation) into space, thereby actualizing it. Rather than perceiving isolated points or instants, we experience "here-nows"—a continuous temporal flow. For Hegel, temporality is marked by a hierarchical sequence of ever enjoining dialectical steps ultimately culminating as absolute Concept. As absolute Concept. But for Concept to truly be absolute it must concretize itself, 'not' itself, as space and then time and then becoming and so on until yet again when we once again arrive at absolute Concept for the first time. Pardon my sarcastic desecration of Hegel's genius. I freely admit that I am not worthy! Oh, did I mention that Hegel is a distant relative of mine, so I guess this is a bit of a family feud (; -).

Catherine Malabou, in her work "The Future of Hegel: Plasticity, Temporality and Dialectic," challenges the prevailing notion that Hegel's philosophy is solely oriented toward the past and lacks a future perspective. Malabou argues that Hegel's concept of "plasticity" plays a crucial role. Plasticity refers to the capacity to receive and produce form, allowing for transformation and adaptability. Hegel's view of humanity, divinity, and philosophy is more flexible than often portrayed. He conceives of these aspects as inherently amenable to future changes. In his discussion of anthropology, Hegel emphasizes habit as liberating the human subject from the constraints of nature. Human beings possess "plastic individuality", implying the ability to develop and differentiate over time. Hegel's concept of plasticity plays a pivotal role. Plasticity refers to the capacity for transformation and adaptability. When we engage with Hegelian concepts, we participate in a dialectical process. Rather than

* Georg Wilhelm Friedrich Hegel, *Philosophy of Nature*, trans. A. V. Miller (Oxford: Clarendon Press, 1970), 36–37. See Hegel's Philosophy of Nature

demonstrating propositions, we allow concepts to unfold and reveal their implications. The dialectical motion is meant to allow us to see how time and space intertwined, affecting our understanding of existence and change.

Hegel considers time as the first mediation of outsideness. It negates space and forever remains related to it. Unlike isolated points or instants, we experience time as a continuous flow—a series of "here-nows". Time's role is not merely chronological; it mediates our perception of the external world. In Hegel's Logic, being itself is temporality. The past, timeless yet involved in actualization, shapes the present. The present unfolds through *Gewordensein* (becoming), revealing inherent possibilities of actuality. Hegel believes his Concept transcends mere abstraction; it becomes a dynamic interplay between time and space. *

This process of abstracting and then concretizing is a technique which is typical of Hegel's entire circular enterprise from Being and nothingness through Concept through solitude overlapping here and now/point to sociality overlapping multiple 'heres' and 'nows' in which a spatial point becomes temporal points arising back again to Being and nothingness. For Hegel, concretizing does not undo the tension of abstraction but completes what is lacking in abstraction. It seems to me that each step along the way might also be thought of as relative uncertainties, abstractions, which get taken up into relative certainties, concretizing, and then evolving through an entire process of such movements which, in an almost Nietzschean eternal recurrence mode, finds a circularity. In any case, these abstract to concrete progressions seem to me to have at least a momentary similarity to classical certainty. In a kind of circular ascending and descending fashion classical certainties in the history of science seem to be eclipsed with paradigm shifts which eventually expand and perhaps give expanded context to prior paradigms. If so, both Hegelian dialectics and knowledge itself might find a kind of cyclical musical harmony and discophony which have an ascendant progressive quality. This is mere speculation, could this kind of move have at least an analogous similarity with the whole classical question of how the anarchical and chaotic get transformed wholly and completely concretizing as origin. Could origin then be abstract and in turn concretized as cause and effect?

In any case, exformation is not willing to play in Hegel's Concept. In quantum physics exformation is severely different than what Hegel and our history of knowledge leads us to believe. What 'is' has a shady association with what is not. "*Coming-to-be*" and "*ceasing-to-be*" are not separate and determinate moments which rise into becoming. Neither is "*coming-to-be*" and "*ceasing-to-be*" the same. Exformation has a hazy middle, an uncertainty, an indeterminacy which can find a limited classical amount of predictability in the branching of probability associated in some hazy fashion with the observer effect. So, the question to be or not to be has a hazy middle. Hegel's notion of becoming obfuscates the current findings of physics and appears to yet again be trapped in a certain philosophical tradition of the classical world.

In Hegelian dialecticism we see a succession of logical oppositions which lead to transformations that unify the oppositions while maintaining their difference. By dialectic, I mean a positive concept

---

* See The Future of Hegel: Plasticity, Temporality and Dialectic
And Temporality and Time in Hegel and Marx
And Understanding Hegel's Theory on Time
And Hegel and the Concepts of Time
And The Concept of Space in Hegel:The Early Jena Years
And SPACE AND TIME IN HEGEL'S PHILOSOPHY

followed by its negation, the negative of that concept, followed by the unification of both concepts through a transformative concept. For Hegel, there is no way to separate knowing and truth. Truth must always be in the context of knowing. For Hegel, as Leibniz, all predicates resided in the subject but for Hegel the subject eventually rises as absolute Concept not the monad. However, exformation as the sense of Hesiod's chaotic, yawning gap cannot simply be the transformation of dialectic. It cannot follow classical notions and cycles of absolute and completeness. It exceeds logical determinations in its necessary indeterminacy. Virtual particles in a Hegelian sense cannot be transformed, lifting up, sublimated (in German *aufhebung)* from particles into Concept (e.g., as mathematics). There is an incomplete excess in quantum physics which knows nothing of rising into Concept.

The mathematics of the wave function is highly incomplete. Hegel's *aufhebung* is considered as complete with regard to fully transforming and accounting for both positive and negative, concrete and abstract, conceptual components. Once a 'leaky', uncertain, spooky middle is acknowledged the entire Hegelian conceptual frame is put into question. Only language in its classical narratives would give us the illusory mirage of completeness (Hegel) or mathematics mirrored in the tautology of identity (Leibniz). 'Knowing' has an uncertainty about it which cannot be fully accounted for or 'taken up'. In my estimation Hegel's dialectic is rooted in a certain custom of language which obliviates the 'ex' in exformation. There is a persuasion in naming and logic with regard to identity that gets taken up into our lived life which gives us a false sense of assurance, even to the point of the extremism of violence and arrogance.* This is not to suggest that 'knowing' has no 'truth'. In the analogous way that classic physics is a subset, a limiting case, of quantum mechanics; knowing is a subset and limited case of an 'ex', an externality, which defies absolute determinacy. Additionally, "knowing' can never be detached from a language, a custom, and a history as it can never fully detach itself from hermeneutics. With regard to the wave function, the practical, applied findings of mathematics give us a certain amount of predictability. One practical effect of predictability is to enable us to manufacture faster, smaller, less power consuming integrated circuits. However, there is nothing in the wave function that would indicate finality. The actual details of the wave function in current science are highly contrary to any sense of completeness, classic certainty, and any equivocations which find agreement with, on the one hand, absolute certainty, determinacy, and on the other hand, truth and knowing.

Hegel's Kant Fix

Kant's thing-in-itself is built on a classical modernist idea that knowledge is internalized to the individual. Any other is a thing which we can know nothing about. This has become the 'it' of the historic abstraction of reality as neuter, as objects, of which we have no direct relation with. In the 19th century G.W. Hegel believed he had unraveled Kant's thing-in-itself.

> However, Kant's solution to Hume's skepticism led to a skeptical conclusion of its own that Hegel rejected. While the intersubjectively shared structure of our reason might allow us to have knowledge of the world from our perspective, so to speak, we cannot get outside of our mental, rational structures to see what the world might be like in itself. As Kant had to admit, according to his theory, there is still a world in itself or "Thing-in-itself" (*Ding an sich*) about

* See Double Consciousness

> which we can know nothing. Hegel rejected Kant's skeptical conclusion that we can know nothing about the world- or Thing-in-itself, and he intended his own philosophy to be a response to this view.*

Hegel, in agreement with Kant, proposed that necessary truth must be imposed by the mind, but he rejected Kant's concept of the "thing-in-itself" as unknowable. According to Hegel, all that exists *is mental* and is consequently accessible to individual consciousness. Kant's notion of the "thing-in-itself" (German: *Ding an sich*) refers to the status of objects as they are, independent of representation and observation. Kant argued that the empirical world—the sum of all objects—is a complex of appearances, whose existence and connection occur only in our representations. The thing-in-itself remains elusive, as we can only know its appearances, not its true nature. Hegel, however, took a different stance. Hegel radicalized Kant's transcendental idealism into absolute idealism, asserting that the concept of things-in-themselves is a contradiction. For Hegel, a thing must be an object of our consciousness to exist at all. He critiqued Kant's incomplete process of abstraction and the absolute distinction between subject and object. While Kant grappled with the unknowable "thing-in-itself," Hegel emphasized the centrality of consciousness and rejected the existence of such hidden entities. The fix for Hegel was to jettison the metaphysical abstraction of neutrality and objectivity altogether. However, for Hegel, this then is ultimately lifted up from subject/object absolute Concept. In this way, Hegel has more of an affinity to Wheeler in that there is no exteriority to concept. The thing-in-itself arises wholly into concept without any residue. For Hegel noumena is phenomena is concept. Hegel was taken with Leibniz and his idea that all predicates reside in the subjects. However, for Hegel there was a difference. † The following will be a brief sketch of Hegel's overall strategy for his two major works, "Phenomenology of Spirit" (1807) and the "Science of Logic" (1812-1818). These works are intensely difficult to understand. Incredibly, there are still many active schools of philosophy which are still researching and publishing Hegel's works. I think this should be read with the backdrop of previous discussions on quantum physics.

Hegel's Start with Subjective Spirit: Consciousness to Self-Consciousness to the State

> Hegel as a philosopher who had introduced this historically reflective dimension into philosophy (and set it on the characteristically romantic path which has predominated in modern continental philosophy) but who had unfortunately still remained bogged down in the remnants of the Platonistic idea of the search for ahistorical truths (Rorty 1982). Those adopting such an approach to Hegel tend to have in mind the (relatively) young author of the

* Robert Stern, *Hegel and the Phenomenology of Spirit* (London: Routledge, 2002), 11.
See Hegel's Dialectics

† See Georg W. F. Hegel (1770–1831) (.pdf)
And Thing-in-itself
And German Idealism
And Understanding Hegel's Mature Critique of Kant
And thing-in-itself

> Phenomenology of Spirit and have tended to dismiss as "metaphysical" later and more systematic works like the Science of Logic. *

Leibniz's focus on predicates residing within subjects aligns with his intricate metaphysical system of monads. Hegel, on the other hand, emphasized the subject's emergence within historical and communal contexts. While Leibniz's theory emphasizes the relationship between predicates and individual monads, Hegel's perspective centers on the subject's connection to the human community in his "Phenomenology of Spirit". In "The Phenomenology of Spirit", Hegel embarks on a profound exploration of the development of human knowledge. Notwithstanding, all of this eventually, through the dialectic of his "Science of Logic," ends up in Concept. † What binds all this together must have something to do with the question of consciousness.

> Hegel describes spirit as sunk in nature, and treats consciousness as largely limited to what now might be described as sentient or phenomenal consciousness alone—the feeling soul. Consciousness in the sense of the modern subject–object opposition only makes its appearance in the following second section, Phenomenology of Spirit, which, reprising key moments from the earlier book of that name, raises a problem for how we are to understand the relation of phenomenology and systematic philosophy: is it a path to it or part of it? ‡

This is where Hegel starts his investigation in his early work the "Phenomenology of Spirit". Hegel starts with unhappy consciousness. Unhappy consciousness has three stages: external beyond, changing individual, and achieved reconciliation. At the outset, consciousness grapples with sensory inputs, attempting to grasp objects in the external world. This initial stage represents a fundamental encounter with the external phenomenal realm. As consciousness evolves, it moves beyond mere sensory perception. It engages in more sophisticated ways of relating to the external world. Here, the individual undergoes transformations, adapting and growing. Finally, Hegel envisions a state where consciousness attains reconciliation. This reconciliation occurs through self-awareness, mutual recognition, and the interplay of self-consciousness with other self-consciousnesses. "Phenomenology of Spirit" grapples with the journey from sensory experience to self-consciousness, weaving together metaphysics, epistemology, ethics, and political philosophy. §

In the "Phenomenology of Spirit" Hegel grappled with the concept of "Concept" and where it encountered challenges. Hegel described the book as an "exposition of the coming to be of knowledge." He meticulously traced the journey of spirit through various stages, from sense perception to

* Paul Redding, "Georg Wilhelm Friedrich Hegel," in *The Stanford Encyclopedia of Philosophy*, ed. Edward N. Zalta (Spring 2024 Edition), sec. 1.
See Georg Wilhelm Friedrich Hegel (Stanford Encyclopedia of Philosophy)

† See Hegel and Recent Analytic Metaphysics
And Hegel: The Idea of the Subject

‡ Paul Redding, "Georg Wilhelm Friedrich Hegel," in *The Stanford Encyclopedia of Philosophy*, ed. Edward N. Zalta (Spring 2024 Edition), sec. 4.
See Georg Wilhelm Friedrich Hegel (Stanford Encyclopedia of Philosophy)

§ See Hegel's Phenomenology of Unhappy Consciousness
And Spirit as the Self-Awareness of Society
And The Phenomenology of Spirit
And Hegel on Self-Consciousness: Desire and Death in the Phenomenology of Spirit

absolute knowing. The concept of Concept emerges as spirit evolves, but it is not a static entity. Instead, it undergoes a necessary process of self-origination and dissolution. This dynamic movement is central to Hegel's dialectical method. Hegel's understanding of how concepts come into being implies a certain instability or insecurity in knowledge. Unlike Kant, who emphasized the mediating role of concepts, Hegel recognized that concepts themselves are subject to change and transformation. This instability arises from the inherent tension between different shapes of spirit as it progresses toward pure knowledge. Hegel's exploration reveals that knowledge is not a fixed endpoint but an ongoing process. In the Preface to the Phenomenology, Hegel rejected the notion of absolute knowledge based on immediate intuition or feeling. Such an approach, he argued, would dissolve the rich differentiation and determination of empirical content into a vague obscurity. Instead, Hegel sought a more rigorous path—one that grappled with contradictions, engaged in dialectical movement, and ultimately arrived at a more coherent system of knowledge. When multiple self-consciouses attain the state subjectivity has attained its highest goal. But how does the state arise to objectivity, to universals? [*]

Hegel's Transition to Objective Spirit: Concept

> In contrast, the British Hegelian movement at the end of the nineteenth century tended to ignore the Phenomenology and the more historicist dimensions of his thought, and found in Hegel a systematic metaphysician whose Logic provided the basis for a definitive philosophical ontology. This latter traditional metaphysical view of Hegel dominated Hegel reception for most of the twentieth century, but from the 1980s came to be challenged by scholars who offered an alternative non-metaphysical, post-Kantian view. By "non-metaphysical" these thinkers had in mind metaphysics in the sense that Kant had been critical of, a point sometimes missed by critics. But in turn, this post-Kantian reading has been challenged by a revised metaphysical view, critical of the purported over-assimilation of Hegel to Kant by the post-Kantians. In the revised metaphysical view, appeal is often made to Aristotelian or Spinozist conceptual realist features of Hegel's thought, as well as to features of recent analytic metaphysics. [†]

Hegel does not want to end at Kant's thing-in-itself, *a priori* categories of knowledge, and a moral categorical imperative. Hegel wants to understand how subjectivity and community can arise into objectivity, the universal, and eventually the absolute as pure Concept.

> Hegel, however, thought that philosophy had to unify theoretical and practical knowledge, and so the Phenomenology has further to go. Again, this is seen differently by traditionalists and

---

* See The Phenomenology of Spirit
And Phenomenology of Spirit: Chapters 1 to 3: Shapes of Consciousness
And Hegel's Preface to the Phenomenology of Spirit
And Hegel and the Phenomenology of Spirit
And A commentary on the preface to Hegel's 'Phenomenology of Spirit'

† Paul Redding, "Georg Wilhelm Friedrich Hegel," in *The Stanford Encyclopedia of Philosophy*, ed. Edward N. Zalta (Spring 2024 Edition), sec. 1.
See Georg Wilhelm Friedrich Hegel (Stanford Encyclopedia of Philosophy)

revisionists. For traditionalists, Chapters 7, Religion and 8, Absolute Knowing, testify to Hegel's disregard for Kant's critical limitation of theoretical knowledge to empirical experience. Revisionists, on the other hand, tend to see Hegel as furthering the Kantian critique into the very coherence of a conception of an in-itself reality that is beyond the limits of our theoretical (but not practical) cognition. Rather than understand absolute knowing as the achievement of some ultimate God's-eye view of everything, the philosophical analogue to the connection with God sought in religion, post-Kantian revisionists see it as the accession to a mode of self-critical thought that has finally abandoned all non-questionable mythical givens, and which will only countenance reason-giving argument as justification. However we understand this, absolute knowing is the standpoint to which Hegel has hoped to bring the reader in this complex work. This is the standpoint of science, the standpoint from which philosophy proper commences, and it commences in Hegel's next book, the Science of Logic. *

How does self-conscious and community end subjective spirit and find a beginning in objective spirit? Hegel starts out the "Science of Logic" with Being and Nothingness. He believes these are dialectical opposites which both depend on each other to mean something. By themselves they mean are contradictory and mean nothing. However, the tension between the two must necessarily lead to becoming. This short synopsis demonstrates Hegel's method throughout the "Logic".

Reading into the first chapter of Book 1, Being, it is quickly seen that the transitions of the Logic broadly repeat those of the first chapters of the Phenomenology, now, however, as between the categories themselves rather than between conceptions of the respective objects of conscious experience. Thus, being is the thought determination with which the work commences because it at first seems to be the most immediate, fundamental determination that characterises any possible thought content at all. (In contrast, being in the Phenomenology's Sense-certainty chapter was described as the known truth of the purported immediate sensory given—the category that it was discovered to instantiate.) Whatever thought is about, that topic must in some sense exist. Like those purported simple sensory givens with which the Phenomenology starts, the category being looks to have no internal structure or constituents, but again in a parallel to the Phenomenology, it is the effort of thought to make this category explicit that both undermines it and brings about new ones. Being seems to be both immediate and simple, but it will show itself to be, in fact, only something in opposition to something else, nothing. The point seems to be that while the categories being and nothing seem both absolutely distinct and opposed, on reflection (and following Leibniz's principle of the identity of indiscernibles) they appear identical as no criterion can be invoked which differentiates them. The only way out of this paradox is to posit a third category within which they can coexist as negated (Aufgehoben) moments. This category is becoming, which saves thinking from paralysis because it accommodates both concepts. Becoming contains being and nothing in the sense that when something becomes it passes, as it were, from nothingness to being. But

* Paul Redding, "Georg Wilhelm Friedrich Hegel," in *The Stanford Encyclopedia of Philosophy*, ed. Edward N. Zalta (Spring 2024 Edition), sec. 4.
See Georg Wilhelm Friedrich Hegel (Stanford Encyclopedia of Philosophy)

> these contents cannot be understood apart from their contributions to the overarching category: this is what it is to be negated (aufgehoben) within the new category.
>
> In general this is how the Logic proceeds: seeking its most basic and universal determination, thought posits a category to be reflected upon, finds then that this collapses due to a contradiction generated, like that generated by the category being, and so then seeks a further category with which to make retrospective sense of those contradictory categories. However, in turn the new category will generate some further contradictory negation and again the demand will arise for a further concept that can reconcile these opposed concepts by incorporating them as moments. *

At this point Becoming embarks upon a quest for answers. Becoming searches for reasons why it is something and not nothing. This is the journey towards truth. This journey is characterized by something we now might call 'language games'. Interestingly, I find this also in Wheeler's 'it from bit' of yes and no judgements. In Hegel's "Science of Logic", the quest for truth takes on a profound dimension. While truth is commonly understood as the agreement of thought with the object, Hegel delves deeper into its philosophical essence. According to him, philosophical truth is the agreement of a content with itself or of an object with its concept. This concept transcends mere correspondence and ventures into the realm of self-consistency. Hegel's "Science of Logic" aims to overcome a flaw he perceives in other systems of logic. These systems often assume a complete separation between the content of cognition (the world of objects, considered independent of thought) and the form of cognition (the thoughts about these objects, which are pliable and dependent on their conformity to the world of objects). Hegel seeks to bridge this gap, emphasizing that reality is shaped by thought and is, in a strong sense, identical to thought. In this pursuit, Hegel explores the dynamic process by which consciousness assimilates objects into mental concepts. His journey traces the movement from basic categories to more complex ones, revealing the intricate interplay between thought and reality. Therefore, the quest for truth in Hegel's "Science of Logic" is not merely about universality; it encompasses the very fabric of thought, weaving together content, form, and self-consistency in a philosophical tapestry. †

> If the concrete object of a de re judgment is effectively what had been under consideration in Chapter 2, Perception, in the Phenomenology (the thing with properties), we now might envisage where Hegel's thought is headed in these sections of the subjective logic. In the Phenomenology it turned out that the capacity for a subject to entertain objects of consciousness such as perceptual ones was that such a subject was capable of self-consciousness. It then turned out that to be capable of self-consciousness the subject had to exist in a world with other embodied subjects whose intentions it could recognize. It is here that we might pick up Robert Brandom's suggestion, following Sellars, that we should think of the existence of infer-

---

* Paul Redding, "Georg Wilhelm Friedrich Hegel," in *The Stanford Encyclopedia of Philosophy*, ed. Edward N. Zalta (Spring 2024 Edition), sec. 4.
See Georg Wilhelm Friedrich Hegel (Stanford Encyclopedia of Philosophy)

† See Hegel's Truth: A Property of Things?
And Science of Logic
And Science of Logic 1
And Hegel: The Science of Logic
And 6 - Hegel, ethics and the logic of universality

ential processes or processes of reasoning as presupposing participation within social communicative interactions in which the making of an assertion is considered as a move in a language-game of the "giving and asking for reasons". In short, we may think of Hegel's syllogism of necessity, which constitutes the ground or "truth" of the earlier formal conception of syllogisms, as a type of inter-subjective practice embodying thought—a type of syllogising practice that is by necessity inter-subjective and recognitive. Formally considered we might think of this syllogism as the logical schematization of the most developed form of recognition in which thinkers acknowledge others as free thinkers. *

Hegel believes he lays out a firm case for subjective and objective spirit. Subjective spirit deals with the problem of 'mind'. In contrast remember the previous discussion in the section "Quantum Mind". I will come back to this soon.

3.2.2 Philosophy of Subjective and Objective Spirit

In the Encyclopaedia, Philosophy of Nature is followed by Philosophy of Spirit (Geist). Hegel's usual triadic pattern when applied here results in sections devoted to the philosophies of subjective spirit, objective spirit, and absolute spirit. Philosophy of subjective spirit constitutes what is closest in Hegel's philosophy to a philosophy of mind in the contemporary sense, while the philosophy of objective spirit concerns those objective patterns of social interaction and the cultural institutions within which spirit is objectified in patterns of human life we have seen at work in Phenomenology of Spirit. Within subjective spirit, we may anticipate that the first division, Anthropology, will follow on from topics with which Philosophy of Nature ends —the animal organism—and so it does. Thus here Hegel is concerned with what he terms "Seele", "soul"—which seems to translate more the ancient Greek term, "psyche"—and hence the mind-body relation:

If soul and body are absolutely opposed to one another as is maintained by the abstractive intellectual consciousness,

Hegel comments,

then there is no possibility of any community between them. The community was, however, recognized by ancient metaphysics as an undeniable fact. †

Finally, Hegel believes he has made the case for absolute spirit.

Phenomenology from shapes of consciousness to shapes of spirit. The internal Phenomenology of Spirit seems to play an important role in setting up this transition from Psychology to Objective Spirit (Williams 2007), but it might also be seen as crucial in relating the more cognitive dimensions of Psychology back to the theme of embodiment prominent in Anthro-

---

* Paul Redding, "Georg Wilhelm Friedrich Hegel," in *The Stanford Encyclopedia of Philosophy*, ed. Edward N. Zalta (Spring 2024 Edition), sec. 4.

See Georg Wilhelm Friedrich Hegel (Stanford Encyclopedia of Philosophy)

† Paul Redding, "Georg Wilhelm Friedrich Hegel," in *The Stanford Encyclopedia of Philosophy*, ed. Edward N. Zalta (Spring 2024 Edition), sec. 4.

See Georg Wilhelm Friedrich Hegel (Stanford Encyclopedia of Philosophy)

pology (Nuzzo 2013a).Thus any naturalistic analysis is ultimately surpassed by a social and historical one, which itself cannot be understood as anti-naturalistic.

The philosophy of subjective spirit passes over into that of objective spirit, which concerns the objective patterns of social interaction and the cultural institutions within which spirit is objectified. The book entitled Elements of the Philosophy of Right, published in 1821 as a textbook to accompany Hegel's lectures at the University of Berlin, essentially corresponds to a more developed version of the philosophy of objective spirit and will be considered here. *

But this is not your father's God. Absolute spirit cannot be what it is without being what it isn't. Absolute carries the requirement of everything with no excess to itself in order to qualify as absolute. So, absolute spirit must start all over again by once again for the first time forgetting itself, differentiating itself as a simple point, this is space, but point must differentiate itself so it must become time and from there we eventually go back through the entire cycle.

> Hegel concludes the lectures with the claim that he has, tried to exhibit their (this series of spiritual configurations) necessary procession out of one another, so that each philosophy necessarily presupposes the one preceding it. Our standpoint is the cognition of spirit, the knowledge of the idea as spirit, as absolute spirit, which as absolute opposes itself to another spirit, to the finite spirit. To recognize that absolute spirit can be for it is this finite spirit's principle and vocation. (LHP 1825–6, III: 212) †

But now, in current physics the notion of 'it' has taken a new turn. Instead of Kant's 'thing-in-itself' conveying no information about noumena or Hegel's pure Concept, quantum computing finds no metaphysical 'thing' or Concept. Instead, it has a spooky way to find exformation in the wave function's branching of the observer effect of quantum branching. Quantum computing tells us that vast sources of quantum information can be utilized in the form of qubits. These vast sources can be thought of as reducing the vastness of what history calls 'it' into orders of magnitudes more powerful practical applications. However, this usefulness does not infer our conventional notions of practicality which subsists in some underlying 'reality'. Neither does it exist as some metaphysical notion of psychic energy. Neither does it lapse into some purely subjective substratum. This technology relies on non-spatial, non-temporal phenomena to perform calculations very differently from today's digital technologies. Quantum computing has vastly improved results for highly complicated problems in many areas of scientific research, encryption, communications, and memory storage. However, the fact still remains that all this relies on quantum entanglement which has no particular persisting state except in our utilization of it.

In some ways Hegel's circularity of object spirit and its absoluteness inclusive of what it is not might remind us of Nietzsche's eternal recurrence of the same. But for Hegel the same never reoccurs

---

* Paul Redding, "Georg Wilhelm Friedrich Hegel," in *The Stanford Encyclopedia of Philosophy*, ed. Edward N. Zalta (Spring 2024 Edition), sec. 4.

See Georg Wilhelm Friedrich Hegel (Stanford Encyclopedia of Philosophy)

† Georg Wilhelm Friedrich Hegel, *Lectures on the History of Philosophy 1825–6*, vol. 3, *Medieval and Modern Philosophy*, ed. K. Ludwig, trans. R. F. Brown and J. M. Stewart (Oxford: Clarendon Press, 2009), 212.

See Georg Wilhelm Friedrich Hegel (Stanford Encyclopedia of Philosophy)

as this would require absolute spirit not forgetting itself altogether. In other ways, this could sound a bit like LaPlace's demon as the universal (or absolute) wave function. However, the wave function has no unifying Concept, no ultimate completion beyond chaos, complexity, emergence and novelty in contradistinction to a unifying ~~universal~~[*] wave function. Even the phrase ~~universal~~[†] wave function is misleading as universal must be essentially thought from classical branching not from quantum physics. Hegel ends and begins again anew in absolute spirit. Quantum physics has no beginning or ending, no origin or *telos,* other than branchings from the observer effect. Quantum physics does not consolidate into classical terms. It remains from the intuitions of the ancients as chaos and from Levinas as radical alterity. Absolute spirit has no necessary appeal to ethics. Ethics is contingent on the development of spirit. Ethics is a higher principle, evolution of spirit, but at the expense of an alterity which can be subsumed into origin, of belonging and finding its home along the way towards absolute spirit. For Levinas, radical alterity of the other remains without the option for an echo in objective spirit.

[*] Jacques Derrida is the primary postmodern philosopher associated with "erasure" or overstriking (sous rature) words like "universal," "truth," or "being." This technique signals that a word is necessary for language yet inadequate or deceptive, challenging the stability of traditional philosophical concepts.

[†] Jacques Derrida is the primary postmodern philosopher associated with "erasure" or overstriking (sous rature) words like "universal," "truth," or "being." This technique signals that a word is necessary for language yet inadequate or deceptive, challenging the stability of traditional philosophical concepts.

4
THE 20TH CENTURY

Certainly, we can no longer relate and understand each other as sense machines. We can no longer be cogs in the machinery of history. Who we are, who I am, who my neighbor is has been brought before us as radical alterities. Throughout this major section of the volume, I want to explore deeply some of the varying implications of how we understand ourselves and 'me', who we are with each other. In my opinion, at the end of all epochs we must not merely bracket, epoque, our collective and individual histories but allow them to find their way back to the oblivion from which they came in order to let severe uncertainties undermine the 'isness' of histories and perhaps guide us towards responsibilities we never imagined.

In this section I want to focus more on the philosophical history which brought us here and how 20th century philosophy has tried to deal with the extreme uncertainties which face us. This story has not been written yet as it may not be writable but perhaps, we shall find it sayable. In any case, it is in its rough draft form at the present time. It took millenniums for the *epos*, the epic tale of our epoch(s) 'reality' to be written. Now I want to accelerate towards the direction of 20th century philosophy with more of a view to questions posed from that time on. We have still not yet begun to find our way to these ideas in popular culture. However, we have felt our way towards them in disruptions, challenges and changes to religious beliefs, cultural norms, the environment, democracies, sexuality, ethnic, gender, and disabled equality. The very power structures which ruled and demanded violent hierarchical order are like Humpty Dumpty. Either the 17th century term Humpty Dumpty applies to a drink of brandy boiled with ale, a short and clumsy person (or both) or,

Humpty Dumpty sat on a wall,
Humpty Dumpty had a great fall;
All the King's horses
And all the King's men,
Couldn't put Humpty together again.*

In my opinion Humpty Dumpty and quantum physics have much in common.

Our environmental problems alone are increasingly going to affect our immune systems, agricultural productivity, economic stability, ocean, and weather sustainability. We have also seen great achievements in medicine, automation, technology, education, robotics, artificial intelligence, quantum computing, and information accessibility. All of these changes have put immense pressures on sociological, cultural, and governmental viability and services. The rate of these changes imposed on communities brings stress and reactionary push back threatening regional and world security. Practicalities which enabled humans to make progress for centuries are increasingly calling for more personal responsibility if we are to effectively respond to species survival threats. The sciences have made enormous progress.

As we have seen in physics and philosophy the scope of these changes has rendered classical and modernistic norms inadequate to a level which challenges us to even be able to be able to make sense;

* See Humpty Dumpty.

exactly what is being rendered senseless. Our ready-made historic settings have not equipped us to even begin to adequately find words and ideas, much less practical actions, for making cultural and individual sense out of the chaos and complexity of the abyss we face. As I discussed, classicism and modernity fitted us with notions of 'reality' which found practicality based on the sensing machine. Our sensing machines are born and die in an absolute space and time universe of cause and effect of matter, things, and substances bumping into the night. In the 20th century the sensing, thinking machine was becoming more and more obsolete in its abstractions. At this point I want to look at some 20th century philosophical schools. This century began to unravel some of the underlying assumptions of classicism. This was partly due to an emerging abyss of insufficiency - quantum mechanics.

Classicism's Exhaustion

The idea of the self has evolved and sustained itself in many ways through Western history. This is a very brief summary of the way the concept of self developed over the centuries. Notice in all the descriptions below how the self always has some relationship, a continuity, to or claim as essence and reality. All these histories were very different from what culminated in our conception of the self, 'me'. But a strain which runs through them is a sense of access, relatedness, and availability to fundamental truth concerning what we think as reality, essence, Being, etc. Analogously this functions a bit like special and general relativity. They stake out domains of narration in a smooth, continuous function of relatedness to light - to reality, essence, Being, etc. The self is always consumed in an economy of relatedness to the absolute speed limit of light, to truth. I will give a brief summary here, but a more extended summary will be indicated in endnotes.

- Plato conceived of the absolute light as the Forms. Analogically, the Forms were the absolute speed limit of light The self has a rational soul (psyche). However, humans have a highly obscured understanding of absolute truth, shadows on the cave walls of the Forms. [1]
- Aristotle believed that intellect and reason were immortal and perpetual. This was the permanence of the self, analogically, the absolute speed limit of light. However, Aristotle believed this permanence was intricately linked to actuality, the materiality of the body. Becoming as a combination of potential and actual was not strictly a being because beings change. The soul is the form of the human body, but intellect and reason are the permanence which remains for change. [2]
- Descartes believed in the mind/body binary dualism called the thinking thing. The rational, thinking soul was absolute truth, analogically, the absolute speed limit of light. The soul was immaterial, immortal, and conscious. The material body was an impermanent substance. [3]
- Leibniz believed the self was an absolute monad with no exteriority. Analogically, the monad was the absolute speed limit of light. [4]
- Isaac Newton was a devout Christian. Analogically, God was the absolute speed limit of light. The self was a creation of God. The self gets its identity from God. [5]

From Christian, religion and metaphysics, the self is creation, a child of God, with divine revelation. Revelation" (Latin *revelatio*) is a translation of the Greek word *apokalypsis*, which means the removal of a veil so that something can be seen. * Epiphany is the knowledge of God. Analogically, God was the absolute speed limit of light.

From Enlightenment, there was a plethora of notions of the self and reality. This ranged from Christian ideas of God, mysticism, humanism, and Eastern philosophy. Dominant themes were reason, knowledge, freedom, and budding ideas of modernity in the British empiricists near the end of Enlightenment. Likewise, analogical notions of the absolute speed of light ranged from God to deism to humanism. Additionally, we see with the empiricists that analogical notions of the absolute speed of light first explicitly taking on the form of neutrality. Many of the empiricists of this era also believed in God but held that empirical observation beheld the light of a neutral universe. Here the self is cast off into agnosticism or merely personal preference. Empiricism challenged traditional notions of an unchanging, immortal soul. Instead, it emphasized the role of sensory experiences and the dynamic nature of our mental lives. The soul, according to many empiricists, is not a fixed essence but a product of our perceptions and consciousness. [6]

Hegel is a very difficult thinker to ascertain an analogous absolute speed of light. The main reason for this is due to the wide variations in the schools of Hegel. The vertical schools would certainly tell us the analogous speed of light would be Hegel's idea of the Concept in his "Logic". However, the horizontal schools would disagree on this point. For them, the absolute speed of light would not be an Idea but a process the process of self-creation, of emergence and organic evolution. This would imply that no kind of absolute could take on any typical traditional meanings including the use of the term 'absolute'. It would also disregard neutrality as absolute. However, the process itself is a dialectic. In this sense there is a logical, smooth transition to different stages of becoming and the 'not' of having become. So, this would still fit into the continuity of a relativistic analogy. [7]

From modernism, gathering together the logic of neutrality wherein all meaning has been condensed and reduced into sheer isness. Neutrality like the *'is'* and its totalizing 'not' recognizes no externality. It is the bankruptcy of classicism into units of exchange. The only meaning to be derived is from transactionalism, the echo of the thinking thing. Even nature, our Roman inheritance, transacts with the currency of cause and effect. The uncaused as metaphysics is relegated to classicism's logic of negation in which no excess can remain. Meaning voids itself in the culminating neutrality of historic mediocrity.

It seems fitting that classicism finally gasps its 'not' as the age-old intuition of the absolute as the speed of light in relativity. Relativity ends the epoch of essences and truth with the silent star of the absolute speed of light. Everything in spacetime, even spacetime, is bent towards the speed of light, disappearing into the light. But the vast majority of spacetime is not light but darkness, the lingering, simmering night of the soul, vanished by the impoverished 'not' of classicism. But imperceptibly exceeding this final totality is quanta and its spooky function. This can no longer be light or its 'not' but an excess having no relation to continuous, smooth, and relative spacetime. Rather, irreconcilable disjuncture, without relation and relativity with some kind of spooky, hazy, chaotic branching, analogic confluences of many-worlds, multiverses – enumerable variations of possible actualizing narratives. But always excess, apart from observer effects, from selves, and yet, already simultaneously within the wave function's spooky externality. This is the gap we face at the end of history's epochs.

---

* See Divine Revelation (Stanford Encyclopedia of Philosophy)

Phenomenological Transitions

In the early 20th century, physics's relativity found a new kind of spacetime from the absolute space and time of classicism. It found the spacetime is curved by the absolute speed limit of light. This spacetime is bent, has stretch, compression, and distortion, changes the experience of temporality and extension. Reality is moldable to experience, to lived experience. Concurrently, phenomenology is born. Phenomenology as well as modernism had its early roots all the way back to the 17th century and perhaps even earlier. And as was the case for modernism, phenomenology also came into its own in the late 1800s with Husserl. In phenomenology, the concurrent findings of special and general relativity are accommodated with the effort to get back to how we find ourselves in actual living experiences. And from there phenomenology wants to investigate how meaning and significance may arise. Phenomenology brought on a banishment of modernist abstractions, the absolutes and essences of history, and back to perspectives of experiences. Modernism was the last gasp of classicism.

Often in the history of physics and philosophy there seems to be an interesting interplay between them in which both complement and expand each other's mutual areas of interest. The 20th century seems to be no different. Einstein's theory of relativity went against a centuries old idea called aether. Even in the 19th century, physics and philosophy held aether in esteem. Aether was both physical and spiritual. Aether was thought to pervade the universe almost like what we have called vacuum energy. It had a kind of unknown mystique perhaps a bit like how we think of dark energy today. Aether did function like a speed limit or smallest possible medium, but Einstein's special theory of relativity made it irrelevant due to his discoveries about the speed of light. Einstein's special theory of relativity in 1905 met with outright critics in physics, philosophy, and religion or he was simply ignored. But for some at the time Einstein's special relativity was just another rendition of the aether. However, the aether would shortly become an unnecessary holdover from the past.

> Eventually, Albert Einstein published in September 1905 what is now called special relativity, which was based on a radical new application of the relativity principle in connection with the constancy of the speed of light. In special relativity, the space and time coordinates depend on the inertial observer's frame of reference, and the luminiferous aether plays no role in the physics. Although this theory was founded on a very different kinematical model, it was experimentally indistinguishable from the aether theory of Lorentz and Poincaré, since both theories satisfy the relativity principle of Poincaré and Einstein, and both employ the Lorentz transformations. After Minkowski's introduction in 1908 of the geometric spacetime model for Einstein's version of relativity, most physicists eventually decided in favor of the Einstein-Minkowski version of relativity with its radical new views of space and time, in which there was no useful role for the aether.*

The 20th century's relativity destroyed the notions of absolute time and space which historically and metaphysically got intertwined with the absolute God of Christianity. In relativity time could no

* Michel Janssen, "Einstein's First Years in Zurich (1909–1911)," in *The Cambridge Companion to Einstein*, eds. Michel Janssen and Christoph Lehner (Cambridge: Cambridge University Press, 2014), 165–166.

See Criticism of the Theory of Relativity

longer be unchanging and absolute. In quantum mechanics time does not exist and neither does the local realism of 'here'. Frank Wilczek, the American theoretical physicist and mathematician and Nobel laureate at MIT I discussed earlier tells us,

> As relativity undermines 'now', quantum theory undermines 'here'.*

When relativity relativized time and space to the inertial frame of the observer, many outspoken Christian evangelists proclaimed it as demonic heresy. While all this was going on against Einstein, modern phenomenology was born which Einstein seemed to favor. Einstein's general theory of relativity in 1915 also met with criticism but was mostly ignored for a number of years. A collection of essays was published in a book in 1931 which detailed criticisms of relativity from many different disciplines. Here is more information about the book.

> A collection of various criticisms can be found in the book ""*Hundert Autoren gegen*" Einstein (One Hundred Authors Against Einstein), published in 1931. It contains very short texts from 28 authors, and excerpts from the publications of another 19 authors. The rest consists of a list that also includes people who only for some time were opposed to relativity. From among Einstein's concepts the most targeted one is space-time followed by the speed of light as a constant and the relativity of simultaneity, with other concepts following. Besides philosophic objections (mostly based on Kantianism), also some alleged elementary failures of the theory were included; however, as some commented, those failures were due to the authors' misunderstanding of relativity. For example, Hans Reichenbach wrote a report in the entertainment section of a newspaper, describing the book as "a magnificent collection of naive mistakes" and as "unintended droll literature." Albert von Brunn interpreted the book as a pamphlet "of such deplorable impotence as occurring elsewhere only in politics" and "a fallback into the 16th and 17th centuries" and concluded "it can only be hoped that German science will not again be embarrassed by such sad scribblings", and Einstein said, in response to the book, that if he were wrong, then one author would have been enough.
>
> According to Goenner, the contributions to the book are a mixture of mathematical–physical incompetence, hubris, and the feelings of the critics of being suppressed by contemporary physicists advocating the new theory. The compilation of the authors show, Goenner continues, that this was not a reaction within the physics community—only one physicist (Karl Strehl) and three mathematicians (Jean-Marie Le Roux, Emanuel Lasker and Hjalmar Mellin) were present—but a reaction of an inadequately educated academic citizenship, which did not know what to do with relativity. As regards the average age of the authors: 57% were substantially older than Einstein, one third was around the same age, and only two persons were substantially younger. Two authors (Reuterdahl, von Mitis) were antisemitic and four others were possibly connected to the Nazi movement. On the other hand, no antisemitic expression can be found in the book, and it also included contributions of some authors of Jewish

* Milic Capek, *The Philosophical Impact of Contemporary Physics* (Princeton: D. Van Nostrand Company, 1961), 373. See When words fail (pdf)

ancestry (Salomo Friedländer, Ludwig Goldschmidt, Hans Israel, Emanuel Lasker, Oskar Kraus, Menyhért Palágyi).*

As mentioned, starting at the same time Einstein faced a lot of scientific and religious opposition early on and with his later general theory of relativity. Anti-Semites attacked his theory and him personally. Einstein's relativity sparked a lot of religious indignation in the early 20th century. Einstein was associated with politically left ideologies including the struggle of civil rights for African Americans. He was well versed in philosophy and wrote about it. Kantians criticized Einstein's spacetime theory as Kant's *a priori* Categories of Understanding seemed to contradict relativity. In 1924 Einstein responded to the Kantians with this,

> Until some time ago, it could be regarded as possible that Kant's system of a priori concepts and norms really could withstand the test of time. This was defensible as long as the content of later science held to be confirmed*) did not violate those norms. This case occurred indisputably only with the theory of relativity. However, if one does not want to assert that relativity theory goes against reason, one cannot retain the a priori concepts and norms of Kant's system.
>
> [Footnote] *) To refute Kant's system it actually suffices to indicate a logically conceivable theory (corresponding to conceivable observational material) that conflicts with Kantian norms. Whether non-Euclidean geometries accomplished this remained controversial.
>
> For starters, this does not exclude, at least, the retention of Kant's way of posing the problem, as Cassirer, for instance, does. I am even of the opinion that this standpoint cannot be strictly refuted by any scientific development. For, one will always be able to say that critical philosophers had hitherto erred in setting up the a priori elements and one will always be able to set up a system of a priori elements that does not conflict with a given physical system. I surely may briefly indicate why I do not find this standpoint natural. Let a physical theory consist of the parts (elements) A, B, C, D, which together form a logical whole that correctly connects the pertinent experiments (sensory experiences). Then the tendency is that less than all four elements, e.g., A, B, D, still say nothing about the experiences, without C; no more so A, B, C, without D. One is then free to regard three of these elements, e.g., A, B, C, as a priori and only D as empirically determined. What always remains unsatisfactory in this is the arbitrariness of the choice of elements to be designated as a priori, even disregarding that the theory could be replaced at some point by another theory that substitutes some of these elements (or all four of them) with others. One could be of the view, though, that through direct analysis of human reason, or thought, we would be in a position to recognize elements that would have to be present in any theory. But most researchers would probably agree that we lack a method for recognizing such elements, even if one were inclined to believe in their existence. Or should one imagine that the search for a priori elements was a kind of asymptotic process that advances along with the development of science?†

* Milada Scharff and Petr Scharff, "The Case of *Hundert Autoren gegen Einstein*," in *Einstein's Relativity and Its Contexts*, eds. Petr Scharff and Milada Scharff (Berlin: Springer, 2011), 45–47.

See Criticism of the Theory of Relativity

† Albert Einstein, "Is Theory of Relativity Compatible with Kantian Philosophy?," in *The Collected Papers of Albert Einstein*,

Effectively, Einstein uses Karl Popper's argument that religion is unfalsifiable on the Kantians. Einstein tells us that Kant's Categories of Understanding as mentioned earlier [quantity (unity, plurality, totality), quality (reality, negation, limitation), relation (inherence and subsistence (substance and accident), causality and dependence (cause and effect), community (reciprocity), and modality (possibility, existence, necessity)] were thought by Kant as pre-conditioning thinking. These categories were *a priori* so there would be no way to stand outside of them to critique them since thinking itself is already conditioned by the categories. This made the categories unfalsifiable. Einstein seems to indicate in the quote above that even if specifics of the categories were changed or eliminated, according to Kant the *a priori* assumption could not be proven false. If something is asserted as prior to existence as merely an assertion, without evidence, how could it possibly be falsifiable? Think about trying to disprove religious dogma. It is ironic that in the case of exformation which Einstein himself was largely responsible for simultaneously recognizing and criticizing (e.g. in search of a unified field theory), we do have a case of information as *a priori*. What is the difference?

Exformation has empirical evidence which has had substantial recent histories which experimentally verify drawing these conclusions. This satisfies the empiricist conditions for knowing. Certainly, there could be future evidence which could de-spook entanglement but more and more this seems unlikely. There are a minority of physicists which are still actively trying to find alternate explanations but for the majority of physicists this seems to be a settled issue. For entanglement to be anything like a 'settled issue' in physics it has to have undergone many decades of verifiable, repeatable, and exhaustive searching for alternative solutions which do not have the 'spook' factor. There are some professors who actively try to teach entanglement by downplaying its implications, restricting their mentoring to more strict mathematical terms. This is certainly understandable for students making their careers in physics. However, in the view of many famous and not so famous physicists who allow themselves to engage in philosophical implications there are solid reasons why exformation implies massive historic, social, and philosophical consequences. Falsifiability must always be a foundation of science for science to be science. Additionally, initial findings in science most often are abstract. If these findings survive the test of time, they tend to become more concrete and simpler like Einstein's famous $e = mc^2$. Philosophy has a fundamentally different basis than science.

On Husserl

Aristotle's "Physics" and "Metaphysics" books show us why philosophy can never be purely empirical as the Positivists claimed. As mentioned earlier, Einstein was fully aware of why philosophy could not be reduced to empiricism. I also think philosophy should never be exempt from falsifiability or more precisely, critical systemic questions and shortcomings in any particular philosophical system. Even as Positivism anchored scientific findings in phenomenal observations which can never be considered complete, Socratic ignorance must be the foundation of philosophy so that it does not lapse into dogmatism. Edmund Husserl (1859-1938) founded phenomenology as an empirical science of science.

---

vol. 14, *The Berlin Years: Writings & Correspondence, April 1923–May 1925*, eds. Diana Kormos Buchwald et al. (Princeton: Princeton University Press, 2015), 346–347.

See Einstein on Kant

His motto was "back to things themselves" not historic abstractions but rigorous and bracketed from historic biases focus on lived phenomenon.

> In the first decade of the 20th century, Husserl considerably refined and modified his method into what he called "transcendental phenomenology". This method has us focus on the essential structures that allow the objects naively taken for granted in the "natural attitude" (which is characteristic of both our everyday life and ordinary science) to "constitute themselves" in consciousness. (Among those who influenced him in this regard are Descartes, Hume and Kant.) As Husserl explains in detail in his second major work, Ideas (1913), the resulting perspective on the realm of intentional consciousness is supposed to enable the phenomenologist to develop a radically unprejudiced justification of his (or her) basic views on the world and himself and explore their rational interconnections...
>
> As a philosopher with a mathematical background, Husserl was interested in developing a general theory of inferential systems, which (following Bolzano) he conceived of as a theory of science, on the ground that every science (including mathematics) can be looked upon as a system of propositions that are interconnected by a set of inferential and grounding relations. Following John S. Mill, he argues in Logical Investigations that the best way to study the nature of such propositional systems is to start with their linguistic manifestations, i.e., (sets of) sentences and (assertive) utterances thereof.
>
> How are we to analyse these sentences and the propositions they express? Husserl's approach is to study the units of consciousness that the respective speaker presents himself as having—that he "gives voice to"—in expressing the proposition in question (for instance, while writing a mathematical textbook or giving a lecture). These units of consciousness he labels intentional acts or intentional experiences, since they always represent something as something—thus exhibiting what Brentano called intentionality. According to Husserl, there are non-intentional units of consciousness as well. (He quotes pain as an example.) What distinguishes intentional from non-intentional experiences is the former's having intentional content. *

There is strong evidence that Einstein was aware of what Husserl was doing in phenomenology. In 1905, the same year as Einstein's special theory of relativity, Edmund Husserl founded the school of modern phenomenology. Much of Einstein's language discussed by the paper below reflects an in-depth knowledge of what Husserl and his student Martin Heidegger were working on. The bold indicates that Einstein had a sympathetic view of the budding philosophies of phenomenology. Also, note in this text that the positivists found Einstein to be an existential threat to science.

> In an unpublished manuscript from the same year as Husserl's Crisis, Einstein seems to turn to this phenomenological problem using the same Husserl's terminology:
>
> Science as something existing, confectioned, is more than something of **objectivized and,**

---

* Christian Beyer, "Edmund Husserl," in *The Stanford Encyclopedia of Philosophy*, ed. Edward N. Zalta (Winter 2022 Edition), sec. 2.

See Husserl (Stanford)

at the same time, impersonalized that we as human beings know in general. Science as something becoming, as destination, is also **equally subjective, psychologically conditioned** as all the other human aspirations. (...) Naturally there are people who affirm that science has produced a too high connection among experiencing **facts** [erlebbaren], so we can deduce **from experienced facts other experiencing facts**. According to the opinion of some positivists, the possible solution of such assignment would be the unique end of science. (...) There is in fact a stronger and, for this reason, also darker impulse behind these assiduous disputes: the will of understanding Being, the Reality. However, it seems that we have avoided using such words, since we are really embarrassed in clarifying what we must properly intend for real [Wirklich] and what we must understand [begreifen] in this general affirmation. All these efforts are founded upon the trust that Being in its structure is incomplete harmony.

Referring himself to Husserl's phenomenology – as I think – Einstein concludes that:

(...) **In our time a new and original thought is beginning to emerge. If this time has produced a progress of the epistemological sphere, so in fact it seems to me that we could not give for granted any reasonable ways that brings us from the Erleben (experiencing) to the conceptual Erfassen (knowledge) of things, since every thought is founded on a free theoretical construction, which systematically derives through sense experiences.** *

The latter quote by Einstein appears to show an in-depth knowledge of Husserl's work. The German words Erleben (experiencing) and Erfassen (knowledge) are at the root of Husserl's phenomenological investigations. Particularly, why are human experiences rejected from the science of 'things'? Husserl's motto was back to the things themselves or the way we intentionally orient ourselves to phenomenon. Einstein seems to validate the effort to reevaluate thought with regard to sense experience.

From Einstein's quote, the Positivists opposed anything *a priori* such as Kant's Categories of Understanding. For the Positivists, this kind of truth was true by definition such as metaphysics or *a posteriori* meaning already 'empirically' determined. What was meant by empiricism for the Positivists? As empirical, science was falsifiable and not mere assertions of dogmatism, unfalsifiable. Additionally, Einstein points out that the Positivists were incapable of understanding any such notion as 'Being'. Being was the main concern of Heidegger's early work "Being and Time". We have previously used the term epoch to indicate an age. Now I would like to introduce the term '*epoché*'. How did we get epoch from epoché?

Epoch comes to us, via Medieval Latin, from Greek epochē, meaning "cessation" or "fixed point." "Epochē," in turn, comes from the Greek verb epechein, meaning "to pause" or "to hold back." When "epoch" was first borrowed into English, it referred to the fixed point used to mark the beginning of a system of chronology. That sense is now obsolete, but today "epoch" is used in some fields (such as astronomy) with the meaning "an instant of time or a date selected

* Christian Beyer, "Edmund Husserl," in *The Stanford Encyclopedia of Philosophy*, ed. Edward N. Zalta (Winter 2022 Edition), sec. 2.

See Phenomenology and Relativity: Husserl, Weyl, Einstein and the concept of Essence

> as a point of reference." The "an event or a time that begins a new period or development" sense first appeared in print in the early 17th century, and "epoch" has been applied to defining moments or periods of time ever since. *

For Husserl *epoché* means,

1. Moment of theoretical suspension of all action.
2. Moment of theoretical suspension of disbelief. †

Epoché refers to a fundamental technique that goes beyond mere abstraction and examination of essences. It serves to highlight consciousness itself by suspending trust in the objectivity of the world. In Hellenistic philosophy, epoché plays a crucial role in Pyrrhonism, a skeptical philosophy. Pyrrhonists use it to suspend judgment about non-evident matters, aiming for a state of ataraxia—freedom from worry and anxiety. Sextus Empiricus, a Pyrrhonist philosopher, defines epoché as a state of intellect where we neither deny nor affirm anything. The Pyrrhonists employ techniques like the Ten Modes of Aenesidemus, the Five Modes of Agrippa, and Pyrrhonist maxims to achieve epoché. In Stoicism, epoché describes withholding assent to Phantasia (impressions). Epictetus explains that epoché involves feeling uncertainty, neither affirming nor denying, but recognizing the indeterminacy of a situation. Edmund Husserl introduced epoché as a systematic procedure in his work Ideas I. Through phenomenological reduction, practitioners suspend judgment regarding the existence of the external world. This allows them to examine phenomena as they are originally given to consciousness. Husserl distinguishes between universal epoché (leaving behind all assumptions of existence) and local epoché (setting aside specific assumptions). The former has a stronger effect than the latter. For Husserl, epoché serves as a tool for exploring consciousness, questioning assumptions, and gaining deeper insights into our experience of reality. ‡ Husserl used the word *epoché* to mean 'bracket'. Husserl asks us to bracket our individual assumptions about our normal everyday life.

> The modern founder of phenomenology is the German philosopher Edmund Husserl (1859–1938), who sought to make philosophy "a rigorous science" by returning its attention "to the things themselves" (zu den Sachen selbst). He does not mean by this that philosophy should become empirical, as if "facts" could be determined objectively and absolutely. Rather, searching for foundations on which philosophers could ground their knowledge with certainty, Husserl proposes that reflection put out of play all unprovable assumptions (about the existence of objects, for example, or about ideal or metaphysical entities) and describe what is given in experience. The road to a presuppositionless philosophy, he argues, begins with suspending the "natural attitude" of everyday knowing, which assumes that things are simply there in the external world. Philosophers should "bracket" the object-world and, in a process he calls

---

* "Epoch," Merriam-Webster.com Dictionary, s.v. "History and Etymology."
See epoch (merriam-webster)

† See epoche

‡ See epoche (Britannica)
And Epoché
And Bracketing (phenomenology)

> epoché, or "reduction," focus their attention on what is immanent in consciousness itself, without presupposing anything about its origins or supports. Pure description of the phenomena given in consciousness would, Husserl believes, give philosophers a foundation of necessary, certain knowledge and thereby justify the claim of philosophy to be more radical and all-encompassing than other disciplines (see Ideas 95–105 and Meditations 11–23). *

Husserl's phenomenology involved not theorizing about what **is** the case in any way. It also meant not simply taking, for example, classical science's findings (e.g. absolute time and space) as relevant. Husserl wanted to get back to the phenomenon itself in perception and experience. This is where lived time and lived space found relevance. In my view of Husserl, the notions of lived body and lifeworld were intentional objects which wanted to downplay the copula, the determinate *is* in experience, in sensuality. **Sensibility is not sensuality.** I will use these words carefully in everything which follows to keep this difference clear. For Husserl, sensuality was how we live as 'affect', as feeling, as immediate lived experience. This is what we mean as sensuality. For Husserl, sensibility arises from sensuality taken up into intentionality.

Husserl delved into the intricate realms of consciousness and perception. Husserl believed that life primarily consists of the immanent mental life of a subject. This life encompasses both conscious and unconscious dimensions. The conscious aspects involve our direct awareness, thoughts, and experiences. The unconscious aspects, however, remain hidden from immediate awareness but play a crucial role in shaping our mental life. In his later works, Husserl emphasized the importance of understanding life as a whole, including its unconscious layers. For Husserl, the unconscious is not limited to mere sensory data (hyletic sensations). It extends beyond that. It encompasses perceptual unities, structures, thoughts, and even memories. In his work entitled "Analyses of Passive Synthesis", Husserl explored various meanings of the unconscious. Husserl's notion of life is closely tied to embodiment. Life is not an abstract concept; it is inherently connected to our bodily existence. He considered life as the pure immanent life of subjectivity or consciousness. Through his genetic phenomenology, Husserl aimed to understand the a priori conception of experiential genesis. Life, according to him, involves both conscious and unconscious elements, and it is through this life that we engage with the world. From a Husserlian perspective, we are part of a universal community of living beings. As self-conscious rational subjects, we can acquire apodictically self-evident ethical insights. These insights guide us toward ethical authenticity and responsibility, even in our interactions with non-human and non-conscious beings. Husserl's exploration of sensuality goes beyond sensory data, touching upon the intricate interplay of consciousness, embodiment, and ethical understanding within the fabric of life. †

> Husserl compares this process of intentional "forming" of sensual matter to the interpretation of a linguistic expression, but this comparison should not mislead us to conclude that he subscribes to a sense-datum theory of perception (see Section 2 above, headword: mental image

* "Phenomenology," Joukowsky Institute for Archaeology & the Ancient World, Brown University, n.d. See Phenomenology (Brown University)

† See Edmund Husserl on the Conscious and Unconscious Aspects of Immanent Life
And Husserl's Layered Concept of the Human Person: Conscious and Unconscious
And 9 - Husserl's theory of meaning

theory). Rather, his view on perception is best characterized as a sophisticated version of direct (i.e., non-representationalist) realism. *

Husserl focused on the lived experience of intentionality.

In Logical Investigations Husserl developed a view according to which conscious acts are primarily intentional, and a mental act is intentional only in case it has an act-quality and an act-matter. Introducing this key distinction, Husserl writes:

The two assertions '2 x 2 = 4' and 'Ibsen is the principal founder of modern dramatic realism', are both, qua assertions, of one kind; each is qualified as an assertion, and their common feature is their judgment-quality. The one, however, judges one content and the other another content. To distinguish such 'contents' from other notions of 'content' we shall speak here of the matter (material) of judgment. We shall draw similar distinctions between quality and matter in the case of all acts.

An additional notion in the Investigations, which grows in importance in Husserl's later work and will be discussed here, is the act-character. Husserl views act-quality, act-matter and act-character as mutually dependent constituents of a concrete particular thought. Just as there cannot be color without saturation, brightness and hue, so for Husserl there cannot be an intentional act without quality, matter and character. The quality of an act (called 'intentional act' above) is the kind of act that it is, whether perceiving, imagining, judging, wishing, and so forth. The matter of an act is what has been called above its intentional content, it is the mode or way in which an object is thought about, for example a house intended from one perspective rather than another, or Napoleon thought of first as "the victor at Jena", then as "the vanquished at Waterloo". The character of an act can be thought of as a contribution of the act-quality that is reflected in the act-matter. Act-character has to do with whether the content of the act, the act-matter, is posited as existing or as merely thought about and with whether the act-matter is taken as given with evidence (fulfillment) or without evidence (emptily intended). †

Modernity assembled objects into an abstract objective world for a subject. Husserl's lifeworld was the science of how human experience can assemble any such thing as modernity's science.

Among the fundamental beliefs thus uncovered by Husserl is the belief (or expectation) that a being that looks and behaves more or less like myself, i.e., displays traits more or less familiar from my own case, will generally perceive things from an egocentric viewpoint similar to my own ("here", "over there", "to my left", "in front of me", etc.), in the sense that I would

* Christian Beyer, "Edmund Husserl," in *The Stanford Encyclopedia of Philosophy*, ed. Edward N. Zalta (Winter 2022 Edition), sec. 3.

See Edmund Husserl (Stanford)).

† Marianne Sawicki, "Husserl, Edmund: Intentionality," in *Internet Encyclopedia of Philosophy*, ed. James Fieser and Bradley Dowden, sec. 2.

See Edmund Husserl: Intentionality and Intentional Content
And Husserl (Stanford)

> roughly look upon things the way he does if I were in his shoes and perceived them from his perspective. This belief allows me to ascribe intentional acts to others immediately or "appresentatively", i.e., without having to draw an inference, say, by analogy with my own case. So the belief in question must lie quite at the bedrock of my belief-system. It forms a part of the already pregiven (and generally unreflected) intentional background, or "lifeworld" (cf. Crisis), against which my practice of act-ascription and all constitutive achievements based upon that practice make sense in the first place, and in terms of which they get their ultimate justification. *

The lifeworld had two senses to it. It had personal beliefs, opinions, or intentions projected into what Husserl calls intentionality. It also had a collective sense where intentionality congeals into society, history, language, etc.

> Husserl's notion of lifeworld is a difficult (and at the same time important) one. It can roughly be thought of in two different (but arguably compatible) ways: (1) in terms of belief and (2) in terms of something like socially, culturally or evolutionarily established (but nevertheless abstract) sense or meaning. †
>
> Here is a brief but more detailed explanation of these two notions.
>
> (1) If we restrict ourselves to a single subject of experience, the lifeworld can be looked upon as the rational structure underlying his (or her) "natural attitude". That is to say: a given subject's lifeworld consists of the beliefs against which his everyday attitude towards himself, the objective world and others receive their ultimate justification. (However, in principle not even beliefs forming part of a subject's lifeworld are immune to revision. Hence, Husserl must not be regarded as an epistemological foundationalist; see Føllesdal 1988.)
>
> (2a) If we consider a single community of subjects, their common lifeworld, or "homeworld", can be looked upon, by first approximation, as the system of senses or meanings constituting their common language, or "form of life" (Wittgenstein), given that they conceive of the world and themselves in the categories provided by this language.
>
> (2b) If we consider subjects belonging to different communities, we can look upon their common lifeworld as the general framework, or "a priori structure", of senses or meanings that allows for the mutual translation of their respective languages (with their different associated "homeworlds") into one another.
>
> The term "lifeworld" thus denotes the way the members of one or more social groups (cultures, linguistic communities) use to structure the world into objects (Husserliana, vol. VI, pp. 126–138, 140–145). The respective lifeworld is claimed to "predelineate" a "world-horizon" of

* Christian Beyer, "Edmund Husserl," in *The Stanford Encyclopedia of Philosophy*, ed. Edward N. Zalta (Winter 2022 Edition), sec. 7.

See Husserl (Stanford)

† Christian Beyer, "Edmund Husserl," in *The Stanford Encyclopedia of Philosophy*, ed. Edward N. Zalta (Winter 2022 Edition), sec. 7.

See Husserl (Stanford)

> potential future experiences that are to be (more or less) expected for a given group member at a given time, under various conditions, where the resulting sequences of anticipated experiences can be looked upon as corresponding to different "possible worlds and environments". *

When worldviews begin to break down, they not only become less functional, but they also become more abstract. That is, they no longer seamlessly allow practicality. Another way to think of Husserl's intentionality is as a movement away from abstraction towards the concrete experience of lived human phenomena. How does experience come to us? Is it as concepts or simple practicality which need no further elaboration? Practical acts are not mere mindless movements but always, already assume specific contexts. The context brought to the fore in 2 + 2 = 4 is different from the context of the 19th century playwright Ibsen's juxtaposition of modernism and the idealism of Hegel contrasted with the iconoclasm of Nietzsche's anti-idealism. † Here Husserl does not want to debate the specific themes, the 'facts of the case', he wants to think about the lived modalities in which these different kinds of contextual relationships occur. Intentional acts have qualities ranging from evidentiary-based to wishful (without evidence). But what is 'evidence' for Husserl? For Plato evidence was the shadows of what is seen emanating from what is not seen, the forms. For Descartes evidence was the thinking thing, the *cogito*, which cannot doubt that it thinks.

> While, in Descartes, the concept of evidence meant "seeing" something without any doubt, in Husserl it does not have a single meaning as apodictic [my comment, absolute] certainty. This is due to the fact that Husserl used evidence in two different but closely interrelated senses. Husserl did not separate experience and evidence. That is why evidence emerged as a form of consciousness, an intentionality, in Husserl. This intentional act resembles the structure of lives and acts of consciousness, the periphery of consciousness, and life with such a structure is determined to be the consciousness of something. Husserl explains such an intentional experience (that is, intentionality) as "being directed towards something aimed at", "being turned to something". Evidence thus is, in fact, a form of consciousness, an intentionality. Intentionality and evidence are, for this reason, two concepts which are, to some extent, concurrent in the philosophy of Husserl. ‡

In this way Husserl is more a psychologist of living phenomenon than a theorist of philosophical or ideological propositions. However, let's note here that experiential observation is not so unlike scientific experimental observation with the exception of the specific context of mathematical repeatability. However, the motifs or contexts which science allows cannot account for such repeatable notions as cultural values, narratives of history, and ways language shapes perception. We do see the effects of all these in science in the forms of competing scientific paradigms which always contextually underlie the

* Christian Beyer, "Edmund Husserl," in *The Stanford Encyclopedia of Philosophy*, ed. Edward N. Zalta (Winter 2022 Edition), sec. 7.
See Husserl (Stanford)

† See The Drama of History: Ibsen, Hegel, Nietzsche

‡ Christian Beyer, "Edmund Husserl," in *The Stanford Encyclopedia of Philosophy*, ed. Edward N. Zalta (Winter 2022 Edition), sec. 4.
See Husserl's Evidence Problem

assumptions wherein phenomenon is allowed to show itself in science. One such example of this is how classic physics allowed scientific discovery to come to the fore. We see this effect in Einstein's belief that quantum physics was lacking due to uncertainty and needed the classic notion of locality to complete it. So, what is it that orients us in the world where science, religion, culture, language, and history orient us? For Husserl it was the body and the 'lifeworld'.

> According to Husserl, it is the body which provides orientation ; man's intentionality is possible thanks to the body. A person's body is, therefore, essentially important in phenomenology. It is the centre for orientation. According to Husserl, who emphasizes the unity of the soul and the body by pointing out that "the body is all I possess", and who claims that the "I" is a whole composed of the soul and the body, there are two features of the physical organism. These two features cannot be thought of apart. The body is, firstly, an organ for perception and demand. I perceive anything through my body, which is an organism that belongs to me. If my body were not a whole made up of individual organs, nothing would exist. My body, for this reason, is a prerequisite for the world to exist. I am, therefore, dependent on a centre of orientation, and thus on my body, in whatever I perform. This centre of orientation is originally bestowed upon me with every turn I take. I am unable to perform without my body. According to Husserl, whatever one perceives, wherever one goes, one experiences everything with respect to the relationship between oneself and one's body. Such dimensions as "over", "below", "right", "left" and so forth derive meaning only in relation to my body. [*]

Husserl's motto of 'back to things in themselves' meant back to our concrete experience as embodied. So, in this sense Husserl looked at phenomenology as archeology. Unearthing artifacts of lived experience was the process of phenomenology. However, even the phenomenologist's observation conditioned what showed itself as lived experience and lifeworld. Is this so unlike quantum physics' complementarity idea where experimental observation has a determinacy which conditions what it observes but says nothing about absolute 'reality' when not being observed? However, just as quantum observation is not some sort of mere subjectivity so phenomenology derives a sense of agreement not necessarily from some kind of 'pure' detached scientific observation, so culture, language, and history have a determinate kind of arena where judgements can more or less find agreement.

> In experience we become conscious of a thing. Evidence, on the other hand, is an act of consciousness which gives (presents) a thing – that is, brings about an experience. Thus, evidence is the name for experiencing which may have various degrees or grades of adequacy. Therefore, according to Husserl, there is no one evidence, but many evidences or experiences. There also are evidences which are relative, which are not certain, allowing for doubt one way or the other, which in fact depend upon other experiences, which are not consistent, and which thus leave some aspects in the dark, or even appear altogether wrong in the end, as a result of which "what" is shown is exposed as "nothing" [†]

---

[*] Christian Beyer, "Edmund Husserl," in *The Stanford Encyclopedia of Philosophy*, ed. Edward N. Zalta (Winter 2022 Edition), sec. 4.

See Husserl's Evidence Problem

[†] Christian Beyer, "Edmund Husserl," in *The Stanford Encyclopedia of Philosophy*, ed. Edward N. Zalta (Winter 2022 Edition),

On Heidegger

Martin Heidegger (1889-1976), a student of Husserl, addresses the matter of presuppositions in "Being and Time". Heidegger would rather propose understanding as "ahead of itself" as *"projecting expectations that interpretation then makes explicit"* from a past, a "forestructure", which has already made a specific understanding possible.

> Later phenomenologists have been skeptical of Husserl's contention that description can occur without presuppositions, in part because of Husserl's own analysis of the structure of knowledge. According to Husserl, consciousness is made up of "intentional acts" correlated to "intentional objects." The "intentionality" of consciousness is its directedness toward objects, which it helps to constitute. Objects are always grasped partially and incompletely, in "aspects" (Abschattungen) that are filled out and synthesized according to the attitudes, interests, and expectations of the perceiver. Every perception includes a "horizon" of potentialities that the observer assumes, on the basis of past experiences with or beliefs about such entities, will be fulfilled by subsequent perceptions (see Meditations 39–46).
>
> Extrapolating from Husserl's description of consciousness, Martin Heidegger(1889–1976) argues that understanding is always "ahead of itself" (sich vorweg), projecting expectations that interpretation then makes explicit. In the section "Understanding and Interpretation" in Being and Time (1927) Heidegger argues that inherent in understanding is a "forestructure" (Vorstruktur) of assumptions and beliefs that guide interpretation. Heidegger's account of the interdependence of understanding and expectations is in part a reformulation of the classic idea that interpretation of texts is fundamentally circular, inasmuch as in interpretation the construal of a textual detail is always necessarily based on assumptions about the whole to which it belongs (see Palmer and hermeneutics). His theory of understanding also reflects his own assumptions about human existence, which he describes as a process of projection whereby we are always outside of and beyond ourselves as we direct ourselves toward the future. Heidegger's conception of the anticipatory structure of understanding is important for later versions of phenomenology that focus on interpretation and reading. Hermeneutic phenomenology (especially as developed by Hans-Georg Gadamer and Paul Ricoeur) further the role of presuppositions in understanding, and phenomenological theories of textual reception (especially the "Constance school," led by Hans Robert Jauss and Wolfgang Iser) investigate how literary works are understood differently by audiences with different interpretive conventions (see reader-response theory and criticism and reception theory). *

Let's recall our discussion on Aristotle's idea of primary and secondary or accidental being. Heidegger thinks that Aristotle was on the right track but mistook being as presence. Heidegger takes Being to be more fundamental than what comes to presence.

---

sec. 4.
See Husserl's Evidence Problem

* "Phenomenology," Joukowsky Institute for Archaeology & the Ancient World, Brown University, n.d.
See Phenomenology (Brown University)

Each of these aspects of Heidegger's framework in Being and Time emerges out of his radical rethinking of Aristotle, a rethinking that finds its fullest and most explicit expression in a 1925–6 lecture course entitled Logik (later renamed Logik (Aristoteles) by Heidegger's student Helene Weiß, in order to distinguish this lecture course from a later one he gave also entitled Logik; see Kisiel 1993, 559, note 23). On Heidegger's interpretation (see Sheehan 1975), Aristotle holds that since every meaningful appearance of beings involves an event in which a human being takes a being as—as, say, a ship in which one can sail or as a god that one should respect—what unites all the different modes of Being is that they realize some form of presence (present-ness) to human beings. This presence-to is expressed in the 'as' of 'taking-as'. Thus the unity of the different modes of Being is grounded in a capacity for taking-as (making-present-to) that Aristotle argues is the essence of human existence. Heidegger's response, in effect, is to suggest that although Aristotle is on the right track, he has misconceived the deep structure of taking-as. For Heidegger, taking-as is grounded not in multiple modes of presence, but rather in a more fundamental temporal unity (remember, it's Being and time, more on this later) that characterizes Being-in-the-world (care). This engagement with Aristotle—the Aristotle, that is, that Heidegger unearths during his early years in Freiburg and Marburg—explains why, as Sheehan (1975, 87) puts it, "Aristotle appears directly or indirectly on virtually every page" of Being and Time. (For more on Heidegger's pre-Being-and-Time period, see e.g., Kisiel 1993, Kisiel and van Buren 1994, and Heidegger's early occasional writings as reproduced in the collection Becoming Heidegger. For more on the philosophical relationship between Husserl and Heidegger, see e.g., Crowell 2001 and the review of Crowell's book by Carman 2002; Dahlstrom 1994; Dostal 1993; Overgaard 2003.) *

What does Heidegger think is more fundamental? Heidegger tells us,

"dasein [the there of being, existent beings] is ontically [in the particular sense of a lived being] distinguished by the fact that, in its very Being, that Being is an issue for it" (Being and Time 4: 32). †

Heidegger's whole early work in 'Being and Time' is to try to uncover an ontological meaning to being. In other words, he wants to try to clarify what it is that makes us, individual beings, have a peculiar concern for Being. Capitalizing 'Being' here is simply to point toward and describe why we are so concerned about 'what's it all about'. Heidegger thinks Aristotle tried to strain this idea through his kind of empirical sense of how beings show themselves in nature. By 'showing themselves in nature' we mean how they appear. This is what he is referring to as presence. Heidegger thinks that there is a circularity which is endemic to thinking about Being. He calls this circularity the hermeneutical circle.

* Michael Wheeler, "Martin Heidegger," in *The Stanford Encyclopedia of Philosophy*, ed. Edward N. Zalta (Fall 2020 Edition), sec. 2.1.

See Heidegger (Stanford)

† Martin Heidegger, *Being and Time*, trans. John Macquarrie and Edward Robinson (New York: Harper & Row, 1962), 32.

Michael Wheeler, "Martin Heidegger," in *The Stanford Encyclopedia of Philosophy*, ed. Edward N. Zalta (Fall 2020 Edition), sec. 2.1.

See Heidegger (Stanford)

Hermes was the interpreter between earth and the underworld. The hermeneutical circle is an inquiry wherein the 'there' of being, us, uncovers more and more about our odd condition where Being is an issue for us. This circularity cannot be taken into presence as some kind of object in nature as present-at-hand as Aristotle tried to do. Below is a quote from a more difficult and technical writing of Heidegger but first I will first try to make it more accessible:

- Notice how being, both capitalized and not, are treated in this text.
- *dasein* is German for the 'there of being' or human being.
- Anxiety is the withdrawal of Being from *Dasein*, the 'there of [a] being'.
- We are thrown as the 'there' of being, as *Dasein*.
- Thrownness is always spread over time from a past in a present towards projection, towards a future.
- Resoluteness, projecting, and taking hold of one's authenticity may have some kinship to Husserl's notion of intentionality and quality although this would be debated among scholars.
- Authenticity is taking hold in resolute-ness of one's being.
- Care is the resolute–ness of taking hold of one's authentic being.

> We have defined "resoluteness" as a projecting of oneself upon one's own Being-guilty—a projecting which is reticent and ready for anxiety.(iv) Resoluteness gains its authenticity as anticipatory resoluteness.(v) In this, Dasein understands itself with regard to its potentiality-for-Being, and it does so in such a manner that it will go right under the eyes of Death in order thus to take over in its thrownness that entity which it is itself, and to take it over wholly. The resolute taking over of one's factical 'there', signifies, at the same time, that the Situation is one which has been resolved upon. In the existential analysis we cannot, in principle, discuss what Dasein factically resolves in any particular case. Our investigation excludes even the existential projection of the factical possibilities of existence. Nevertheless, we must ask whence, in general, Dasein can draw those possibilities upon which it factically projects itself. One's anticipatory projection of oneself on that possibility of existence which is not to be outstripped—on death—guarantees only the totality and authenticity of one's resoluteness. But those possibilities of existence which have been factically disclosed are not to be gathered from death. And this is still less the case when one's anticipation of this possibility does not signify that one is speculating about it, but signifies precisely that one is coming back to one's factical "there". Will taking over the thrownness of the Self into its world perhaps disclose a horizon from which existence snatches its factical possibilities away?(1) Have we not said in addition that Dasein never comes back behind its thrownness?(vi) Before we decide too quickly whether Dasein draws it authentic possibilities of existence from thrownness or not, we must assure ourselves that we have a full conception of thrownness as a basic attribute of care. *

I am not going to try to decipher Being and Time in this volume so I will try to give a more relat-

* Martin Heidegger, *Being and Time*, trans. John Macquarrie and Edward Robinson (New York: Harper & Row, 1962), 434–435.

able account of this quote to folks who are not acclimated to academic philosophy. When Being becomes reality and practicality (the 'there' of Being) it becomes everydayness in Heidegger's terminology. It becomes what I refer to as absolute. It has a kind of certainty which Heidegger refers to as inauthentic. I would add that it simultaneously relieves existential tension and exhausts existential tension to the point of fatigue and exasperation [anxiety]. Contrarily, when space is abstracted as three dimensions of length, width and height traveling through one dimension of time there is no messy concrete or lived middle which makes these relationships more dubious. Time and space have been mechanized. In practical effect, they become absolutely apparent [this is more akin to inauthenticity for Heidegger or ontic (the particular of ontological) everydayness of the 'there' of being]. This makes the notions of local realism self-evident. The cosmos is thought to be ordered and chaos has been vanquished. Such historic certainties imply an endless repetition of 'mere' ritual reenactments without the possibility of contextual reflection (everydayness). In my estimation this creates a facade where the rules are known, and the game becomes game theory.

Whoever negotiates the rule-scape better is promised to reap the benefits. The only value becomes transactionalism or playing the game better than others. Benefit is at the expense of others. Fatigue is the end result of repetition where the rules have no possibility for excess. The facade of excess is given at the expense of the other. This is the definition of transactional mechanism, survival of the fittest machine. Note that I started this discussion with the *meta* of mechanism and ended it with mechanism defining being, in particular, the sense of 'my being'. I have become the victim of a certain history which gets reduced to mere repetition. The result of this lived desperation becomes the source of social power structures and their inevitable violences. Philosophy is the historic questioning of our contextual assumptions. These assumptions walk hand in hand through our history with assumptions built into the history of science as well.

When Einstein questioned the relativity of absolute time and space, philosophy was simultaneously questioning the certainty of the absoluteness of the lived context in which 'reality' was implicitly taken as ready-to-hand, as a tool which blindly validated or invalidated behavior. As a student of Husserl, Martin Heidegger exemplified the thought of ready to hand as a context in which we are related to being as a tool, a hammer.

> The less we just stare at the hammer-thing, and the more we seize hold of it and use it, the more primordial does our relationship to it become, and the more unveiledly is it encountered as that which it is—as equipment. The hammering itself uncovers the specific 'manipulability' of the hammer. The kind of Being which equipment possesses—in which it manifests itself in its own right—we call 'readiness-to-hand'. (Being and Time 15: 98) *

Heidegger refers to this mode of living embodiment with the other as ready-to-hand, *Zuhandenheit,* in which the other has already disappeared into use value. Here is how Heidegger introduces the idea of ready-to-hand in "Being and Time",

* Martin Heidegger, *Being and Time*, trans. John Macquarrie and Edward Robinson (New York: Harper & Row, 1962), 98.
Michael Wheeler, "Martin Heidegger," in *The Stanford Encyclopedia of Philosophy*, ed. Edward N. Zalta (Fall 2020 Edition), sec. 2.2.2.
See Heidegger (Stanford)

> The ready-to-hand is not grasped theoretically at all, nor is it itself the sort of thing that circumspection takes proximally as a circumspective theme [my clarification i.e, with explicit contextuality]. The peculiarity of what is proximally ready-to-hand is that, in its readiness-to-hand, it must, as it were, withdraw [zurückzuziehen] in order to be ready-to-hand quite authentically. That with which our everyday dealings proximally dwell is not the tools themselves [die Werkzeuge selbst]. On the contrary, that with which we concern ourselves primarily is the work—that which is to be produced at the time; and this is accordingly ready-to-hand too. The work bears with it that referential totality within which the equipment is encountered.
>
> The work to be produced, as the "towards-which" of such things as the hammer, the plane, and the needle, likewise has the kind of Being that belongs to equipment. The shoe which is to be produced is for wearing (footgear) [Schuhzeug]; the clock is manufactured for telling the time. The work which we chiefly encounter in our concernful dealings—the work that is to be found when one is "at work" on something [das in Arbeit befindliche]—has a usability [my clarification i.e, practicality] which belongs to it essentially; in this usability it lets us encounter already the "towards-which" [my clarification i.e, Huserl's intentionality] for which it is usable . A work that someone has ordered [das bestellte Werk] is only by reason of its use and the assignment-context of entities which is discovered in using it.
>
> But the work to be produced is not merely usable for something. The production itself is a using of something for something. In the work there is also a reference or assignment to 'materials': the work is dependent on [angewiesen auf] leather, thread, needles, and the like. Leather, moreover is produced from hides. These are taken from animals, which someone else has raised. Animals also occur within the world without having been raised at all; and, in a way, these entities still produce themselves even when they have been raised. So in the environment certain entities become accessible which are always ready-to-hand, but which, in themselves, do not need to be produced. Hammer, tongs, and needle, refer in themselves to steel, iron, metal, mineral, wood, in that they consist of these. In equipment that is used, 'Nature' is discovered along with it by that use—the 'Nature' we find in natural products.
>
> Here, however, "Nature" is not to be understood as that which is just present-at-hand, nor as the power of Nature. The wood is a forest of timber, the mountain a quarry of rock; the river is water-power, the wind is wind 'in the sails'. As the 'environment' is discovered, the 'Nature' thus discovered is encountered too. If its kind of Being as ready-to-hand is disregarded, this 'Nature' itself can be discovered and defined simply in its pure presence-at-hand. But when this happens, the Nature which 'stirs and strives', which assails us and enthralls us as landscape, remains hidden. The botanist's plants are not the flowers of the hedgerow; the 'source' which the geographer establishes for a river is not the 'springhead in the dale'. *

When we are using the hammer, the hammer disappears in its use value while we are nailing boards to build a house. When we are using a hammer, we are unaware of the hammer. We are purely focused

* Martin Heidegger, *Being and Time*, trans. John Macquarrie and Edward Robinson (New York: Harper & Row, 1962), 99–100.

Michael Wheeler, "Martin Heidegger," in *The Stanford Encyclopedia of Philosophy*, ed. Edward N. Zalta (Fall 2020 Edition), sec. 2.2.2

on using the hammer to accomplish a task, an intention. Heidegger tells us in this modality, the hammer disappears in use. The hammer is not present before us as a hammer but disappears in its usefulness to complete a task. When the hammer is in use, the lived space we inhabit with the hammer is the disappearance of the hammer. Rather, we are intent on the goal we intend to accomplish such as framing a house. Our lived context is the immersion in the in-order-to, to frame a house. When reality is taken in this mode its use value comes to the fore.

The negative modality of instrumentality is exhibited in the example of when the hammer breaks while in use. But if the hammer breaks and we hit our finger, we curse the hammer and our smashed finger. In this case, the lived mode we inhabit with the hammer has changed, the hammer has become conspicuous as a damn 'thing'. Now, the hammer comes to the fore as what Heidegger calls present-at-hand. Heidegger calls this lived mode of conspicuous presence where things or beings stand before us 'sitting right in front of us' as 'present-at-hand', *Vorhandenheit*. Now the hammer appears from its invisibility in use, and we inhabit the space of the hammer as showing itself as an object. It no longer disappears in use but becomes present before us. When the hammer breaks the intention to build the house recedes away as the space of a broken hammer and a painful finger come to the fore of our contextual lived modality. Here the quality of our lived experience changes from in-order-to, e.g. frame the house, and transforms to what is simply before me, e.g. what is wrong with this stupid hammer. The quality and context of our lived experience has changed, not just what we are doing. This example shows that space is experienced as lived regionality of different contexts. Here we inhabit a different lived context or horizon in which the tool has become an object. Our lived modality of ready to hand becomes present at hand; becomes conspicuous in its failure to function as a tool. Here the value is no longer in its usefulness but in its stark presence before us. In this example, the lived mode of presence-at-hand shows itself negatively. In a positive lived modality of presence-at-hand we might think of this as the lived mode of empiricism in science. Let's also note the relativity of time and space which changes during the transition from ready-to-hand to present-at-hand.

Likewise, when we experience the being of things in the mode of use value, beings become neutral. Neutrality, 'my' historic branch hides and becomes the face of the other. The other becomes information for our use – to fortify our retreat from the face of the other. I have called this lived modality transactionalism. Transactionalism is the mode of relating to others and the environment as standing reserve, waiting to be used, *Bestand*. Heidegger thinks technology is an example of how we comport ourselves to the environment in this lived modality. In this case, the saying converted as origin to the said is pure information which comports us to the other as neutrality. Here, the light as submerged presence has calcified from chaotic intuitions of privation into sense data thought in terms of commonsense. Neutrality is the dominant modality in which modernism orients our day-to-day comportment to beings.

Note that the abstract notion of space is linear extension, e.g. feet, inches, etc., of equally divided increments. The lived experience is much different. It is more concrete, less abstract as linear extension. The question this should bring to the fore is, "Why do we privilege abstract space over lived space?" 'Deseverance' is a spatial ability of *Dasein*, the 'there' of being, to sever regions of lived space, bringing them closer and further away. For example, when we listen to a lecturer, we are more proximally located by attention to the region of the speaker than the people sitting around us. We spatially bring the speaker closer to us than the folks around us. This lived spatiality is called deseverance. Deseverance is how Heidegger describes the way we live spatiality.

Heidegger uses the term de-severance to describe the "opening up of a space in which things can be near and far". Consider a person wearing glasses and looking at a painting. Physically, the glasses are closer to the viewer, but for the person absorbed in the world and contemplating the painting, the painting becomes closer. The objective distances of things present-at-hand do not necessarily coincide with the remoteness and closeness of what is ready-to-hand within our lived experience. When we are listening to a lecture there is a region of the lectern and the professor which we bring closer to us than the desk we are sitting at. This region also brings with it certain social guidelines of how we comport ourselves in the space of a lecture, e.g. not obtrusive, etc. Below is how Heidegger describes deseverance,

> When we speak of deseverance as a kind of Being which Dasein has with regard to its Being-in-the-world, we do not understand by it any such thing as remoteness (or closeness) or even a distance.(1) We use the expression "deseverance"* in a signification which is both active and transitive. It stands for a constitutive state of Dasein's Being—a state with regard to which removing something in the sense of putting it away is only a determinate factical mode. "De-severing"* amounts to making the farness vanish—that is, making the remoteness of something disappear, bringing it close.(2) Dasein is essentially de-severant: it lets any entity be encountered close by as the entity which it is. De-severance discovers remoteness; and remoteness, like distance, is a determinate categorial characteristic of entities whose nature is not that of Dasein. De-severance*, however, is an existentiale [my comment i.e., a particularity of our existence]; this must be kept in mind. Only to the extent that entities are revealed for Dasein in their deseveredness [Entferntheit], do 'remotenesses' ["Entfernungen"] and distances with regard to other things become accessible in entities within-the-world themselves. Two points are just as little desevered from one another as two Things, for neither of these types of entity has the kind of Being which would make it capable of desevering. They merely have a measurable distance between them, which we can come across in our de-severing.
>
> Proximally and for the most part, de-severing(3) is a circumspective [my comment i.e., contextual] bringing-close—bringing something close by, in the sense of procuring it, putting it in readiness, having it to hand. But certain ways in which entities are discovered in a purely cognitive manner also have the character of bringing them close. In Dasein there lies an essential tendency towards closeness. All the ways in which we speed things up, as we are more or less compelled to do today, push us on towards the conquest of remoteness. With the 'radio', for example, Dasein has so expanded its everyday environment that it has accomplished a de-severance of the 'world'—a de-severance which, in its meaning for Dasein, cannot yet be visualized [my comment, perhaps virtual reality brings a little more acuity]. (Heidegger, 1959, pp. 207-208)*

Another example of phenomenology is that of lived time. We think of time as linear, now moments equally separated as seconds, minutes, hours, etc. However, we experience time qualitatively. When we

* Martin Heidegger, *Being and Time*, trans. John Macquarrie and Edward Robinson (New York: Harper & Row, 1962), 139–140.

Michael Wheeler, "Martin Heidegger," in *The Stanford Encyclopedia of Philosophy*, ed. Edward N. Zalta (Fall 2020 Edition), sec. 2.2.3.

are having fun, time flies by quickly. When we are bored, time drags on slowly. There is a varying stretch of time from a past to a future in the way we experience it. Our abstractions of time are thought of as more concrete than our experience of lived time. In "Being and Time" Heidegger writes of an inauthentic modality of lived time as a succession of 'now' moments. He writes,

> The concern which awaits, retains, and makes present, is one which 'allows itself' so much time; and it assigns itself this time concernfully, even without determining the time by any specific reckoning, and before any such reckoning has been done. Here time dates itself in one's current mode of allowing oneself time concernfully; and it does so in terms of those very matters with which one concerns oneself environmentally, and which have been disclosed in the understanding with its accompanying state-of-mind—in terms of what one does 'all day long'. The more Dasein is awaitingly absorbed in the object of its concern and forgets itself in not awaiting itself, the more does even the time which it 'allows' itself remain covered up by this way of 'allowing'. When Dasein is 'living along' in an everyday concernful manner, it just never understands itself as running along in a Continuously enduring sequence of pure 'nows'. By reason of this covering up, the time which Dasein allows itself has gaps in it, as it were. Often we do not bring a 'day' together again when we come back to the time which we have 'used'. But the time which has gaps in it does not go to pieces in this lack-of-togetherness, which is rather a mode of that temporality which has already been disclosed and stretched along ecstatically. The manner in which the time we have 'allowed' 'runs its course', and the way in which concern more or less explicitly assigns itself that time, can be properly explained as phenomena only if, on the one hand, we avoid the theoretical 'representation' of a Continuous stream of "nows", and if, on the other hand, the possible ways in which Dasein assigns itself time and allows itself time are to be conceived of as determined primarily in terms of how Dasein, in a manner corresponding to its current existence, 'has' its time.
>
> In an earlier passage authentic and inauthentic existing have been characterized with regard to those modes of the temporalizing of temporality upon which such existing is founded. According to that characterization, the irresoluteness of inauthentic existence temporalizes itself in the mode of a making-present which does not await but forgets. He who is irresolute understands himself in terms of those very closest events and be-fallings which he encounters in such a making-present and which thrust themselves upon him in varying ways. Busily losing himself in the object of his concern, he loses his time in it too. Hence his characteristic way of talking—'I have no time'. But just as he who exists inauthentically is constantly losing time and never 'has' any, the temporality of authentic existence remains distinctive in that such existence, in its resoluteness, never loses time and 'always has time'. For the temporality of resoluteness has, with relation to its Present, the character of a moment of vision. When such a moment makes the Situation authentically present, this making-present does not itself take the lead, but is held in that future which is in the process of having-been. One's existence in the moment of vision temporalizes itself as something that has been stretched along in a way which is fatefully whole in the sense of the authentic historical constancy of the Self. This kind of temporal existence has its time for what the Situation demands of it, and it has it 'constantly'. But resoluteness discloses the "there" in this way only as a Situation. So if he who is resolute encounters

> anything that has been disclosed, he can never do so in such a way as to lose his time on it irresolutely.
>
> The "there" is disclosed in a way which is grounded in Dasein's own temporality as ecstatically stretched along, and with this disclosure a 'time' is allotted to Dasein; only because of this can Dasein, as factically thrown, 'take' its time and lose it. H. 411
>
> As something disclosed, Dasein exists factically in the way of Being with Others. It maintains itself in an intelligibility which is public and average. When the 'now that...' and the 'then when...' have been interpreted and expressed in our everyday Being with one another, they will be understood in principle, even though their dating is unequivocal only within certain limits. In the 'most intimate' Being-with-one-another of several people, they can say 'now' and say it 'together', though each of them gives a different date to the 'now' which he is saying: "now that this or that has come to pass..." The 'now' which anyone expresses is always said in the publicness of Being-in-the-world with one another. Thus the time which any Dasein has currently interpreted and expressed has as such already been given a public character on the basis of that Dasein's ecstatical Being-in-the-world. In so far, then, as everyday concern understands itself in terms of the 'world' of its concern and takes its 'time', it does not know this 'time' as its own, but concernfully utilizes the time which 'there is' ["es gibt"]—the time with which "they" reckon. Indeed the publicness of 'time' is all the more compelling, the more explicitly factical Dasein concerns itself with time in specifically taking it into its reckoning. *

In this rather long passage Heidegger wants to expose what he thinks is an inauthentic modality of time as living on the momentary succession of practical concern. For example, we find ourselves losing ourselves in the object of concern in order to accomplish a task which we never have enough time for, and we lose ourselves in forgetfulness. He states this in a way which he wants to simultaneously show us that this is only one modality of time. Earlier he showed how different moods such as elation, fear, anxiety, etc. alter our lived temporality. He later goes on to show how temporality is constituent of Being and being in the world. In terms of this volume, we should see a kind of relativity coming to the fore from classicism to the relativity of modernism.

Heidegger questions the arising of Being from beings. How does the lived temporality of care (*Sorge*) rise into authenticity as resolution, Heidegger criticizes the vulgar conception of time, which views it as a linear sequence of discrete "now-points." According to Heidegger, the past is the "no-longer-now," the future is the "not-yet-now," and the present flows from future to past. Additionally, he rejects any distinction between time and eternity, where temporality is derived from a non-temporal state of eternity. Heidegger contends that we are time. Temporality is not a mere external framework; it is intrinsic to our existence. The past as no-longer-now shapes our understanding of history and memory. The present as the now is where our existence unfolds. The future is the not-yet-now, revealed through anticipation. Lived time in this sense has a dynamic stretch. Temporality is not a linear progression of abstract 'now' moments; it stretches beyond mere incremental succession. Heidegger describes this as "being-ahead-of-myself-already-in-the-world." Our existence is always projected toward the future, and this anticipation defines our being. Temporality is integral to care

* Martin Heidegger, *Being and Time*, trans. John Macquarrie and Edward Robinson (New York: Harper & Row, 1962), 410–411, 464.

(*Sorge*). Care involves our fundamental indebtedness and guilt, which we take over from the future. It holds together the totality of our existence, intertwining facticity, falling, and being. Heidegger's temporality of care reveals that time is not an external container but an essential aspect of our being-in-the-world. *

In his later works Heidegger adopts a mythopoetic style to his writing. *Ereignis* is a central concept in Heidegger's later philosophy emerging during the 1930s. *Ereignis*, often translated as the "event of appropriation". *Ereignis* refers to how being manifests itself in its truth—a profound understanding of existence. Heidegger's writings on *Ereignis* were not initially intended for public consumption. These texts were written without didactic considerations, aiming for an originary (poietic) language—a language that might be called "experimental" or "esoteric." *Ereignis* becomes a lens through which we can explore the essence of being and its unfolding. In his later period, Heidegger's thinking undergoes another shift (though less extreme than his famous "turn"). Even during this phase, Ereignis remains significant, and its articulation deserves special consideration. Ereignis represents a profound encounter with being—an event that transcends conventional language and opens up new dimensions of understanding. †

In his work, "Contributions to Philosophy: From Enowning", Heidegger explored the idea of self-revelation and appropriation. *Ereignis* represents a profound event where humans encounter meaningful presence and determine the meaning of their lives. Heidegger's unique way of "crafting a novel genre of philosophical writing" is characterized by an interventionist style aimed at subverting prevailing neo-Kantian thought. His focus on aesthetics extended beyond World War II, encompassing works from the 1930s to the 1950s.

The 1940s brought on Heidegger's famous 'turn', also known as "*die Kehre*", refers to a significant shift in Martin Heidegger's philosophical thought. The turn involves a reversal of perspective. Instead of focusing solely on human existence (*Dasein* – the 'there of being'), Heidegger turns toward Being itself. This shift aims to understand the fundamental nature of being beyond human consciousness. It's a move away from anthropocentrism and toward a broader consideration of existence. The turn also signifies a transformation in Heidegger's style and approach to philosophy. His later works exhibit a different tone, vocabulary, and emphasis compared to his earlier writings. Heidegger delves into historical texts, poetry, architecture, technology, and other subjects, seeking new avenues for philosophical exploration. Heidegger's concept of *aletheia* (disclosure) pertains to how things in the world appear to human beings. It signifies an "unclosedness" or "unconcealedness"—a way of understanding the world. Initially, Heidegger associated *aletheia* with a redefined notion of truth, but he later corrected this association. An apophantic statement is an assertion that covers up meaning and presents something as "present-at-hand." For instance, sentences like "The President is on vacation" or "Salt is Sodium Chloride" are apophantic because they can be easily repeated in news and gossip, but the deeper context may be lost. Heidegger replaced terms like subject, object, consciousness, and world

* See Heidegger's Being and Time, part 8: Temporality
And Care, Death, and Time in Heidegger and Frankfurt
And Temporality
And 2 - Originary Temporality
And 15 - Temporality as the Ontological Sense of Care

† See 10 - Ereignis: the event of appropriation
And HEIDEGGER AND THE APPROPRIATION OF METAPHYSICS
And On Appropriation

with "Being-in-the-world" (*In-der-Welt-sein*). This move transcends the traditional split between subject and object found in Western thought. Instead, Heidegger emphasizes that consciousness is always consciousness of something, and objects exist in relation to consciousness. At the most fundamental level, moods play a crucial role in our encounter with the world. A mood arises neither from the outside nor the inside; it emerges from our being-in-the-world. Only through a mood can we genuinely engage with things around us. Dasein (the 'there of being'), a term synonymous with being-in-the-world, signifies our openness to the world. It involves projecting ourselves onto possibilities, interpreting the world through those possibilities, and understanding it in terms of potentialities. Dasein is not detached from the world; it is a "thrown projection" that engages with the hidden and revealed possibilities before it. Although Heidegger rarely used the term *Kehre* himself, commentators refer to "the turn" as a significant change in his writings. Themes characterizing this turn include poetry and technology. It marks a shift in perspective—from Dasein's being to being itself—and a transformation in Heidegger's philosophical style. The most profound aspect of the turn is a fundamental change in Being's relation to us. It leads to a new inception that overcomes traditional metaphysics. By examining Being itself, Heidegger seeks to uncover its hidden dimensions and challenge conventional assumptions about reality. The 'turn' in later Heidegger's philosophy, where he shifts from a focus on human existence to a broader exploration of being and its significance. [*]

Heidegger's 'turn' was based on the idea that great works of art were resolutely populist but also had revolutionary aspirations. He thought the greatest works of art "grounds history" by "allowing truth to spring forth". [†] In Heidegger's later writings he dispensed with the words ontic being and ontological Being all together in favor of the event of appropriation or the German word *Ereignis*. Here, the author of this journal article tells us,

> I choose to see Heidegger's critique as anticipating the appropriation of metaphysics where his description of turning supercedes overcoming. Turning is what constitutes Ereignis, the event of being coming into its own, or what Heidegger also refers to as 'appropriation'. Appropriation designates the moment in which the presencing of being is not taken up according to any specific determination but is allowed to presence as itself. [‡]

Ereignis

Ereignis is German for "event" or "appropriation". The term *Ereignis* plays on its similarity to the German word *"eigen"* (meaning own or proper). Thus, *Ereignis* can be understood as the "event of appropriation". *Ereignis* involves the "revealing", "grasping", and "owning" of existence, allowing humans to engage with their own being and the world. *Ereignis* is not merely an event; it is the unveiling of meaningful presence that grants humans the possibility of determining any such thing as

* See Martin Heidegger (1889—1976)
And Heideggerian terminology
And 211. - Turn (Kehre)
And 6 - The turn

† See Heidegger's Aesthetics

‡ Thomas Sheehan, "The Turn," in *The Cambridge Heidegger Lexicon*, ed. Mark A. Wrathall (Cambridge: Cambridge University Press, 2021), 215.
See HEIDEGGER AND THE APPROPRIATION OF METAPHYSICS

meaning. *Ereignis* transcends ordinary events; it is a "self-vibrating realm" where *Dasein* and Being intersect. Through *Ereignis*, *Dasein* and Being achieve their "active nature" by shedding the qualities attributed to them by metaphysics. It is a "dynamic encounter" that opens up possibilities for understanding existence beyond conventional boundaries.

In his book "Poetry, Language, Thought" Heidegger plays on Heraclitus' notion of what I have been terming binary dualities with an imaginary notion of an ancient Greek temple. *

> It is the temple-work that first joins together and simultaneously gathers around itself the unity of those paths and relations in which birth and death, disaster and blessing, victory and disgrace, endurance and decline obtain the form of destiny for human being. ...The temple first gives to things their look and to humanity their outlook on themselves. †

Ereignis is "circularity and appropriation" of "earth, sky, divinities and mortals". In a 1954 essay entitled, "Building Dwelling Thinking" in his book "Poetry, Language, Thought", Heidegger writes of his four-fold, earth, sky, divinities and mortals,

> [H]uman being consists in dwelling and, indeed, dwelling in the sense of the stay of mortals on the earth.
>
> But 'on the earth' already means 'under the sky.' Both of these also mean 'remaining before the divinities' and include a 'belonging to men's being with one another.' By a primal oneness the four—earth and sky, divinities and mortals—belong together in one. ‡

Mortals represent human beings—the earthly inhabitants. They are the ones who dwell in the world, engage with it, and experience existence. Mortals are the recipients of both life and death, and their existence is intertwined with the other elements of the fourfold. The divinities are not God or gods in the traditional sense. Instead, they are hinting messengers of godhood. These entities have approached and come closer to the divine, but they are distinct from the ultimate divine presence. Their role is to reveal glimpses of the sacred and the transcendent within the mortal realm. The earthly temple of mortals gathers together the life and death of mortals and pitches to the highest aspirations of divinities. The earth juts through world without regard to human worlding as 'natural materials'. The earth holds a mystery which the work of art in its pure materiality and worlding as in van Gogh's multiple paintings of a woman's peasant shoes bring together the rich content of the paintings,

> A pair of farmer's shoes and nothing more. And yet. [Ein Paar Bauernschuhe und nichts weiter. Und dennoch.]
>
> From out of the dark opening of the worn insides of the shoes the toilsome tread of the

---

* See Heidegger's Aesthetics

† Martin Heidegger, "The Origin of the Work of Art," in *Off the Beaten Track*, eds. and trans. Julian Young and Kenneth Haynes (Cambridge: Cambridge University Press, 2002), 21.

Poetry, Language, Thought. A. Hofstadter, trans. New York: Harper & Row, 1971.

‡ Martin Heidegger, "Building Dwelling Thinking," in *Poetry, Language, Thought*, trans. Albert Hofstadter (New York: Harper & Row, 1971), 149.

On Amazon, Poetry, Language, Thought

> worker stares forth... The shoes vibrate with the silent call of the earth, its quiet gift of the ripening grain [i.e., "earth" makes "world" possible by inconspicuously giving itself to the world] and the earth's unexplained self-refusal in the fallow desolation of the wintry field [i.e., it is also constitutive of earth that it resists this world by receding back into itself]. (Poetry, Language, Thought. 33-4, A. Hofstadter, trans. New York: Harper & Row, 1971.) *

Heidegger pits the disclosure of the peasant women's shoes with the earth's refusal of disclosure in the painting's brute materiality.

The inconspicuous thing withdraws itself from thought most stubbornly. Or can it be that this self-refusal of the mere thing, this self-contained refusal to be pushed around, belongs precisely to the essential nature of things? (Poetry, Language, Thought. 31-2, A. Hofstadter, trans. New York: Harper & Row, 1971.) †

The Stanford article notes,

> It is worth pausing here to comment on the fact that, in his 1935 essay The Origin of the Work of Art, Heidegger writes of a conflict between earth and world. This idea may seem to sit unhappily alongside the simple oneness of the four. The essay in question is notoriously difficult, but the notion of the mystery may help. Perhaps the pivotal thought is as follows: Natural materials (the earth), as used in artworks, enter into intelligibility by establishing certain culturally codified meanings—a world in the sense of Being and Time. Simultaneously, however, those natural materials suggest the existence of a vast range of other possible, but to us unintelligible, meanings, by virtue of the fact that they could have been used to realize those alternative meanings. The conflict, then, turns on the way in which, in the midst of a world, the earth suggests the presence of the mystery. This is one way to hear passages such as the following: "The world, in resting upon the earth, strives to surmount it. As self-opening it cannot endure anything closed. The earth, however, as sheltering and concealing, tends always to draw the world into itself and keep it there" (Origin of the Work of Art 174). ‡

As a phenomenologist Heidegger dispenses with preconceived notions inherited from the history of philosophy by seeking to concretize philosophy in lived experience. Heidegger wants to chronicle the way appropriation concretizes itself in event, the struggle to allow whatever comes of bringing to presence an essence to lived experience. So, the event of gathering appropriates in order to concretize as clearing.

---

* Martin Heidegger, "The Origin of the Work of Art," in *Poetry, Language, Thought*, trans. Albert Hofstadter (New York: Harper & Row, 1971), 33–34.

See Heidegger's Aesthetics

† Martin Heidegger, "The Origin of the Work of Art," in *Poetry, Language, Thought*, trans. Albert Hofstadter (New York: Harper & Row, 1971), 33–34.

See Heidegger's Aesthetics

‡ Michael Wheeler, "Martin Heidegger," in *The Stanford Encyclopedia of Philosophy*, ed. Edward N. Zalta (Fall 2020 Edition), sec. 3.2.

See Heidegger (Stanford)

> At the heart of Martin Heidegger's philosophy of being was his notion of the "clearing". The clearing is much more than just a space where something has been cleared away. It is an opening through which entities other than ourselves can emerge out of hiddenness, or are made visible by a bringing into the light. In one sense the clearing is the place or site where such unconcealment occurs, in the presence of the human form of being that Heidegger calls Dasein. In another sense Dasein is the clearing. *

Here, we see that the temporality of presence both hides and brings to presence as clearing. For Heidegger there is one epoch which subsumes all others as the history of 'absenting' and 'presenting' of appropriating event, both as the individual subject and as lived essence.

> In Heidegger's lexicon 'being' usually designates what, in this or that historical epoch, it means for any entity to be. Hence, it is not to be confused with a term designating any entity or set of entities, though it necessarily stands in an essential relation to human beings, as creatures uniquely capable of differentiating beings from what gives them meaning. But the meaning of being, so construed, must also be distinguished from what grounds or constitutes its essential correlation with human beings. Heidegger labels this ground the Ereignis. He also refers to it as Seynsgeschichte to signal the fact that, as part of this Ereignis, the history of interpretations of being constitutes and, in that sense, underlies our way of being and understanding being. In the process, this still-unfolding history takes hold of us in the ways we make this destiny our own, mindlessly or not. Indeed, in our preoccupation with particular beings (including the metaphysical preoccupation with them insofar as they exist, i.e., with the being of beings), this history easily escapes our notice. †

As embodied encounter of the other, the other in transactionalism disappears in use value. When history as presence is shown in pre-understandings of 'common sense', the submergence of histories as possible ways of being-in-the-world is forgotten as 'common sense', as a different kind of presence, it is in the mode of what Heidegger calls circumspection (*Umsicht*).

Circumspection can refer to our practical, unreflective engagement with the world. Circumspection can be inauthentic or authentic. In inauthentic circumspection, we navigate life without deep introspection or awareness. Inauthentic circumspection involves a kind of prudent foresight—we act based on habit, social norms, and practical considerations. It's a mode of existence where we blend into the collective, following societal conventions and norms. Inauthentic circumspection characterizes our everyday, mundane interactions with the world. Here, we lose sight of our individuality. We become part of the "they," the anonymous crowd, and conform to societal expectations. Our actions lack depth, and we remain dispersed into the neutrality of the world and existence. For Heidegger, authentic circumspection involves a profound self-awareness. It's the recognition of our unique exis-

---

* W. J. Korab-Karpowicz, "Martin Heidegger," in *The Internet Encyclopedia of Philosophy*, ed. James Fieser and Bradley Dowden, sec. 2.

See THE CLEARING: HEIDEGGER AND EXCAVATION

† Daniel Dahlstrom, "Heidegger's *Being and Time*," in *The Oxford Handbook of the History of Phenomenology*, ed. Dan Zahavi (Oxford: Oxford University Press, 2018), 282.

See Being at the Beginning: Heidegger's Interpretation of Heraclitus (.pdf)

tence, our *Dasein* (being-there). Authenticity requires facing existential realities: mortality, anxiety, and conscience. When we confront these moments head-on, we transcend mere inauthentic circumspection. Authentic existence emerges when we embrace our individuality, acknowledge our finitude, and make choices aligned with our true selves. *

A similar idea to authentic circumspection was thought of by Heidegger's mentor Edmund Husserl as a 'phenomenological reduction' of epoché, or bracketing assumptions made by what Heidegger may have referred to as inauthentic circumspection. Husserl referred to this as "*phenomenological reduction*".

> This deep-structure of intentional consciousness comes to light in the course of what Husserl calls the "phenomenological reduction" (Husserliana, vol. XIII, pp. 432 ff), which uses the method of epoché in order to make coherent sense, in terms of the essential horizon-structure of consciousness, of the transcendence of objective reality. The most global form of epoché is employed when this reality in total is bracketed. There is still something left at this point, though, which must not, and cannot, be bracketed: the temporal flow of one's "present" experience, constituted by current retentions and original impressions. These recurrent temporal features of the horizon-structure of consciousness cannot be meaningfully doubted. They provide a kind of hýle for "inner perception" and corresponding reflective judgements, but it is a very special kind of hýle: one that is a proper part of the "perceived" item and does not get conceptually "formed" in the course of perception (reflecting the fact that unlike spatio-temporal objects, lived experiences "do not adumbrate themselves"; cf. Husserliana, vol. III/1, p. 88). Hence, there is no epistemically problematic gap between experience and object in this case, which therefore provides an adequate starting point for the phenomenological reduction, that may now proceed further by using holistic justification strategies. After all, intentional consciousness has now been shown to be coherently structured at its phenomenologically deepest level. †

In dismantling vague preconceptions and assumptions, Husserl thinks we can start again from rudimentary raw materials (*hýle)* in order to reconstruct our circumspectual comportment to intentional consciousness more authentically. For Husserl in this nexus of physics and philosophy, we stand at the door of another possibility – a new epoch [related to Greek *epoché*].‡ This paper tells us that our word epoch came from the Greek word *epoché* which...

---

* See Heidegger on the Self, Authenticity and Inauthenticity
And 34. - Circumspection (Umsicht)
And The Confluence of Authenticity and Inauthenticity in Heidegger's Being and Time
And 17. - Authenticity (Eigentlichkeit)

† Christian Beyer, "Edmund Husserl," in *The Stanford Encyclopedia of Philosophy*, ed. Edward N. Zalta (Winter 2022 Edition), sec. 4.
See Husserl

‡ See Epoche
And Epoch
Also ἐποχή

> inferred the need for total suspension of judgment (epoché) on things." ... "Epoché (ἐποχή, epokhē "suspension") is an ancient Greek term which, in its philosophical usage, describes the theoretical moment where all judgments about the existence of the external world, and consequently all action in the world, is suspended.*

Our English word 'epoch' denotes a fixed point of time, an age or a long duration of time. We use the word as demarcated by some common era such as age of Enlightenment, classical or modern age, etc. In Husserl's foundational work "Logical Investigations" (1900-1901) Husserlian scholars have argued that he may have used the ambivalence of *epoché* and epoch to imply that the phenomenological reduction would usher in a new epoch where science would lose the primacy of its historic abstractions and found itself on principles which go back to the things themselves, how we intentionally experience things. † It is interesting that Husserl's intuitions only prefigured Einstein's work in special relativity and eventually quantum entanglement by only a few years. The storm clouds signaling the end of modernism's supremacy were gathering.

However, in all this one fact of history interrupts Heidegger's work – the violence of Nazism.

> Heidegger joined the Nazi Party (NSDAP) on May 1, 1933, ten days after being elected Rector of the University of Freiburg. A year later, in April 1934, he resigned the Rectorship and stopped taking part in Nazi Party meetings, but remained a member of the Nazi Party until its dismantling at the end of World War II. The denazification hearings immediately after World War II led to Heidegger's dismissal from Freiburg, banning him from teaching. In 1949, after several years of investigation, the French military finally classified Heidegger as a Mitläufer[1] or "fellow traveller."[2] The teaching ban was lifted in 1951, and Heidegger was granted emeritus status in 1953, but he was never allowed to resume his philosophy chairmanship. ‡

Edmund Gustav Albrecht Husserl was born into a Jewish family in Prostějov (Prossnitz), Moravia, which was part of the Austrian Empire at the time. Although he had converted to Protestantism in his youth, the Nazi regime classified him as a Jew. As a result, he faced discrimination, was deprived of many honors, and was expelled from the library of the University of Freiburg due to his Jewish background. Despite these challenges, Husserl's groundbreaking work in phenomenology profoundly influenced 20th-century philosophy and continues to resonate in contemporary thought. Heidegger, who acknowledged his debt to Husserl, later took a political stance that was offensive to Husserl after the Nazis came to power. Heidegger, infamously a Nazi proponent, was himself of Jewish origin. §

---

* "Sextus Empiricus," in *The Stanford Encyclopedia of Philosophy*, ed. Edward N. Zalta (Fall 2021 Edition), sec. 1.
See EPOCH AND EPOCHE

† See Epoché: Meaning, Object, and Existence in Husserl's Phenomenology
And The Role of Husserl's Epoche for Science
And The Problem of the Epoché in Husserl's Philosophy
And Bracketing (phenomenology)

‡ "Martin Heidegger," in *The Stanford Encyclopedia of Philosophy*, ed. Edward N. Zalta (Fall 2020 Edition), sec. 1.
See Martin Heidegger and Nazism

§ See Edmund Husserl
And Edmund Husserl (1859—1938)
And Edmund Husserl (Wikipedia)
And Husserl, Edmund

Levinas, a Lithuanian Jew, became a naturalized French citizen in 1931. He fought with the French and was captured by the Nazis where he remained a prisoner of war until the end of the war in 1945. His father and brothers died at the hands of the Nazi SS in Lithuania. Maurice Blanchot helped Levinas' wife and daughter spend the war in a monastery (Emm). Both men understood the horror of war and made brilliant strides to wrestle with the absolute need for meaning in what appeared to be a meaningless world. They both described in very different ways the pitfalls of humanity and articulated with painful integrity and brilliance an avenue of hope in a hopeless world. Levinas is not so easy to pin down with a political philosophy. Levinas warns us of the insidious nature of totalitarianism. In this way, his forebodings about the state have a kinship to Orwell's critique of nationalism.

In 1988, Levinas wrote an article titled "As If Consenting to Horror," where he revealed that he learned early, perhaps even before 1933, about Heidegger's sympathy toward National Socialism (Nazism). This insight sheds light on the complex relationship between Heidegger's philosophy and his political inclinations. While Levinas criticized Heidegger's ontology, he also acknowledged that Heidegger's reflections on death led him to view *Dasein* (human existence) as anti-social rather than overly immersed in the socio-historical context. *

Confluences of Phenomenology and Relativity

The dance of philosophy and physics continued in phenomenology and relativistic physics. Physics relativized absolutes of classical physics while concurrently, phenomenology relativized absolutes of classical philosophy. In relativity space and time are relative to the speed of light. Space and time can be twisted and distorted in extreme fashions in which absolutes like extension and linear clock time can no longer be 'real' but are subjected to 'actual' distortions by the speed of light. Relativity informs us that 'reality' is relative to the frame of the observer. Therefore, classical scientific absolutes are 'absolutely' relative. However, relativity still could be said to posit an absolute as the speed of light. If it were not for quantum physics which does not recognize temporality at all e.g., quantum entanglement, relativity might indeed be yet another rehash of the classical absolute in terms of the speed of light. Similarly, phenomenology wants to get away from classical abstractions about spatiality and temporality by,

---

And Edmund Husserl (Britannica)

* See Hannah Arendt and the Fragility of Human
And Heidegger's Aesthetics
And Heidegger (Stanford)
And The Origin of the Work of Art (Wikipedia).
And Martin Heidegger and Nazism
And Otherwise than Being-with: Levinas on Heidegger and Community (.pdf)
And Rethinking Levinas on Heidegger on Death
And Martin Heidegger (Wikipedia)
And Heidegger's Style: On Philosophical Anthropology and Aesthetics
And 9 - The work of art
And 2 - German Aesthetics after World War II
And Heidegger and the Aesthetics of Rhetoric
And Antisubjectivism and the End of Art: Heidegger on Hegel
And Heidegger, Wagner, and the History of Aesthetics
And AESTHETICS OF DEFAMILIARIZATION IN HEIDEGGER, DUCHAMP, AND PONGE (.pdf)
And 10 - Ereignis: the event of appropriation
And On Appropriation
And Contributions to Philosophy (Of the Event)

for example, relativizing them to lived regions and lived temporality. The old idea that relativity and phenomenology would simply make everything subjective no longer finds relevance in the 20th century. The reactionary push back to 20th century neo-physics and neo-philosophy came from 'institutionalists of the previous era of physics and philosophy. In the early to middle 20th century the overwhelming evidence in relativistic physics and the appeal to concrete lived experience of space and time won out over modernist notions of subjectivity. But what happened in this dynamic reappraisal of language and meaning was nothing new to history and classicism.

Attention to the future, the yet to be determined, places a requirement that tired old truisms must renew themselves. Science has an advantage over philosophy in that mathematics can largely escape the emotive investment of historic terms in language. However, philosophy must work in terms of historic languages. Classical philosophy takes the route of endlessly assigning new and more adaptive meanings to previous ideas. This is also the reason that philosophy tends to be relegated more easily to the ethereal junk yard of absurdity in common perceptions. However, both science and philosophy must be true to the future no matter what the reactionary push back might be. This volume has brought to the fore examples of religion bogged down in meanings of the past, tradition, and institutional conservatism. But this happens in secular history as well. However, neither can resist necessities imposed by the future. In science, mathematics allows a kind of unencumbered freedom to strike out in radically new directions. However, even mathematics is rewarded economically when new mathematics can assimilate or explain the dominant mathematics of its day. In spite of this, the most pervasive revolutions in the history of science tend to be represented by the adventurers who threw all caution to the wind, e.g., Einstein. Certainly, the contemporaneous budding of phenomenology was also a revolutionary step which made previous histories abstract and less concrete. But before phenomenology there was Hegel.

In the 19th century Hegel was caricatured in his day as an 'idealist'. Idealism in that sense carried with it the intonations of subjective, not objective. Again, the binary dualisms of history's narration and search for a continuum of meaning pressed on from the past. But Hegel recognized the failure of classicism which preceded him. Hegel understood that the problem was not so much with language but the assignment of meanings to the language. This was quite insightful and also in keeping with the scientific discoveries of his day, e.g., evolution, germ theory of disease, periodic table of elements, electromagnetism, thermodynamics, major advances in spectral analysis leading to electromagnetic fields, X-rays, and radioactive elements, and in mathematics complex variables gave rise to new geometries, series convergences, and hypercomplex numbers and let's not forget the new mathematics gave way to the more concrete embodiment of the atom.

In the 19th century Hegel saw all these scientific massive changes. He also understood how these very important breakthroughs also rhymed with more repressed echoes from the past. Hegel understood that commonly accepted definitions and meanings of language were not fundamentally static but dynamic. The problem with language and classicism was that it tried to force meanings into essential static forms. Hegel called these static forms of language 'abstract'. Hegel, ahead of his time, understood that abstract meanings, meanings which had lost the day in history, were the problem. The answer as in science was to radically change the way we understood common vernacular. Hegel determined that logical contradictions were not absolutes but semantic vehicles for phenomenology. In other words, contradictions need not end in absolute dualities like Descartes' mind-body or Kant's categories of understanding and the impenetrable thing-in-itself but could find more concrete meanings by transformation. By 'transformations' I mean undermining abstract and conservative trends for static binary dualisms. Hegel found in historically muted echoes of Leibniz the idea that all predicates are contained within the subject. The clearest and also hardest to understand example of this is Hegel's radical approach to the meaning of 'concept'.

Classical logic was dominated by noncontradiction as the exclusive right of epistemology or truth. Hegel's logic nurtured contradiction in transformation. For Hegel, Concept was not our abstract semantic of a thought, an idea, which takes place in the 'mind' of a subject, and which fermented into the mid/body split of Hegel's predecessors. Concept through a process of dialectical transformations was built in *phusis* organically in a similar way in which we think of biology, or what *is*. The previously assumed classical notion of *is* was the problem. The reactionaries wanted to reduce Concept in Hegel to 'idealism' but the power of Hegel's thinking eventually turned the momentum of history towards another shore, a future in which absolute time and space could be made relative, lived time and space could find meaning without being 'idealized' as mere subjectivity. It may also be argued that Hegel's philosophy aided communism and perhaps some versions of nationalist fascism (although that is very problematic in his writings of the state). However, Hegelians would argue these critiques as a demented form of Hegelianism. But in terms of dialectical materialism and leftist schools of Hegel, the historic record of communism in Karl Marx was an explicit outgrowth from Hegelian philosophy. Additionally, even today there are many leftist Hegelian schools of Christian dogma which adhere to the teachings of Hegel e.g., Hegelian Catholic schools. The leftest schools take Hegel's notion of Concept into a theological framework. There are also more centrist schools of Hegel which focus on Hegel's dialectic in the Logic, the "Science of Logic". Hegel is one example of how philosophy in the 19th century danced with major transformations in science heeding the emptying out of that period's contemporaneous tiring of the 'same ol' song' in previous classical language. Once again, the future beckoned.

The early Heidegger in "Being and Time", somewhat akin to Hegel's Concept, tried to think about Being in a phenomenological novel fashion. And just as Hegel's Concept was superficially criticized as 'ideal' so Heidegger's notion of Being was superficially criticized as what Heidegger termed 'onto-theological'. By 'onto-theological' Heidegger was referring to the tendency to make 'Being' into a kind of substance (onto) or thing separate from beings, from us. Likewise, Heidegger's early work was not an effort to reiterate the history of Being in terms of a god-like notion of Being. Heidegger, like Hegel, wanted to find a more organic relationship to beings, to us, in the very elusive and taken-for-granted notion we have Being. And also, like Hegel, Heidegger's novel and complex notion involved a radical rethinking of what we can possibly mean by our hazy idea of Being. The tendency to take an onto-theological approach to Heidegger's notion of Being was a partial reason for Heidegger giving up his use of the term 'Being' altogether in his latter writings. Instead, Heidegger's 'turn' went towards a more mythopoetic style and his notion of the German word '*Ereignis*'. *Ereignis* in German has the idea of an 'event of appropriation'. Again, *Ereignis* is a very technical term not appropriate for this volume. But perhaps the idea might have some interesting analogical intonations to what I have discussed concerning branching, falling, collapsing into classicism from the observer effect of the wave function. In any case, in Heidegger's early writings some have taken the notion of Being as a more novel approach to intuitions from classicism.

This general approach may also have some superficial similarities to what Hegel did with his notion of Concept. From a general technique perspective, perhaps similar to finding more relevant reassignments of meanings from classicism's worn-out ideas of concept, being, and constituents of existence. The early Heidegger and Hegel follow a tendency to reappropriate classically understood terms in novel and more nuanced and relevant ways. However, It seems to me that these techniques have an analogical similarity not unlike classical chaos theory which tries to maintain classical notions of causality by reassigning causality to asymptotic infinitesimals. However, the inability of this type of undertaking still cannot face the bizarre philosophical implications of quantum chaos theory. To be fair, it is arguable that in my understanding both Hegel and Heidegger still had not reckoned with the extreme ramifications of the void which faces us in quantum chaos and radical externality. It is also arguable that this kind of refitting classicism from worn out meaning-bestowing aspects of historic language implicitly takes on a kind of backseat approach to responsibility and ethics. Mine and Levinas' contention is that the reckoning of the abyss we face in the radical externality cannot find a home in the incessant wearing out and ever more reassigning of novel ontological and epistemological classical notions. These techniques provide a pseudo-ethical basis for responsibility to the other which always seem to get reappropriated in terms of force and violence. In any case, certainly Hegel and Heidegger were not the first re-interpreters of classicism. Progressive classicism entertains an ever more linguistic refitting to rescue the lagging meaning bestowing ability of language.

Classicism itself was permeated with revolutions in science and math all throughout its history. However, the institutionalized momentum of tradition and commonsense in popular culture was always a dampening effect to change and transformations. Our classical past, our 'nature', wants to sterilize the future, find meaning which links the past to the future. I call this meaning-bestowing. This dynamic continues to this day. Eventually the future presses on to upcoming generations, making change and transformations inevitable. History is the flow of past and future, reactionaries and revolutionaries. More concerning, all the while classicism retains the signature of force and violence throughout its course, underscoring a more pervasive and dominant theme at any cost – the perception and appeal of continuity.

All major shifts in scientific or philosophical paradigms must eventually be recaptured by narrative. Language must ready itself to accommodate extreme shifts which are forward looking. History is the accommodation of continuity. Continuity desires origin. Origin is the guarantee of meaning for classicism. The desire for classicism is the quest for meaning. Meaninglessness is not an option for species survival. The binary duality of meaning and meaninglessness forces continuity as history and narrative. Is there an exterior to this desire to conserve, retain, find meaning in binary dualisms? This is the question which looms before us now. In the 20th century, the last vestiges of classicism hang on in phenomenology. As I will soon discuss, postmodern philosophy is intently and painfully aware of our inability to be able; to find yet another classical hope for a novel refitting of philosophy to unbridgeable exteriority. Likewise, the last vestiges of classicism hang on in relativity. In spite of the most successful physics in human history, underneath relativity we find a troublesome void. In quantum chaos theory physics faces a spooky chasm papered over in the veneer of complementarity.

Quantum physics finds no home in the past. Certainly, the desire of quantum physicists to maintain continuity with relativity is, to a degree, found in the successes of quantum field theory. But even that is no done deal in spite of vigorous work to make it so. Bohr's desire for complementary relations between quantum physics and classical science attempts continuity with classical physics as a limited case of quantum physics. As such, it has drawn much attention and funding from working physicists who have practical concerns about their own livelihood. But even they are fully aware of the radical rupture quantum physics brings to the fore which vastly exceeds complementary aspirations. The previous volumes have been an effort to bring quantum physics' implications out of mathematics and into the purview of those of us who cannot be conversant in the highly specialized language of quantum mathematics. And I am not the only one to attempt this. As I have detailed, the greatest, and to this day, even the not so well-known geniuses of quantum physics well understand its radical implications. They wrote unapologetically about it in philosophical terms. Quantum physics really is a radical departure from history, our collective narratives. Language does not easily or novelly accommodate the abyss which looms in quantum physics. We cannot easily adapt meanings to its findings. This is not because of its admitted difficulty for non-mathematical folks but because of the awkwardness and inability for language to bring it to 'sense', to reawaken it once again to history's narrative.

Quantum physics represents a gap which makes it difficult to find words. Perhaps it is better to feel its effect as an unbridgeable gap. I believe we see this affective dis-ease in the forebodings of fanatical and extremist anti-woke reactionaries. This radical disruption of history can find no revolutionaries. We are venturing beyond the domain of narrative which cannot resurrect classical notions as postmodernism demonstrates. I find in the philosophy of Levinas, similar to quantum physics, a radical departure from classicism and its perennial offspring. Both physics and philosophy bring us to an abyss which cannot be bridged by novel and rigorous shifts in meaning and definition. Both leave us speechless, unable to be able. However, what quantum physics cannot capture, Levinas brings before our face – responsibility.

The radical rupture of quantum physics is not some concept which can be fashioned from the scrap pile of history. Rather, quantum physics brings us to an abyss, chaos, which finds no dominant home in history. Furthermore, in Levinas we are not merely left dangling without recourse as in quantum physics. Levinas brings us face to face with the other. The other here is not my assumed, comfortable, and meaning-bestowing certainties from language and history. The other here is he, she, or binary gender ambivalences who not only contest us but call us to responsibility. There is no home in Levinas' radical alterity of the other. Our inability to even be able in the radical alterity of the other is whom history retreats. More importantly we, me, self, as the constitution of isness is itself retreat. There is no underlying basis to mine-ness. Mine-ness retreating from radical alterity becomes 'me', a self (might we think analogically branching from the observer effect?). Mine-ness as falling away from classical re-foundationalism has no recourse to classical novelty before unbridgeable interruption. Mine-ness only makes possible echo and transactionalism and even the very basis of commonsense. Commonsense assures, assumes, and is rooted in the mystical incantation of 'I am that I am'. In the face of radical alterity, the self is created as retreat. The self as retreat finds mytho-metaphysical 'reality' in its attempt to flee radical externality. In the drying up of classical novelty the pitched desperation of terror is the end of the very possibility of classically fueled, meaning-bestowing ability. This is the horror of death. Death twists and distorts mythopoetic intuitions of radical externality in which the 'my' of origin finds no home. More on that later. For now, I want to take a deeper look into postmodernism and other attempts reckoning with unrecoverable devastations to classicism in the 20th century.

Structuralism

In the lead up to postmodernism in the 20th century, the heir of existentialism was called structuralism.

* Existentialism was a philosophical movement, dominant in Europe during the 1940s and 1950s which emphasized individual experience, choice, freedom, and responsibility. Existentialists, led by thinkers like Jean-Paul Sartre (1905-1980), maintained the primacy of lived experience. Interestingly, existentialism made contributions to structuralism and furthered both poststructuralism and hermeneutics.

> The most popular voices of this movement were French, most notably Jean-Paul Sartre and Simone de Beauvoir [1908-1986]], as well as compatriots such as Albert Camus [1913-1960], Gabriel Marcel [1889-1973], and Maurice Merleau-Ponty [1908-1961], the conceptual groundwork of the movement was laid much earlier in the nineteenth century by pioneers like Søren Kierkegaard [1813-1855] and Friedrich Nietzsche [1844-1900] and twentieth-century German philosophers like Edmund Husserl [1859-1938], Martin Heidegger [1889-1976], and Karl Jaspers [1883-1969] as well as prominent Spanish intellectuals José Ortega y Gasset [1883-1955] and Miguel de Unamuno [1864-1936]. †

Structuralism emerged as an intellectual current and methodological approach primarily in the social sciences. It interprets elements of human culture by examining their relationship to a broader system. In essence, structuralism seeks to uncover the underlying structural patterns that shape human actions, thoughts, perceptions, and emotions. Philosopher Simon Blackburn succinctly captures this idea: "The belief that phenomena of human life are not intelligible except through their interrelations. These relations constitute a structure, and behind local variations in the surface phenomena, there are constant laws of abstract structure".

Structuralism was influential in various fields, including linguistics, anthropology, and literary theory. Structuralists believed that underlying structures (such as language, culture, or society) shape human experience and meaning. They focused on binary oppositions, patterns, and deep structures. Structuralism was criticized for its rigidity and tendency to oversimplify complex phenomena. It often ignored historical context and individual agency.

The roots of structuralism can be traced back to early 20th-century Europe, particularly in France and the Russian Empire. Structuralism goes back to Ferdinand de Saussure (1857-1913) who developed a theory of structural linguistics. Wilhelm Wundt, a German psychologist, is often considered one of the founders of structuralism. He established the first psychology laboratory in Leipzig, Germany, in 1879. Edward B. Titchener, a student of Wundt, further developed structuralism in psychology. He introduced it to the United States and emphasized introspection as a method for understanding mental processes. In psychology, structuralism aimed to break down mental experiences into their basic components (such as sensations, feelings, and thoughts). Researchers used introspection to analyze conscious experiences. Structural linguistics, influenced by Swiss linguist Ferdinand de Saussure, focused on the underlying structure of language. Saussure introduced the concepts of "langue" (the abstract system of language) and "parole" (actual speech). Structuralism influenced these fields by examining cultural practices, rituals, and social institutions as interconnected systems.

---

* See Structuralism

† Thomas Flynn, "Existentialism," in *The Stanford Encyclopedia of Philosophy*, ed. Edward N. Zalta (Fall 2021 Edition), sec. 1. See Existentialism

Psychology emphasized introspection. In psychology, participants were asked to describe their conscious experiences (such as sensations or emotions) in detail. Researchers aimed to identify recurring patterns. Structuralists analyzed elements by contrasting them with their binary opposites (e.g., light vs. dark, male vs. female, etc.). Language analysis was important to structuralists. Saussure's approach involved studying linguistic signs (signifier and signified) and their relationships within a language system.

Prominent structuralists include Ferdinand de Saussure and Claude Lévi-Strauss. Structuralism gained prominence through the structural linguistics work of Ferdinand de Saussure and subsequent schools in Prague, Moscow, and Copenhagen. Interestingly, structuralism inherited some of its intellectual momentum from existentialism. After World War II, scholars across various disciplines adopted Saussure's concepts, including French anthropologist Claude Lévi-Strauss, who played a pivotal role in popularizing structuralism. Linguist Roman Jakobson and psychoanalyst Jacques Lacan were also prominent figures associated with this movement.

Critics argued that introspection was subjective and unreliable. Additionally, structuralism focused too narrowly on conscious experiences. The rise of functionalism (which emphasized the purpose and function of mental processes) led to a decline in structuralism's popularity. Despite its decline, structuralism laid the groundwork for later developments in psychology, linguistics, and cultural studies. Concepts like binary oppositions and the study of underlying structures continue to influence various disciplines. Structuralism sought to uncover the fundamental building blocks of human experience and culture, emphasizing systematic analysis and interconnections. *

Saussure argued that there is a difference between an idealized abstraction of language he called *langue* and language actually used in daily life he called *parole.* † *Langue*, abstract 'meanings' we ascribe to words we actually use, *parole*, have no 'meaning' if by 'meaning' we assume they inform us of concepts which are immanent externalities to language. By immanent externalities I mean something which gives our concepts a direct meaning and validity based on an assumed externality like our word 'reality'. For structuralism the notion of immanent meaning was a kind of metaphysic imagined as immanent, as the direct embodiment of an idea. For Saussure our words have no sense of meaning like this. Instead, words are merely signs which point to signifiers, syntax which points to semantic. Saussure tells us that since many different words can apply to the same object or concept, words exhibit a system of signs which endlessly refer to other signs. The sign only has meaning in relationship to the other signs to which it points. Saussure tells us, "*...in language, there are only differences 'without positive terms'*" ‡

Many thinkers in diverse areas such as anthropology, sociology, psychology, literary criticism, economics, and architecture have utilized Saussure's methodology in very interesting ways. This approach to language and philosophy is synchronic meaning how it is used in the present moment without resorting to history, a temporality. This is contrasted to diachronic which refers to a history, a progression through time. If language is purely relational it is synchronic meaning it refers to nothing

---

* See Structuralism (Wikipedia)
And What Is Structuralism In Psychology?
And Structuralism

† See Langue and parole

‡ de Saussure, Ferdinand. [1916] 1959. Course in General Linguistics, translated by W. Baskin. New York: Philosophical Library. p. 120.

other than itself. If it is diachronic language brings with it a history of ideas which finds some meaning in normal usage. Saussure's metaphor was to imagine language as a series of static points (like frames in a film), each representing a specific stage. The film is a diachronic account of language. But language change occurs between these static points as synchronic. It is interesting that physics has a similar situation between general relativity and quantum mechanics.

While existentialism and structuralism have distinct emphases, they intersected in intellectual history. Structuralism rose to prominence in France following the existentialist wave, particularly during the 1960s. The initial popularity of structuralism in France led to its global influence, offering a unified approach to understanding human life across various fields such as anthropology, sociology, psychology, literary criticism, economics, and architecture. Although existentialism and structuralism diverged in focus, their intertwined legacy continues to shape contemporary thought. *

In non-relativistic quantum mechanics, time is regarded as universal and absolute. By using the terms 'universal' and 'absolute' quantum physics does not mean it gives some sort of 'physical' meaning to time. The use in quantum physics means it is simply used as a classic background parameter which is external to the system itself and merely a utilitarian tool for research and experimentation. It is more like snapshots in a film where we have one state here and another here and physics asks what, if any, is the relation between the states. Or in other words, what are the synchronic changes. The notion of 'time' and history in this case is more like frames in the film or diachronic. However, the synchronic change or lack of change is more interesting as it is purely relational and does not try to investigate for example what time 'actually is'. In thinking of time as universal and absolute there is no philosophical importance attached to the concept of time itself. In early science and philosophy time was believed to be an actual phenomenon based in 'reality'. But modern science realized that time had no basis in anything 'absolute' but was stochastic or statistical, discontinuous, quantum 'jumps' between observed stationary states. In the time dependent Schrödinger wave equation time dependence is only understood in superposition as a relationship between quantum wave interference states. This is called perturbation theory or how things react when they are perturbed. In contrast, the observables which means collapse or branching into classical physics is not a continuum of time into classicism but a rupture, kinematic meaning simply mechanical points without any association to forces, motions, or continuums which might link to points. What we think of as historical time, diachronous time, is not at all the common classical notion we have of time as some sort of glue between, causal relation between, two states. Like structuralism which thinks of language as purely relational, quantum physics thinks of spacetime as stochastic, totally separate statistical, states with no on-going relation or history filling the gap that unifies and makes continuous like our common notion of time. † If you saw the recent movie "Everything Everywhere All at Once" that is the idea in quantum mechanics.

However, in general relativity, time is considered continuous, malleable, and relative. Quantum mechanics regards time as an independent variable, not a meaningful factor in the phenomenon it studies. As an independent variable it is discontinuous, utilitarian, where differences of state are purely referential and synchronic. This chasm creates a conceptual conflict between quantum physics and general relativity, known as the problem of time. The problem of time raises questions about the

---

* See Structuralism
And 6 p. 104Existentialism in the 21st century
And Life Of A Human As Seen By Existentialism And Structuralism

† See Time in Quantum Theory (.pdf)

nature of time in a physical sense and whether it is truly a real, distinct phenomenon. It also involves the related question of why time seems to flow in a single direction, despite the fact that no known physical laws at the microscopic level seem to require a single direction. The problem of time arises from this conceptual conflict between the two theories and raises questions about the nature of time in a physical sense. So, for structuralism time is not like general relativity or evolutionary. It is interesting to note that Aristotle was more synchronic than diachronic.

Postmodernism

Postmodernism is an intellectual stance or mode of discourse characterized by skepticism towards elements of the Enlightenment worldview. In the 1940s postmodernism really began to address the abstractions of modernism and the relativistic trend of phenomenology to found dominant themes emerging from lived experience. At its height, postmodernism generally ranges from 1970 to 1990. Postmodernism questions the "grand narratives" of modernity, which include Enlightenment ideals such as reason, science, and progress. It rejects the certainty of knowledge and stable meaning, acknowledging that ideology plays a role in maintaining political power. It acknowledges the influence of ideology in maintaining political power. Postmodernism embraces self-referentiality, epistemological relativism, moral relativism, pluralism, irony, irreverence, and eclecticism. It opposes the "universal validity" of binary oppositions, stable identity, hierarchy, and categorization. Emerging in the mid-twentieth century as a reaction against modernism, postmodernism has permeated various disciplines and is linked to critical theory, deconstruction, and post-structuralism. Critics argue that postmodernism promotes obscurantism, abandons Enlightenment's rationalism and scientific rigor, and contributes little to analytical or empirical knowledge. *

Postmodernism is a multifaceted artistic movement that emerged in the mid-20th century as a reaction against the modernist ethos. Postmodernism is interested in conceptual art. Conceptual artists shifted the focus from the physical object to the idea behind it. They explored concepts, language, and context, often challenging traditional artistic forms. Postmodernism values minimalism. Minimalists embraced simplicity, reducing art to its essential elements. Their works featured clean lines, geometric shapes, and a sense of austerity. Postmodern artists harnessed video technology to create dynamic and time-based artworks. Video art challenged traditional notions of static visual representation. Performance art arises from postmodernism as performers engaged with their bodies, time, and space to create live, ephemeral art experiences. These performances often blurred the boundaries between art and life. Postmodernism is critical of institutions. Artists critiqued art institutions, questioning their authority and power structures. They explored how museums, galleries, and curatorial practices shape artistic meaning. Postmodernism questions identity. Postmodernism emphasized diverse identities and perspectives. Feminist art, LGBTQ+ art, and art by marginalized communities challenged dominant narratives. Postmodernism embraced irony, humor, and playfulness. Artists subverted conventions, creating fragmented and self-referential works. Postmodernists blurred distinctions between high art (traditionally associated with elite culture) and low art (popular culture). Elements from mass media,

* See Postmodernism
And Postmodernism Philosophy (Britannica)
And What is Postmodernism?

advertising, and consumer culture found their way into artworks. Postmodernism critically analyzes authenticity and originality. Postmodernism questioned the idea of a single, inherent meaning in art. Viewers became active participants, shaping interpretations. Concepts like appropriation and ready-mades challenged notions of authenticity. Postmodernists rejected grand narratives and embraced the local, contingent, and temporary. They resisted totalizing theories and celebrated diversity.

Postmodernism is a broader cultural and artistic movement that emerged in the late 20th century. Postmodernism rejects absolute truths, celebrates diversity, and challenges authority. It questions meta-narratives (universal explanations) and embraces hybridity. Irony, intertextuality, and pastiche characterize postmodern works. Postmodernism draws from post-structuralist ideas. Both emphasize language, subjectivity, and the constructed nature of reality. Postmodernism extends beyond academia to influence architecture, literature, film, and popular culture. In summary, while structuralism laid the groundwork, post-structuralism critiqued and expanded upon it. Postmodernism, in turn, absorbed elements from both, creating a rich tapestry of thought and creativity. *

Poststructuralism

Poststructuralism is a philosophical movement that emerged in France during the late 1960s. It challenges the objectivity and stability of the interpretive structures proposed by structuralism. Unlike structuralism, which posits that human culture can be understood through a structure modeled on language, poststructuralism asserts that these structures are constituted by broader systems of power. In particular, it finds systems of power that organize themselves around a narrative of binary oppositions. As we saw with Hegel binary oppositions were necessary at every step along the way with the added difference that binary oppositions gave way to higher binary opposition which unified hierarchically lower binary oppositions which gave rise to higher unities. Even absolute spirit had to 'not' itself to earn the ultimate title - 'absolute'.

Post-structuralism emerged as a critical response to structuralism. It gained prominence in the 1960s and 1970s. Post-structuralists rejected the idea of fixed, universal structures. They emphasized the role of language, discourse, and power in shaping meaning. Jacques Derrida, Michel Foucault, and Roland Barthes are notable post-structuralist thinkers. Post-structuralists questioned binary oppositions and sought to deconstruct them. They highlighted the instability of meaning and the impossibility of pure objectivity. Context, contingency, and subjectivity became central concerns. Post-structuralism blurred boundaries between disciplines. Post-structuralism is a precursor to postmodernism. While not identical, they share themes like skepticism toward grand narratives, playfulness, and a focus on representation. Post-structuralism influenced postmodern art, literature, and philosophy.

Poststructuralists critique the binary oppositions that form the basis of structuralist structures. They reject the idea that media or the world can be interpreted within pre-established, socially constructed frameworks. Structures require a kind of self-sufficiency, a unifying and organizing sameness to qualify as 'structure'. Poststructuralism discards the notion that structuralist definitions of signs are both valid and fixed. It questions the assumption that authors employing structuralist theory

* See Postmodernism (Wikipedia)
And Post-postmodernism (Wikipedia)
And Postmodern art (Wikipedia)
And History of modernism and postmodernism

are somehow above and apart from the structures they describe. Notable figures associated with post-structuralism include Roland Barthes, Jacques Lacan, Julia Kristeva, and Michel Foucault. Poststructuralism challenges the rigidity of structuralist thinking and emphasizes the role of power dynamics in shaping meaning and interpretation.

Structuralism, simply put, was based on the idea that language is purely relational and built up from a system or structure layer upon layer, syntax upon syntax, sign upon sign which points to nothing other than the dynamics of the structure. This is synchronic. Poststructuralism began to question these assumptions. Poststructuralism criticized the notions of signs as a definite pointer to other signs. Signs do not build up a structure layer upon layer, a synchronic center. There never is a meaningful structure extant from the arbitrary and random play of signs pointing to signifiers which always point to other signs. Signs do not always point consistently to another sign which conforms to a linguistic structure. The whole notion of 'structure' cannot substantiate itself from the random play and lack of consistency within any particular sign. Poststructuralism then gives way to deconstruction. *

Deconstruction

Deconstruction is a term that has different meanings in different contexts. In philosophy and literary criticism, deconstruction is a method of analysis that challenges the assumptions and oppositions that underlie Western thought. It was developed by the French philosopher Jacques Derrida and his followers. Deconstruction examines how language, logic, and meaning are constructed and destabilized by texts. In art, architecture, and design, deconstruction is a style or movement that breaks down conventional forms and structures into fragmented and distorted elements. It often creates a sense of disorientation, complexity, and unpredictability. Some examples of deconstructivist artists and architects are Pablo Picasso, Marcel Duchamp, Frank Gehry, and Zaha Hadid. In building and construction, deconstruction is the process of dismantling a structure in a way that preserves and reuses its components, rather than demolishing it and sending it to landfills. Deconstruction is considered a more environmentally friendly and cost-effective alternative to demolition. It also reduces waste, saves energy, and creates jobs.

In deconstruction within any system of signs, any system must always and necessarily deconstruct any structure from within its own terms. The endless and random play of signs must undo itself any notion of structure. It de-centers itself of its own accord. Deconstruction wants to show that any system must necessarily imply its own negation. Any structure necessarily implies its own deconstruction by its own terms e.g., without any external negation but in its own hidden and suppressed affirmative declarations (margins of the text). Any systemic organization cannot help but undo itself. This

* See Post Structuralism
And Poststructuralism (Britannica)
And Post-Structuralism (Oxford)
And Poststructuralism and Postmodernism in International Relations
And Post-structuralism (Wikipedia)
And Post-Structuralism (Oxford)
And Postmodernism and Poststructuralism (Oxford)
And Poststructuralism and Postmodernism in International Relations

brings me back to the problem I have referred to as binary dualisms. At the center of binary dualisms is logic itself. *

Deconstruction is a philosophical and critical method introduced by the French philosopher Jacques Derrida. It challenges traditional notions of meaning, hierarchy, and stability within texts and language. Deconstruction examines the relationship between text and meaning. It questions the idea of fixed, absolute truths and instead focuses on the fluidity and complexity of interpretation. Jacques Derrida's influential work, including his book Of Grammatology, laid the foundation for deconstruction. He emphasized that meaning arises from contrasts between signs and words, rather than from self-sufficient definitions. Derrida drew inspiration from linguist Ferdinand de Saussure. He argued that language gains meaning through the tension between opposing terms. For example, the word "being" only makes sense in contrast to "nothing." Meaning is deferred, never directly present. Deconstruction seeks to identify and overturn hierarchical oppositions within texts. These include signified over signifier, intelligible over sensible, speech over writing, and activity over passivity. However, deconstruction doesn't aim to eliminate all oppositions; they are structurally necessary for creating meaning. Deconstruction challenged fixed meanings, revealing the complexities inherent in language and thought. It has influenced fields such as philosophy, literary criticism, and social sciences. †

From Physics and Postmodernism

Let's tie this back to what was happening in physics around the same time as Husserl and Heidegger were describing lived time and lived space. Leonard Susskind tells us in the "The Cosmological Principle",

> Earlier we discussed the fact that in Einstein's theory, space is stretchable and deformable like the surface of a balloon. It can be stretched flat and smooth or it can be all wrinkled and bumpy. Combine this idea with quantum mechanics, and space becomes very unfamiliar. According to the principles of quantum mechanics, everything that can fluctuate does fluctuate. If space is deformable, then even it has the "quantum jitters." If we could look through a very high-powered microscope, we would see space fluctuating, shaking and shimmering, bulging out in knots, and forming donut holes. It would be like a piece of cloth or paper. On the whole it looks flat and smooth, but if you look at it microscopically, the surface is full of pits, bumps, fibers, and holes. Space is like that but worse. It would appear not only full of texture but of texture that fluctuates incredibly rapidly. ‡

---

* See Deconstruction - Wikipedia
And Deconstruction (Britannica)
And Deconstruction (building)
And Deconstruction (simple Wikipedia)

† See Deconstruction (Wikipedia)
And Deconstruction (Simple Wikipedia)
And Deconstruction (Merriam-Webster)
And Deconstruction (New World Encyclopedia)
And Deconstruction (Britannica)

‡ Brian Greene, *The Fabric of the Cosmos: Space, Time, and the Texture of Reality* (New York: Alfred A. Knopf, 2004), 332.
Leonard Susskind, *The Black Hole War: My Battle with Stephen Hawking to Make the World Safe for Quantum Mechanics* (New York: Little, Brown and Company, 2008), 86.

If space is "stretchable" and "jitters", this does not just apply to the very large. This applies to the quantum level as well. We know from Einstein's relativity that spacetime has relative frames. These frames of reference can be drawn on any scale. On smaller scales they have smaller effects, but they still have effects. Likewise, we know that the distortion of spacetime is the definition of gravity. We also know that space is time and time is space. Additionally, Einstein's most famous formula $e = mc^2$ tells us energy times the square of the speed of light is equal to mass. Or mass is energy and energy is mass. We have mass. This means we have energy. Our mass will make minute changes in spacetime around us. Effectively, this means that what phenomenology calls the stretches of lived time and regionalities of lived space have a physics component. Relativity finds its philosophical expression in phenomenology. Phenomenology is not merely philosophy; it is physics as well. This should give us another indication of how historic assumptions of absolute time and space, linear time and linear extension of space, show us something about ourselves. We do not just learn history; we are history and not just our personal history. With relativity and quantum entanglement a new history is developing which we can welcome with wonder and awe and attention, or we can try to force a history which is rapidly fading.

What I am trying to show here is how deeply entrenched ways our lived historicity abstractly holds sway over our concrete lived experiences. This should indicate a certain kind of way in which we are, a capacity we have which simultaneously defines us and imprisons us. When these phenomenological capacities start to break down our social norms become less functional, we face a precipice. We can either decide to violently reinstate our conventional historical narratives to conserve a supposed world-view. This appears to reduce consternation and guarantee order but only becomes a reactionary case of diminishing returns. Or we can reflect on the causes for these breakdowns and consider the inadequacies of our historic certainties. Another practical outcome of doing the latter has been termed adaptation.

In the later 20th century postmodernism became popular across Europe especially in Italy and France with Jean-François Lyotard, Jacques Derrida, and Jean Baudrillard. Derrida finds recurrent textual narratives running through modernity which undermine themselves from within. He tells us that these narratives undermine themselves from within in what Derrida calls 'logocentrism'. Baudrillard looks at modernity as a hyperreality which is merely simulacrum. By simulacrum he means that historic repetition has made so many copies of themselves that there is no original anymore. There are only endless copies of values and meanings which can only replenish themselves with more and more extremism and fanaticism. Baudrillard tells us,

> Baudrillard presents hyperreality as the terminal stage of simulation, where a sign or image has no relation to any reality whatsoever, but is "its own pure simulacrum". The real, he says, has become an operational effect of symbolic processes, just as images are technologically generated and coded before we actually perceive them. This means technological mediation has usurped the productive role of the Kantian subject, the locus of an original synthesis of concepts and intuitions, as well as the Marxian worker, the producer of capital though labor, and the Freudian unconscious, the mechanism of repression and desire. "From now on," says Baudrillard, "signs are exchanged against each other rather than against the real", so production now means signs producing other signs. The system of symbolic exchange is therefore no longer real but "hyperreal." Where the real is "that of which it is possible to provide an equiva-

> lent reproduction," the hyperreal, says Baudrillard, is "that which is always already reproduced". The hyperreal is a system of simulation simulating itself. [*]

Bereft of meaning, an epoch ends in a whimper while extremism only appends this end with violence and authoritarianism. What is meant by the 'death of an age' seems to me to have an analogous relationship with the difference between Heidegger and Levinas on death. Both find a phenomenological experience of death as 'my death'. Heidegger tells us that my death is the possibility of the absolute impossibility of '*Dasein*' or the 'there' of being, my being. This absolute impossibility of my 'there of being' is more than a mere contradiction. Death recoils, overflows itself, on our lived experience from my lived impossible future to my present. Therefore, Heidegger's notion of death reaches back to me as my ownmost possibility for authenticity. Levinas tells us that death is my most severe experience of passivity, as my suffering of death. Levinas tells us death is not something which grants power as the possibility for authenticity reaching back from 'my death' but an externality which comes to us passively in which we are no longer 'able to be able' in the suffering of death.

> Levinas characterizes Heidegger's view as one that emphasizes the assumption of a position of virility, power, and mastery in the face of death. In contrast, Levinas argues that death is the most radical experience of passivity and impossibility, in which one is 'no longer able to be able'—emphasizing that death is not something assumed, but rather something that comes. The futurity of death as something to come (à venir) means that the encounter with death is an encounter with something other, a conclusion that Heidegger does not make in relation to his own understanding of the futurity of death. [†]

I see echoes of this phenomenological reduction as what marks the end of an age, an epoch. However, when the death of an age is thought in terms of virility and possibility, Nietzsche's overman, *Ubersmensch*, more colloquially superman, it must desperately refresh itself on its ownmost, age-wise understanding of authenticity. This means it must perpetually return to a copy, yet another simulacrum, of its narrative which it gives itself - 'the old days were blah, blah, blah'. As the heroic old days become harder and harder to prop up in its disappearance, it must revert to desperation, exasperation, fanaticism, and finally, violence. The end of our current epoch can no longer thrive on tired narratives of absolute 'values'. 'Absolute' itself has submerged in a sea of relativity where uncertainty is not a suspension until we can reestablish yet another simulation of certainty. This is not a hermeneutical pause which will find its footing yet again in an underlying 'reality'. This is either annihilation from climate change or nukes, whichever one comes first, or it is a step into the abyss. 'Exteriority' is no longer opposed to interiority, object to subject, other to me to be raised up into a transformative unity which binds them both in their oppositions. 'Exteriority' is not a *meta* to *phusis*, yet another metaphysics. It is analogously exhibited and hinted at by the previous discussion on exformation. Exforma-

---

[*] Gary Aylesworth, "Postmodernism," in *The Stanford Encyclopedia of Philosophy*, ed. Edward N. Zalta (Fall 2021 Edition), sec. 4.

See Postmodernism (Stanford

[†] Bettina Bergo, "Emmanuel Levinas," in *The Stanford Encyclopedia of Philosophy*, ed. Edward N. Zalta (Winter 2024 Edition), sec. 3.

See The Otherness of Death: The Creative Translation of Being and Time

tion leaves no room for any of the categories of modernity. Absolute time and space, cause and effect, 'reality', practicality, and the 'common' in sense now marks the tombstone of modernity in which we live and move and have our being.

For the sake of this volume, let's think of the inevitable breakdown of worldview, of classicism to modernity and modernity's budding inadequacies as the disorientation of chaos, perhaps *apeiron* in the Greek sense. The breakdown of order is brought on by the uncertainty of what is no longer working (like the breakdown of the hammer). The pull of chaos is now focusing on the totalitarian death of the leveling off of 'reality' into the same. There is no new age of Aquarius which will squelch uncertainty and usher in a certain euphoria. Uncertainty remains in exteriority. Exteriority has no name but has a face - the face of the other, the he or she which practical response to responsibility cannot get subsumed yet once again in totalitarian simulacrum. This practical responsibility in the face of an other cannot be synchronized to my lived time. This is what Levinas calls Ethics. Ethics here has nothing to do with altruism which was firmly rooted in modernism from Descartes' 'thinking thing' to the monadology of Leibniz to the categories of human understanding in Kant to the Concept which rises from the dialectic of Hegel to the senses of British Enlightenment are all sealed in the certainty of modernity's tombstone.

Remember in general relativity, time is considered malleable and relative. In quantum mechanics time does not arise from phenomenon and is discrete, an auxiliary, as an independent variable which the quantum wave function uses as an operator to distinguish between two discrete states. However, there is no implication of any physicality to temporality as general relativity assumes. This dissonance creates a conceptual conflict between the two theories, known as the problem of time. The problem of time raises questions about the nature of time in a physical sense and whether it is truly a real, distinct phenomenon. It also involves the related question of why time seems to flow in a single direction, despite the fact that no known physical laws at the microscopic level seem to require a single direction. Quantum mechanics regards time as universal and absolute, while general relativity regards time as malleable and relative. So analogously, for postmodernism time is not like general relativity or evolutionary, smooth and continuous, as a glue which binds narratives together. For postmodernism, hierarchies and logic are dispositions of power structures. Postmodernism finds these structures nothing other than historic narratives which only have an appearance of stability and continuity. At bottom, they perpetuate violence and force by establishing edicts in which identity is privileged. All this sounds a bit esoteric for many folks but its insidiousness seeps into simple commonsense and pre-understandings of 'the way it is'.

In the case of postmodernism, binary dualisms are succeeded not by logical negations but by its own logical explicit and implicit narratives. Derrida calls the appeal to apparent universal truths of logic logocentrism (*logos-centric*). Ok, this is too much of a narrow explanation, but the full explanation would be another volume of this size. The bottom line is that both structuralism and poststructuralism are superseded by deconstruction. And so goes the history of philosophy. There is a movement in all these historic narratives towards symmetry and broken symmetry leading to (eventually in classicism) yet again symmetry. However, with postmodernism what has come later seems to me to be a somewhat stochastic proliferation of a variety of ideas. Of course, this may sound a bit like Schrodinger's idea of statistics which I mentioned earlier. However, in terms of postmodernity Humpty-Dumpty cannot be put together again. This is the malaise of contemporary philosophy. And this is why I think quantum physics may be a bigger help than philosophy in understanding the abyss in which we face. It is interesting that the failures of classical philosophy and classical physics are now both pointed towards an abyss, primordial chaos, if you will. I do not think either philosophy or physics will be able to resolve themselves and that is the point. There will be no classical resolution into a higher order or more perfect logic as classicism leads us to believe. We are at the brink of alterity which cannot ever be reincorporated. So, back to complexity and emergence and novelty...radical alterity. From the alterity intuited in the dawn of *phusis* to the disjunctive avenues opened by the void we face in quantum physics, I want to turn to how these findings necessarily must lead us to a new understanding of each other.

5

THE RADICAL ALTERITY OF THE OTHER

Rather than think subjectivity in the historic garb of Being, Levinas wants to denude the historic machinery of assumptions which level the alterity of the other to the same, judgements which cover over the other with language and history that pre-understands the other as the same as my idea of the other. Levinas does this by focusing on the relational approach to the other in a very different way. In physics we might think of the relational approach as concerning boundary conditions, the collapsing of infinites in a way which is useful for assessing probabilities (think of point particles which have string-like behaviors given from extended dimensions). Over the course of these volumes, I have discussed the relational approach to quantum computing, relativity, Schrödinger's equation and its unitary approach to information updating. Also, recall the paper Jianhao Yang published in 2018 * concerning relational quantum mechanics. If one single idea could simplify quantum physics it would be how *phusis* can no longer be thought of in terms of absolutes of time and space, completely independent and separate quantum states but as a spooky, relational approach.

Hard boundaries no longer exclude quanta from each other, subjects from objects, me from things, me from others, observers from observed. Quantum systems intermingle in reversible, bidirectional probabilistic co-determining quantum states which branches, collapses, as the observer effect simultaneously along all possible informational pathways without regard to our abstractions of a classical 'past' moving towards a 'future'. All of this highly interactivity gurgles up into our sensual, lived experience. And our branching as historic/me narratives are captured by the history of ideas into what we simply think of as plain 'ol common sense. In quantum physics boundary conditions are far removed from our abstract ideas of 'hard and clear' limits. Rather, boundaries are patterns of flux, dimly recognized patterns of waves/fields popping in and out of existence. Even in this case, boundaries are not solid, rigid particles or merely waves, fields, and patterns. Boundaries are dependent on the observer effect, complementarity and the collapse of the wave function, the branching of worlds within worlds as informational updating. Boundaries are ways we think about infinites of interactions and possibilities in order to create sense out of them. In using the phrase 'create sense' I mean truncate all these spooky, infinitely varying possibilities into some form of predictable behavior. For example, when we interact with another person, we are completely oblivious to them and us bidirectionally entangling and co-determining each other in unpredictable ways as entangled systems. We have evolved in such a way as to filter (e.g., senses) so we are not murkily changing together interaction with as barely existent globs. Our biology and our history of ideas have created practical boundaries which enable us to function in the day-to-day world. These spooky relationships defy logic, the history of physics, even of ontology, the history of Being, existence, reality. So, relational boundaries are highly contingent and impose a necessity to be thought-felt-lived under these terms. Boundaries are an odd sort of imagined membranes which can no longer be thought of in terms of cause and effect and some monologue of

* See A Relational Formulation of Quantum Mechanics

evolutionary notion of progress. Boundaries are fluxes of arbitrary reference points which terminate in infinities.

For Levinas, the other infinitely exceeds even these fragile boundary conditions, but history congealed as 'my commonsense' terminate the other as reductions of the same. Leveling the other off to an idea of the other sets the other up for transactionalism, marketing, political rhetoric. The market of ideas even requires professional physicists to think long and hard about how to describe spookiness, so their careers are not endangered. Philosophers are also plagued with concerns about careers in their specific fields of study.

Early on in his youth Levinas focused on the sensuous way we encounter the other. The face of the other is not an interface of a thinking object-machine but sensuously affective. Affectivity gets sifted through history as 'sensible', modernity's reduction of affect to the machinery of the sense organs. The five senses crystallize the affective encounter with the other as ready for processing, for categorization, predetermination, and idea. But lived affectivity always exceeds this reduction to the same. The look from the eyes of the other, their gesture, instantaneously interrupts our projected idea of the other. Our embodiment does not occur at the level of idea, thinking, and reflection.

Early on Levinas explores enjoyment, desire, touch, and the caress which does not know what it seeks in terms of an object or a thing we hold in our hands. Love and fecundity sensually draw us out towards an infinity where 'idea' will not take us. The sensual living embrace and vulnerability do not live in transactionalism. The sensual experience is also of fear, horror, indolence, obsession, insomnia. These fears arise from indeterminateness, the exhaustion off the reductionary enterprise to succeed, the hollowing out of excess to identity of the same. The sensuous experience of insomnia is the absolute inability to slip away, to distract oneself. Insomnia is an analogous example of a primal experience Levinas calls the '*il y a*'.

Il y a is a common French expression which simply means 'there is'. Levinas borrows this idea from his friend Blanchot. Leibniz in monadology writes, "Why is there something rather than nothing at all?". Levinas tells us there is no 'no-thing at all' and at the most fundamental level of awareness there is no ego, no subject and neither is there simply nothingness. The common notion of nothingness may be thinkable, but it cannot be experienced. Absolutely indeterminate, nocturnal being is still something. As mentioned perhaps we might think analogically of the unescapable, stark, and bare awareness of chaotic unassembled exformation in which the possibility of retreat is impossible. It is the absolute destitution of externality into pure chaotic presence. The 'there is' is void, empty of determination, without ego, empty of all content but still aware. The 'there is' has no 'other'. The 'there is' simply and interminable simultaneity awareness without the bare self, ego, or self-consciousness meaning awareness of awareness. It is like insomnia as the 'there' of sheer isness except sleep is never a possibility. In the history of Being, light is opposed to darkness. Darkness is the absence of light, but history's darkness remains defined by light as its other. The 'there' of sheer isness has no counter opposed to light, no contrary, to which it can appeal. It has no oppositional determinations, only atemporal awareness of the 'presencing' of nothingness. The mythologies of Hell, Hades, the underworld, etc. are sitting on a warm tropical beach sipping on a margarita compared to what Levinas is telling us.

Sheer awareness, the experience of interminable nothingness, is the paralyzing inability to be able and therefore, a persecution without object, without hope or even a notion of termination. The 'there is' has no past, present, or future. It has no memory. Levinas writes of *il y a* in the quote below,

> The oneself cannot form itself; it is already formed with absolute passivity. In this sense it is the victim of a persecution that paralyzes any assumption that could awaken in it, so that it would posit itself for itself. This passivity is that of an attachment that has already been made, as something irreversibly past, prior to all memory and all recall. It was made in an irrecuperable time which the present, represented in recall, does not equal, in a time of birth or creation, of which nature or creation retains a trace, unconvertible into a memory. Recurrence is more past than any rememberable past, any past convertible into a present. *

The il y a cannot form a self. The il y a is not awareness of awareness. It is sheer awareness without content. Levinas considers the il y a as a formless void, an ambiguous and formidable 'present absence'. It represents a field of impersonal existence, devoid of meaning or purpose. This is unquenchable and interminable nothingness which finds no echo of a self. Levinas draws a parallel between the impersonal phenomenon of *il y a* and the condition of insomnia. Levinas describes the *il y a* as an impersonal, indifferent force that pervades existence. It represents the sheer facticity of existence—the inexorable presence of things, events, and beings in the world. Imagine a desolate landscape where objects, people, and events barely exist absolutely devoid of meaning. Insomnia is a sleep disorder characterized by the inability to fall asleep or stay asleep. It disrupts the natural rhythm of rest and wakefulness. Now, consider how Levinas' relates insomnia to *il y a.* Just as insomnia prevents restful sleep, *il y a* disrupts any content as sheer existence. It is like being awake, metaphorically speaking, in a state of perpetual vigilance which persists relentlessly, filling our existence with indifferent presence. Insomnia often leads to anxiety, restlessness, and a sense of unease. But *il y a* is like insomnia with the exception that it finds perpetual and interminable inability to sleep or even the inability to experience nothingness. It must stare relentlessly without content. *Il y a* has no self. Contrarily, life is the impossibility of all our possibility to retreat from *il y a*. Might we compare life to the impossibility of all our possibility to quantum tunnelling?

In life, repulsion from the nocturnal horrific awareness of *il y a* retreats from the 'there' of sheer isness as ipseity. Ipseity is from Latin *ipse*, bare self, and the suffice '-ity' taken from *itatem*, referring to conditions or quality of being. Ipseity is the pre-reflective basic or core self. For Levinas, ipseity is recoil from the '*il y a*', the 'there' of sheer isness into selfsameness. Ipseity is horrific retreat from *il y a* to the bare determination of an 'I'. It is the escape from *il y a* which makes self-awareness possible, the birth of self-consciousness. The self does not create itself. By the necessity of life, ipseity escapes from sheer isness not from its own power but from its inability, absolute passivity, to form itself. Levinas thinks of ipseity as the raw and bare beginning of the self which must retreat from *il y a.*

> A sphere of pure sensibility was first disengaged, sphere of the nocturnal horror of the indeterminate, and of ipseity taking form as a relationship of immersion in the sensuous element, determining itself as pleasure or contentment. The relationship with things was carefully distinguished from this antecedent relationship of sensuous enjoyment and alimentation. It begins with inhabitation, the establishing of a dwelling, a zone of the intimate closed from the sphere of the alien. Things are apprehended as solid substances and movable goods, furnishings

* Emmanuel Levinas, *Otherwise Than Being or Beyond Essence*, trans. Alphonso Lingis (The Hague: Martinus Nijhoff, 1981), 202.

> for a dwelling, before they are means and implements in a practical field. And there is a new form of ipseity identifying itself in these acts of inhabitation and appropriation, existence for oneself in the form of being at home with oneself, "être chez soi. There was already a relationship with the other at this level, a relationship through cohabitation, with the other as complementary and "feminine" presence. But the relationship with the other in his alterity, the ethical relationship, breaks out in the face to face position in which language takes place. The ethical nature of the relationship with a face constituted the center and principal originality of Levinas's analyses. But Levinas described also an erotic sphere, relationship with the carnal and with the child that comes in the carnal, relationship of voluptuousness and fecundity, as "beyond" the face, a sphere over and beyond the ethical. [*]

Let's see if I can bring the *il y a* and ipseity a little closer. I am going to cite a quantum analogy here simply as an analogical way to arrive at where I think Levinas is trying to bring us with *il y a*. Please indulge me in a 'quantum mythology' to try to make the point. *Il y a* is along the lines of the chaotic (in classically framed terms 'universal') quantum wave function. Use your imagination to think of sheer awareness within the wave function. Not self-awareness but simply awareness which cannot collapse or branch into classicism. I would think the phenomenon of awareness must have some rudimentary chaotic exformation within the wave function which upon branching has the capacity to become classical self-consciousness. However, when the wave function is not observed, its state is purely and wholly indeterminate. Also, remember that temporality and spatiality are all effects of the observer effect and irrelevant to the wave function. The closest we can imagine the wave function is as our notion of simultaneity. If awareness, not awareness of awareness, has some rudimentary chaotic basis in the wave function, it would have no being, existence, reality, truth, or any 'thing', any coherent thought. It would not be able to gather itself - only perpetual vigilance which persists relentlessly and interminably, filling the void with absolute interminability which must nevertheless be aware. Now, imagine that observation, collapses or branches the wave function and you are born (initially in the superposition state). We know that regardless of our perception of branching the wave function in its component of sheer awareness continues on as chaotic awareness. But in this state 'you' are in an impossibility unable to escape as being, classicism, life. Levinas might call this the impossibility of possibility.

Following our discussion in Volume 1 of the quantum bath, we can visualize stages where a highly isolated quantum experiment is increasingly perturbed by its environment. This progressive exposure —or partial environmental immersion—moves the system toward complexity until a chaotic limit is reached. At this threshold of complexity, might we meaningfully think of a ~~universal~~[†] wave function? But the transition in which the wave function branches might be like ipseity, the bare self that Levinas might have in mind. In this analogy, *il y a* is not the 'origin' of the bare self but a kind of simultaneity of layering as bare self and also as wave function. Origin is simply something we call wave function

---

[*] Alphonso Lingis, "Preface," in *Otherwise Than Being or Beyond Essence*, by Emmanuel Levinas, trans. Alphonso Lingis (The Hague: Martinus Nijhoff, 1981), x–xi.

See The Undoing of the Subject Levinas Thought on Ipseity

[†] Jacques Derrida is the primary postmodern philosopher associated with "erasure" or overstriking (sous rature) words like "universal," "truth," or "being." This technique signals that a word is necessary for language yet inadequate or deceptive, challenging the stability of traditional philosophical concepts.

branch, collapse from the observer effect. A quick note here, ipseity born as life is not the observer. The radical alterity of the other would be the observer in this analogy (of course mom and dad also figure into this observational fecundity).

There is no temporality or spatiality, classicism, for the *il y a.* The bare self, retreating from sheer chaotic awareness, is the transition from the spooky, indeterminacy of the wave function to the branch of classicism in which the bare self retreats from il y a as ego for Levinas. It might be tempting to call this an 'origin' but that connotes a cause and effect-like situation. The wave function continues on having already incorporated the quantum state of the observer into its atemporal, aspatial chaos. The wave function does not know anything about a collapse, branch, classicism, etc. It is the simultaneity of all possibilities chaotically. So, the bare self, the ego, is more of a layering effect from *il y a*, chaotic awareness. For *il y a*, this takes the stare of Orpheus to a whole new level. Orpheus' last fixated stare is not on Eurydice but on chaos, not the underworld, but oblivious awareness which cannot awaken. In *il y a,* the echo of Narcissus has died out eternally without even a memory of abyssal echoing.

The retreat as ipseity to oneself takes on absolute compulsion to escape *il y a*, the separation of a bare, unreflective 'me' as form from void. It is passivity as any such thing as before ipseity had no possible ability to even be passive. Escape before there was ipseity, self-sameness, is given from what Levinas calls a "*passivity more passive than all passivity*". Exteriority first speaks collapsing as ipseity, to sheer terror, awareness of inescapable nothingness; ipseity retreats void as sojourner. The "knot of ipseity" forces retreat and collapse into ego. The hither side of ipseity is the ego.

> Obsession as non-reciprocity itself does not relieve any possibility of suffering in common. It is a one-way irreversible being affected, like the diachrony of time that flows between the fingers of Mnemosyne. It is tied into an ego that states itself in the first person, escaping the concept of an ego in an ipseity — not in an ipseity in general, but in me. The knot of subjectivity consists in going to the other without concerning oneself with his movement toward me. Or, more exactly, it consists in approaching in such a way that, over and beyond all the reciprocal relations that do not fail to get set up between me and the neighbor, I have always taken one step more toward him — which is possible only if this step is responsibility. In the responsibility which we have for one another, I have always one response more to give, I have to answer for his very responsibility. *

The ego, this infantile selfsameness is the echo of Narcissus. It is the echo chamber where Being and history play on the stage of totality, the lure into phantasms of distractions alluring oneself in endless deferments as anonymous others which come to presence. Ego teases a break from self-sameness without tearing it from its embryonic habitat. Levinas calls this infantile auto-fascination/affection the 'said'. The said is the production of faux essence. The said is the echo of the 'saying' from, as we will see, the radical externality of the other. It is the pacifier which amazes and delights in pure self-interest, auto-arousal. It is a self-indulgent game where Being hides from ego like peek-a-boo only to jump out in the throes of time to recapture the ecstatic immanence of the moment, of the self. In this gleeful re-presencing of itself to itself from the horrific retreat of ipseity, the ego desires to replace itself

* Emmanuel Levinas, *Otherwise Than Being or Beyond Essence*, trans. Alphonso Lingis (The Hague: Martinus Nijhoff, 1981), 174.

with another version of ipseity, alter-ipseity. In Freud's terms ipseity might be thought of as id. In that case, the bare self of ipseity Freud might call ego desires to replace itself from terror and retreat of *il y a* by echoing itself, replacing itself with a kinder, gentler version we call the ego too or Freud might call super ego. Levinas writes,

> The recurrence of the oneself refers to the hither side of the present in which every identity identified in the said is constituted. It is already constituted when the act of constitution first originates. But in order that there be produced in the drawing out of essence, coming out like a colorless thread from the distaff of the Parques, a break in the same, the nostalgia for return, the hunt for the same and the recoveries, and the clarity in which consciousness plays, in order that this divergency from self and this recapture be produced, the retention and protention by which every present is a re-presentation — behind all the articulations of these movements there must be the recurrence of the oneself. The disclosure of being to itself lurks there. Otherwise essence, exonerated by itself, constituted in immanent time, will posit only indiscernible points, which would, to be sure, be together, but which would neither block nor fulfill any fate. Nothing would make itself. The breakup of "eternal rest" by time, in which being becomes consciousness and self-consciousness by equalling itself after the breakup, presuppose the oneself. To present the knot of ipseity in the straight thread of essence according to the model of the intentionality of the for-itself, or as the openness of reflection upon oneself, is to posit a new ipseity behind the ipseity one would like to reduce. [*]

Typical of Levinas, he packs a lot of his experiences and knowledge of the history of philosophy into a few sentences. I will unpack some of this. In the quote above Levinas thinks analogously of essence like the game of Parques.[†] Parques looks for a break with the same (jail) only in nostalgia to return to the same (jail) if captured by another player. Levinas likens this game to the narcissistic thrill of essence, delighting and disappointing itself in its victories and defeats. This game is where consciousness plays in Husserl's terms as retention and protention. For Husserl, consciousness is an **intentional** stretch upon a field spanned by an immediate present, building on the living past moment he calls retention and projecting towards a future. This is a living present moment to come he calls protention.[‡] Retention is analogously like a previous classical analog state that is not like a particular thought but an accumulation of each past intentional moment into the present intentional moment. It might be thought of as a living intentional field which accumulates the past analog state into its current analog state. At the same present moment, protention is a living intentional moment which projects upon a future moment which opens up towards possibility. It might be thought of as a living intentional field which anticipates the probable future analog state from its current analog state. A musical

---

* Emmanuel Levinas, *Otherwise Than Being or Beyond Essence*, trans. Alphonso Lingis (The Hague: Martinus Nijhoff, 1981), 202–203.

† Parques is a "random thinking game" similar to Pachisi. It is descendant from an ancient game created in India around 500 BC. The game is often subtitled Royal Game of India because Pachisi used red, yellow, blue and green pawns as dancers on palace grounds. Parqués is played with two dice; two to eight players can compete in the same match, depending on what type of game board is used. Each player is given four pieces and uses a specific color. That color is useful to identify the pieces, the jails and the arrival squares of each player. The start is a jail where players start and can be captured by another player only to return to jail. See Parqués

‡ See Retention and protention

song is a bit like this as it builds from previous harmonic constructions which anticipate future harmonic constructions. Maurice Merleau-Ponty (1908-1961) has a good description of retention and protension below,

> Husserl uses the terms protentions and retentions for the intentionalities which anchor me to an environment. They do not run from a central I, but from my perceptual field itself, so to speak, which draws along in its wake its own horizon of retentions, and bites into the future with its protentions. I do not pass through a series of instances of now, the images of which I preserve and which, placed end to end, make a line. With the arrival of every moment, its predecessor undergoes a change: I still have it in hand and it is still there, but already it is sinking away below the level of presents; in order to retain it, I need to reach through a thin layer of time. It is still the preceding moment, and I have the power to rejoin it as it was just now; I am not cut off from it, but still it would not belong to the past unless something had altered, unless it were beginning to outline itself against, or project itself upon, my present, whereas a moment ago it was my present. When a third moment arrives, the second undergoes a new modification; from being a retention it becomes the retention of a retention, and the layer of time between it and me thickens.[*]

In Levinas' previous quote "*behind all the articulations of these movements there must be the recurrence of the oneself*", he critiques Nietzsche's metaphysics of eternal return of the same self which is always arises from the classical metaphysics of essence. The great iconoclast himself uses the most sacred truth of Western religion and philosophy – the essence of self which must always return to the same, to distinguish his early Greek heights of the Übermensch. I have no doubt Nietzsche was well aware of his humorous and ingenious folly (at least in his mind) play upon classicism with Zarathustra's eternal recurrence of the same. Nietzsche thought the highest aspiration was to play with truth. In "Beyond Good and Evil" he writes,

> The Will to Truth, which is to tempt us to many a hazardous enterprise ... what questions has this Will to Truth not laid before us! What strange, perplexing, questionable questions! It is already a long story; yet it seems as if it were hardly commenced. Is it any wonder if we at last grow distrustful, lose patience, and turn impatiently away? ... We inquired about the value of this Will. Granted that we want the truth: why not rather untruth? And uncertainty? Even ignorance? [†]

Imagine if everything in everyone's life was to be repeated over and over again in exactly the same way forever. Nietzsche called this truth the great nausea. I think Nietzsche believed that the endless repeat of the same life over and over again would neutralize the instinct for classical truth and certainty. Yet, in this void Nietzsche would fill it with his heroic and unabashed elitism; with a moment of joyful bliss in nevertheless affirming life. In "Thus Spake Zarathustra", the section entitled "The Vision and

---

* Maurice Merleau-Ponty, *Phenomenology of Perception*, trans. Colin Smith (London: Routledge & Kegan Paul, 1962), 416.

† Friedrich Nietzsche, *Beyond Good and Evil*, trans. Helen Zimmern (Edinburgh: T. N. Foulis, 1907), 5.
See BEYOND GOOD AND EVIL

the Enigma" he writes about a creature half-dwarf and half-mole which reveals the doctrine of eternal recurrence to Zarathustra. Though the creature's discourse on eternal recurrence a snake crawls down the throat of Zarathustra. This is the great nausea. The creature tells Zarathustra to bite the head off the serpent. When Zarathustra bites the head of the serpent, he has affirmed life and Nietzsche writes of Zarathustra, "*No longer shepherd, no longer man—a transfigured being, a light-surrounded being, that laughed! Never on earth laughed a man as he laughed! O my brethren, I heard a laughter which was no human laughter,—and now gnaweth a thirst at me, a longing that is never allayed. My longing for that laughter gnaweth at me: oh, how can I still endure to live! And how could I endure to die at present!*" Nietzsche understood that humans stood at the abyss. Yet, his evocation was to revel in the bliss of his own warrior spirit of conquest and triumph over every pompous truth he despised in classicism. Instead of radical alterity Nietzsche chose radical selfhood. Nietzsche chose jubilant echo of his prowess over an ethical possibility towards the other.

Permeating ancient Greece from the earliest to the latter citizens and thinkers and very important to Aristotle, was the notion of *kairos* meaning 'the opportune moment'.* The temporality of Nietzsche's triumph moment of affirmation in the face of eternal recurrence of the same is a good way to think about the Greek notion of time as *kairos. Kairos* is the opportune moment. It is the temporality of quality as I discussed earlier. *Kairos* is the moment of culmination, completion, ultimate fulfillment captured by *telos*. This is the moment when all tragedy of life is conquered by only the few, the brave, the warrior soul. Nietzsche's moment of *kairos* elevates himself to the rare mountainous peaks of grandeur. Radical alterity is finally and completely effaced by the heroic and elite self. This is the legacy of Nietzsche's echo of himself to himself. This is everything classical essence was supposed to be but failed. In Nietzsche's heroic identity, there is a continuum of the self in an analogous way like relativity. The self is an essence, e.g., spacetime, which continues smoothly in curving and bending but never breaks or gets interrupted by any exteriority. Even in the massive bending of spacetime in a blackhole, there is no break but perhaps a metaphysical-like leap as a singularity (or Zarathustra). For Nietzsche, the essence of the self is preserved at all costs in the eternal triumph of heroic *kairos*. I am not going to take up a lot of space in this volume to shed more light on Nietzsche's obsession with his grandeur and particularly nasty comments on women.

Levinas criticizes Hegel along the same lines of eviscerating any notion of radical alterity when he writes that, "*behind all the articulations of these movements there must be the recurrence of the oneself. The disclosure of being to itself lurks there. Otherwise essence, exonerated by itself, constituted in immanent time, will posit only indiscernible points, which would, to be sure, be together, but which would neither block nor fulfill any fate. Nothing would make itself. The breakup of "eternal rest" by time, in which being becomes consciousness and self-consciousness by equalling itself after the breakup, presuppose the oneself.*" Radical alterity cannot exist or even rise up into Concept for Hegel as it will always be recuperable. All the way from Being and nothingness to the self and other to individuality and collectivity to Concept to space and time and back to Being and nothingness all over again *ad infinitum* there are no breaks, no discontinuities. Everything is anticipated and retained smoothly, logically, continuously as many continuous convolutions. Perhaps similar analogically to the spacetime of relativity. Classicism retains continuity in identity throughout in one way or another. Incomprehensible breaks must always be reappropriated in the return to essence. "*Eternal rest*" is the absolute immanence of Concept. "*Indis-*

* See A Detour of Time

cernible points" cloaks both an allusion to Leibniz and the Hegel's concept of space and time. Concept forgetting itself as indiscernible points can "*neither block nor fulfill any fate*" as mortals can do. For Hegel, the "*breakup of "eternal rest" by time*" is ultimately when Being becomes "*consciousness and self-consciousness* by *equalling itself after the breakup*" in the presupposition of oneself. Levinas' high-density ridicule at the expense of essence and Being is typical of Levinas' fun-poking derision and genius of the history of philosophy. But Levinas does not want to just be a Nietzschean styled iconoclast. He wants to propose a radical alterity which cannot be found by identity, by logic, by Being, by Concept, or by essence. In a way Levinas' technique might be compared analogously with the breakup of relativity by the spooky quantum wave function and its odd observer effect which does not overcome or paper over externality but leaves it dangling without completion, without fulfillment, incapable of being answered classically as essence in all its facades. I am reminded of Levinas' words below,

> To conceive the otherwise than being we must try to articulate the breakup of a fate that reigns in essence, in that its fragments and modalities, despite their diversity, belong to one another, that is, do not escape the same order, do not escape Order, as though the bits of the thread cut by the Parque were then knotted together again. This effort will look beyond freedom. Freedom, an interruption of the determinism of war and matter, does not escape the fate in essence and takes place in time and in the history which assembles events into an epos and synchronizes them, revealing their immanence and their order.
>
> The task is to conceive of the possibility of a break out of essence. To go where? Toward what region? To stay on what ontological plane? But the extraction from essence contests the unconditional privilege of the question "where?"; it signifies a null-site [non-lieu]. The essence claims to recover and cover over every ex-ception — negativity, nihilation, and, already since Plato, non-being, which "in a certain sense is." It will then be necessary to show that the exception of the "other than being," beyond not-being, signifies subjectivity or humanity, the oneself which repels the annexations by essence. *

Let's get back to the retreat of ipseity from the horror of *il y a*. While this may sound dramatic, what Levinas is getting at is the historic inadequacy and inability of the ego to be Superman, to be eternal essence, to be the heroic overcomer, to immortalize itself. Instead, we see the ego as dying, as getting old, as failing to complete itself. The ego approaches its history, its 'me-ness' in the practical everydayness as dismal. It is unable to rise to the heights of immortality, of greatness, of godhood in its inexhaustible echo of itself from and in and through history into the where of 'me'. Levinas continues the quote from above,

> The ego is an incomparable unicity; it is outside of the community of genus and form, and does not find any rest in itself either, unquiet, not coinciding with itself. The outside of itself, the difference from oneself of this unicity is non-indifference itself, and the extra-ordinary recurrence of the pronominal or the reflexive, the self (se) — which no longer surprises us because it enters into the current flow of language in which things show themselves, suitcases

* Emmanuel Levinas, *Otherwise Than Being or Beyond Essence*, trans. Alphonso Lingis (The Hague: Martinus Nijhoff, 1981), 62–63.

> fold and ideas are understood (les choses se montrent, les bagages se plient et les idées se comprennent). A unicity that has no site, without the ideal identity a being derives from the kerygma that identifies the innumerable aspects of its manifestation, without the identity of the ego that coincides with itself, a unicity withdrawing from essence — such is man. *

But as I have introduced throughout this volume with the notion of intuition, Levinas also finds a threat of intuition which only, always must be reappropriated into the said (classicism), into the same which does not allow and cannot follow the thread of the externality of the other.

> The history of philosophy, during some flashes, has known this subjectivity that, as in an extreme youth, breaks with essence. From Plato's One without being to Husserl's pure Ego, transcendent in immanence, it has known the metaphysical extraction from being, even if, betrayed by the said, as by the effect of an oracle, the exception restored to the essence and to fate immediately fell back into the rules and led only to worlds behind the scenes. The Nietzschean man above all was such a moment. For Husserl's transcendental reduction will a putting between parentheses suffice — a type of writing, of commiting [sic] oneself with the world, which sticks like ink to the hands that push it off? One should have to go all the way to the nihilism of Nietzsche's poetic writing, reversing irreversible time in vortices, to the laughter which refuses language.
>
> The philosopher finds language again in the abuses of language of the history of philosophy, in which the unsayable and what is beyond being are conveyed before us. But negativity, still correlative with being, will not be enough to signify the other than being. †

As mentioned, the alter-ipseity, the impossible ego of classical histories, of oneself is not rooted in its own agency, in its illusory play as Idea in the midst of Being. Levinas terms this illusory play 'freedom'. It is the freedom played in the arena of self-creation/subsistence as alter-ipseity. It is the freedom of transactionally echoing oneself to oneself in which ego pays its losing hand to the bitter end. The anarchy of ego, of oneself, cannot be reduced to the 'not' of origin. It cannot play in that historic arena which is doomed to lose. Why? Because it only dreams in echoes of itself in the face of its lack of self-origin. Levinas tells us,

> The oneself has not issued from its own initiative, as it claims in the plays and figures of consciousness on the way to the unity of an Idea. In that Idea, coinciding with itself, free inasmuch as it is a totality which leaves nothing outside, and thus, fully reasonable, the oneself posits itself as an always convertible term in a relation, a self-consciousness. But the oneself is hypostasized in another way. It is bound in a knot that cannot be undone in a responsibility for others. This is an anarchic plot, for it is neither the underside of a freedom, a free commitment undertaken in a present or a past that could be remembered, nor slave's alienation, despite the gestation of the other in the same, which this responsibility for the other signifies. In the expo-

* Emmanuel Levinas, *Otherwise Than Being or Beyond Essence*, trans. Alphonso Lingis (The Hague: Martinus Nijhoff, 1981), 62–63.

† Emmanuel Levinas, *Otherwise Than Being or Beyond Essence*, trans. Alphonso Lingis (The Hague: Martinus Nijhoff, 1981), 62–63.

> sure to wounds and outrages, in the feeling proper to responsibility, the oneself is provoked as irreplaceable, as devoted to the others, without being able to resign, and thus as incarnated in order to offer itself, to suffer and to give. It is thus one and unique, in passivity from the start, having nothing at its disposal that would enable it to not yield to the provocation. It is one, reduced to itself and as it were contracted, expelled into itself outside of being. The exile or refuge in itself is without conditions or support, far from the abundant covers and excuses which the essence exhibited in the said offers. In responsibility as one assigned or elected from the outside, assigned as irreplaceable, the subject is accused in its skin, too tight for its skin. Cutting across every relation, it is an individual unlike an entity that can be designated as τόδε τι ['this' as primary essence in Aristotle]. Unless, that is, the said derives from the uniqueness of the oneself assigned in responsibility the ideal unity necessary for identification of the diverse, by which, in the amphibology of being and entities, an entity signifies. *

In "Otherwise than Being or Beyond Essence" the infinitesimal moment of saying, of the other, to me is the moment before sensual encounter which has not yet crystalized into the said of Being, of thought, of idea. All of this should point us toward the radical alterity of the other which refuses my ideas, judgments, reduction to our repetitive accustomization (common sense) from history leveling off the encounter of the other to the same thought of as the identity of my idea of the other. Levinas writes of the other, the face-to-face encounter with him or her (or more recently, nonaligned with a specific gender) as a barely perceptible boundary of sensuality infinitely exceeded by the radical alterity of the other. The saying of the other which dissipates prior to the temporality of the said, must always be captured and recuperated into the synchrony or the classic temporality of history and Being. Levinas refers to this dissipation or lapse prior to temporality as diachrony and proximity. Diachrony here denotes an abrupt disjunction of temporality between myself and the other. By diachrony Levinas refers to an anarchic excess to temporality with the other which is not contained as 'my time', 'my past, present, or future'.

Analogously perhaps we could associate diachrony to the quantum wave function which can never take hold of some objective other of classical ontology. In this sense, synchrony reflects my branching into informational updates which must always re-imagine itself as the said. However, Levinas' lapse prior to 'informational updating' is not informational at all. It is a call for responsibility to the other in the form of an accusation which must be responded to either from the retreat to ego or the persecution of self. The encounter with the other is not like two computers sharing data. The other is not in my 'informational updating', my temporalizing embodiment. I am not the container of the other. Even more so, the other is on the hither side of my being and Being, which cannot come into presence and its equivocations. The encounter with the other is a diachronous lapse of moments of our freedom of choice which in its pure signifyingness without any object imposes responsibility. This responsibility is not a choice of ethics but an extreme difference which cannot simply fall into traditional categories of ethics and altruistic concern. This is Levinas' distinction as Ethics will be discussed a little further down. Previously, in these volumes I have used 'proximity' in a conventional sense to mean nearness. However, Levinas uses proximity in a special way to indicate disjunction between myself and the other

* Emmanuel Levinas, *Otherwise Than Being or Beyond Essence*, trans. Alphonso Lingis (The Hague: Martinus Nijhoff, 1981), 203–204.

which cannot be bridged, an anarchic, non-spatiotemporal encounter of the other which is more along the lines of spooky entanglement.

For exformation we would think of the spooky quantum wave function which fails us for words, for easy answers provided from classical history. We might like to call it ideal, potential, mathematical, etc. but its 'isness' *is* precisely the question which finds no answer. However, Levinas' alterity of the other is similarly unable to find voice in the said of language and thought. But instead of the spooky we have responsibility. I have referred to entanglement as emerging spacetime. For Levinas, the emergence of me, of subjectivity, is not branching of the wave function but the other who speaks (shall I say observes). Maybe, some would be tempted to think that there is some spooky *phusis* with the encounter of the other which has something to do with the wave function and entanglement. I have also found both ideas to have intriguing similarities but there is quite a distance as well between these two notions in Levinas.

Levinas wants us to once and for all put away these ideas of information, of some kind of mere physics or philosophy which finds its avenues in physics's journals and magazines. What I have tried to do in these volumes is present physics and philosophy more like ancient intuitions of *phusis* which recognized no chasm between physics and philosophy. I have tried to present this material as best I can with clarity and important relevance as a departure from the academically secure ways of guaranteeing one's economic security. Of course, there are professional physicists and philosophers who, like Levinas, are willing to step away from these considerations like Susskind, Everett, Einstein, Bohr and numerous others. I want to make sure that what is at stake in physics today is not taken by others as business as usual. It has affected and will affect more and more commonsense, cultural values and *mores.* * I want to present this material as a precipice upon which history and culture cannot bridge, as a step into the abyss. This is why Levinas is so important to this discussion. In Levinas there is a confrontation with the sleep of common culture fueled by the history of Being. Levinas is not just another philosophy in the ongoing saga of Being. The departure of what Levinas is telling us which physics is unable to articulate, is what the alterity of the other calls to us, to our accusation, and responsibility to the other. Physics certainly does bring us to a departure from history which makes the escape attempt back into business as usual in working physics. Levinas goes further in telling us radical alterity does not end in spookiness but in undeniable responsibility, to the other from which all Being, history, day to day indifference, even in our retreat.

Levinas has a knowledge of philosophy which puts him in the highest echelons of scholars which have ever existed. His writing is very dense and displays skills intimately acquainted with the history of philosophy and contemporary philosophical problems. However, unlike many philosophers his writing is not oriented towards academic security. He is more than willing to speak distinctly about the historic avenues into which Being falls in statements like this,

> The way of thinking proposed here does not fail to recognize being or treat it, ridiculously and pretentiously, with disdain, as the fall from a higher order or disorder. On the contrary, it is on the basis of proximity that being takes on its just meaning. In the indirect ways of illeity, in the

* from Latin mōrēs ['mo:re:s], plural form of singular mōs, meaning "manner, custom, usage, or habit" are social norms, See Mores

anarchical provocation which ordains me to the other, is imposed the way which leads to thematization, and to an act of consciousness. *

Essence: The Violence of Origin Over Radical Alterity

I want to preface this section with the remark that Levinas does have an account for how our histories of essence and Being come about vis-à-vis the third other. So, his account does not simply end in a skepticism of all this. This will be explored in the section entitled, "The Third Other".

Levinas wants to demonstrate how two of the most important philosophers in more recent history, Hegel and Heidegger, still perpetuate the absorption of thematization as essence and consciousness as ontic subjectivity in the identification with Concept and ontology in Being. They do this by introducing and sustaining their own unique understanding of temporality which I discussed earlier ("On Hegel" and "On Heidegger").

In early Heideggerian thought, falling into inauthenticity reduces the individual to the conformist they-self (*das Man*), stripping care (*Sorge*) of its temporal depth. However, a resolute embrace of the 'there' rescues *Dasein* from existential anxiety and insignificance. This authenticity manifests as a commitment to future possibilities, grounding the individual in a nurturing, active care (*Fürsorge*) for both others and the world. † This 'there of being' carries both ontic and ontological weight, linking the individual human being to the historical enigma of Being. Rather than viewing these terms as static nouns, Heidegger frames them as verbs. Existence becomes a dynamic process—an ontic 'there-ing' of *Dasein* that allows the broader, ontological happening of Being to disclose and conceal itself through history.

Sorge is a central concept in Heidegger's philosophy and plays a crucial role in understanding human existence as authentic. As already discussed, Heidegger's philosophy focuses on *Dasein*, which refers to human existence. *Dasein* is not just an abstract concept; it represents our actual being in the world. *Sorge* (often translated as "care") is a fundamental structure of *Dasein*'s existence. It encompasses both an inner state and an external cause. *Sorge* involves a complex interplay of anxiety, worry, and concern. It's not merely about practical care or everyday worries; it delves into our existential condition. When we care, we are attuned to the world, anticipating the future, and apprehensive about our possibilities. Heidegger describes *Sorge* as having three dimensions of time:

- Past: Concern for our history, memories, and past experiences.
- Present: Attentiveness to the current moment and our ongoing existence.
- Future: Anxiety about what lies ahead, our potential, and our choices.

Authenticity in Heidegger's thought involves embracing our unique possibilities and living in accordance with our true self. *Sorge* is essential for authenticity because it reveals our being-as-an-issue—our constant questioning and engagement with existence. *Sorge* is more than mere care; it's authentic, is the existential fabric that shapes choices as existentiality, fears as facticity, and aspirations as fallen-

* Emmanuel Levinas, *Otherwise Than Being or Beyond Essence*, trans. Alphonso Lingis (The Hague: Martinus Nijhoff, 1981), 74.

† See Classic Article: "History of the Notion of Care"

ness. In "Being and Time", Heidegger gives us his notion of *Dasein*'s temporality as "ahead-of-itself", "Being-already-in", and "Being-alongside". *

> Through the unity of the items which are constitutive for care—existentiality, facticity, and fallenness—it has become possible to give the first ontological definition for the totality of Dasein's structural whole. We have given an existential formula for the structure of care as "ahead-of-itself—Being-already-in (a world) as Being-alongside (entities encountered within-the-world)".(2) We have seen that the care-structure does not first arise from a coupling together, but is articulated all the same.(x) In assessing this ontological result, we have had to estimate how well it satisfies the requirements for a primordial Interpretation of Dasein.(xi) The upshot of these considerations has been that neither the whole of Dasein nor its authentic potentiality-for-Being has ever been made a theme. The structure of care, however, seems to be precisely where the attempt to grasp the whole of Dasein as a phenomenon has foundered. The "ahead-of-itself" presented itself as a "not-yet". But when the "ahead-of-itself" which had been characterized as something still outstanding, was considered in genuinely existential manner, it revealed itself as Being-towards-the-end—something which, in the depths of its Being, every Dasein is. We made it plain at the same time that in the call of conscience care summons Dasein towards its ownmost potentiality-for-Being. When we came to understand in a primordial manner how this appeal is understood, we saw that the understanding of it manifests itself as anticipatory resoluteness, which includes an authentic potentiality-for-Being-a-whole—a potentiality of Dasein. Thus the care-structure does not speak against the possibility of Being-a-whole but is the condition for the possibility of such an existentiell potentiality-for-Being. In the course of these analyses, it became plain that the existential phenomena of death, conscience, and guilt are anchored in the phenomenon of care. The totality of the structural whole has become even more richly articulated; and because of this, the existential question of the unity of this totality has become still more urgent. †

In its anticipatory resoluteness, Dasein has now been made phenomenally visible with regard to its possible authenticity and totality. The hermeneutical Situation(iv) which was previously inadequate for interpreting the meaning of the Being of care, now has the required primordiality. Dasein has been put into that which we have in advance, and this has been done primordially—that is to say, this has been done with regard to its authentic potentiality-for-Being-a-whole; the idea of existence, which guides us as that which we see in advance, has been made definite by the clarification of our ownmost potentiality-for-Being; and, now that we have concretely worked out the structure of Dasein's Being, its peculiar ontological character has become so plain as compared with everything present-at-hand [as in the lived mode of science which phenomenally looks in the mode of present at hand or experimenta-

* See What is Sorge (Care) in Heidegger's thought?
And Sorge
And Heideggerian terminology (Wikipedia)
And Sorge (Ohio State University)

† Martin Heidegger, *Being and Time*, trans. John Macquarrie and Edward Robinson (New York: Harper & Row, 1962), 317–318.

tion], that Dasein's existentiality has been grasped in advance with sufficient Articulation to give sure guidance for working out the existentialia conceptually.

The way which we have so far pursued in the analytic of Dasein has led us to a concrete demonstration of the thesis(v) which was put forward just casually at the beginning—that the entity which in every case we ourselves are, is ontologically that which is farthest. The reason for this lies in care itself. Our Being alongside the things with which we concern ourselves most closely in the 'world'—a Being which is falling—guides the everyday way in which Dasein is interpreted, and covers up ontically Dasein's authentic Being, so that the ontology which is directed towards this entity is denied an appropriate basis. Therefore the primordial way in which this entity is presented as a phenomenon is anything but obvious, if even ontology proximally follows the course of the everyday interpretation of Dasein. The laying-bare of Dasein's primordial Being must rather be wrested from Dasein by following the opposite course from that taken by the falling ontico-ontological tendency of interpretation.

Not only in exhibiting the most elemental structures of Being-in-the-world, in delimiting the concept of the world, in clarifying the average "who" of this entity (the "who" which is closest to us—the they-self), in Interpreting the 'there', but also, above all, in analysing care, death, conscience, and guilt—in all these ways we have shown how in Dasein itself concernful common sense has taken control of Dasein's potentiality-for-Being and the disclosure of that potentiality—that is to say, the closing of it off. (Heidegger, 1959, pp. 562-563)

Sorge as the lived, authentic temporality of *Dasein* underlies the inauthentic lived mode of everydayness as *das man*, the they-self. The temporality of 'being there' as *Sorge* is simply assumed in the verb to-be, the 'is', 'are', 'was', etc. Heidegger has a 20th century understanding of temporality not as static 'now' moments but as having a lived-stretch from past (*Vergangenheit*, thrownness, disposedness) through a present (*Gegenwart*, projection, understanding) and towards a future (*Zukunft*, fallenness, fascination, anticipation, ahead-of-itself as not-yet, Being-towards-the-end, death). * Heidegger's idea of temporality has nothing to do with the abstract, linear 'now' moments of classicism except perhaps in the mode of inauthentic being-in-the-world. You might think that his notion of being as for inauthenticity is submerged in everydayness or in authenticity is the temporality of *Sorge* as similar to Levinas' ego in retreat from radical alterity or self in persecution from the radical alterity of the other as resembling each other. To some superficial extent, it does. But, in the mode of authenticity there is the possibility for recovering one's "ownmost possibility". For Levinas there is no recovery of the radical alterity of the other. For Levinas there is never a possibility of recovering saying from the radical alterity of the other.

For Hegel, thinking rises up above, and from Being into Concept. Being is only a moment in the dialectic which in Concept becomes the subject of all predicates and verbs (to be). There is no disjunctive break in Hegel's dialectic. Each stage consumes and preserves as evolutionary transformations and eventually devolution all over again. Perhaps some in the Hegelian tradition might deem this to be simultaneous as well. The 'all over again" is echoed throughout history. This is the temporality of life, death, and rebirth, the endless cycle of yen/yang, Shiva as creation and destruction, self-awareness as "I am Shiva", atman (soul, self, interiority) and Brahman (god, not self, exteriority, growth, *rhyzo-phusis to chaogenesis*). This is Nietzsche's eternal recurrence of the same self. But for Hegel there is a progressiveness which does not abandon the prior but holds all together in higher evolutions which must end

* See Martin Heidegger (Stanford Encyclopedia of Philosophy)

as necessity, as absolute which recovers and exceeds subjectivity as essence. But Levinas wants to make an abrupt distinction which cannot find essence and resolve. The reason that ethics can never be fundamental in any branching, classical system, is because the echo must always find itself back to itself. The echo must always recover itself. The other is recoverable in Being and Concept. There can be no basis for Ethics. Ethics is not ethics if it must always resound back from 'x'. Ethics proceeds from responsibility which has no possibility of return, no seat of sovereignty, rule, and origin or *telos*. In this way the quantum wave function provides a better pedagogical analogy for how to distinguish between other traditions which want to diminish the void into recoverability. Concerning ethics, Levinas writes,

> The ethical language we have resorted to does not arise out of a special moral experience, independent of the description hitherto elaborated. The ethical situation of responsibility is not comprehensible on the basis of ethics. It does indeed arise from what Alphonse de Waelhens called nonphilosophical experiences, which are ethically independent. The constraint that does not presuppose the will, nor even the core of being from which the will arises (or which it breaks up), and that we have described starting with persecution, has its place between the necessity of 'what cannot be otherwise' (Aristotle, Metaphysics, E), of what today we call eidetic necessity, and the constraint imposed on a will by the situation in which it finds itself, or by other wills and desires, or by the wills and desires of others. The tropes of ethical language are found to be adequate for certain structures of the description: for the sense of the approach in its contrast with knowing, the face in its contrast with a phenomenon. *

Here Levinas is telling us that Ethics is not ethics as we normally think about it. What Levinas means by Ethics "*does not arise out of a special moral experience*". What Levinas refers to as responsibility "*is not comprehensible on the basis of [what we normally think of as] ethics*". Levinas' ethics is not about the 'ought' of ethics. Neither does it arrive to us philosophically. Ethics is non-philosophical experience which is independent from the traditional sense of the 'ought' of ethics. Ethics is not about free will to decide to be ethical. Neither is it based on the authenticity of being or essence. Levinas says it is the between of "*what cannot be otherwise*" and what is imposed on us by "*other wills and desires, or by the wills and desires of others*". Levinas wants to draw a sharp contrast with "knowing" and the irreducibility of the he or she we face. This reinforces what I was referring to earlier concerning the 'yawning gap', the incompleteness of knowledge in its historic grounding.

Think about the quantum wave function and the abyss which makes ontology and epistemology unable to give or find an answer. Similarly, the radical alterity of the other which we face prohibits any possibility for recovery back into classicism, essence, Being, logic, etc. only leaving us with a debt which cannot be repaid. The self for Levinas does not arise from some whole, some absolute, some essence. Analogously, if we, me, the 'I' arises from branching of the quantum wave function, it arises from an unbridgeable abyss. 'Me' is retreat from an abyss, a yawning gap, from chaos, radical alterity which cannot be consumed in binary dualism but must always exceed classical narratives. The other, the he or she, is not an answerable to me or even has no way of being made commensurate to me as a product of me, or history, or narrative. The other we face is the demand of chasm in saying. In facing the *il y a* as sheer isness we face radical alterity which I cannot take hold of, cannot temporalize, cannot conceive,

* Emmanuel Levinas, *The Levinas Reader*, ed. Sean Hand (Oxford: Blackwell, 1989), 110.

cannot 'be' or not 'be'. It would be as if we imagined we could just crawl back into the wave function. The retreat from *il y a* as the Gordian knot of ipseity arising into egoity is not a 'taking hold of the quantum wave function' as if to own it and neither can ipseity's retreat arising into self find a 'not' of taking hold of exteriority. The spooky relation is in saying, the face to face in which origin or the 'not' of origin can arrive to me is nonsensical except in the 'hallowed' echoes from the hall of historic narrative which I mistakenly think of as mine, truth, reality, plain ol' common sense. This is where Levinas leaves the spookiness of the quantum wave function as the proximity of the other as "an immemorial past". Levinas writes,

> But is subjectivity thus conceived in what is irreducibly its own? Hegel and Heidegger try to empty the distinction between the subject and being of its meaning. In reintroducing time into being they denounce the idea of a subjectivity irreducible to essence, and, starting with the object inseparable from the subject, go on to reduce their correlation, and the anthropological order understood in these terms, to a modality of being. In the Introduction to the Phenomenology of Mind, in treating as a "pure pre-supposition" the thesis that knowing is an instrument to take hold of the Absolute (a technological metaphor) or a medium through which the light of truth penetrates the knower (a dioptic metaphor), Hegel denies that there is a radical break between subjectivity and the knowable. It is in the midst of the Absolute that the beyond takes on meaning; essence, understood as the immanence of a knowing, is taken to account for subjectivity, which is reduced to a moment of the concept, of thought or of absolute essence. Heidegger says, in a remark at the end of his Nietzsche (Vol. II, p. 451), that the "current term subjectivity immediately and too obstinately burdens thought with deceptive opinions that take as a destruction of objective being any reference from Being to man and especially to his egoity." Heidegger tries to conceive subjectivity in function of Being, of which it expresses an "epoque": subjectivity, consciousness, the ego presupposes Dasein, which belongs to essence as the mode in which essence manifests itself. But the manifestation of essence is what is essential in essence; experience and the subject having the experience constitute the very manner in which at a given "epoque" of essence, essence is accomplished, that is, is manifested. Every overcoming as well as every revaluing of Being in the subject would still be a case of Being's essence. *

Perhaps Aristotle called this event *ousai* which is Greek for pure actuality, awareness, and as a moving *telos*, goal, culmination which is not moved itself. Heidegger actually criticizes the latter Greek priority of presencing taking up by Latin Christianity into the God of light in which darkness is no more. Heidegger thinks privation gets lost as co-arising from Heraclitus and the earlier Greeks.

> When Heidegger observes that Plato's interpretation of the beingness of beings rests on the experience of ov as ousia, Heidegger has in mind the constancy and presence of beings, emerging on their own (vom ihm selbst her), where 'emerging' precisely means coming out from being closed off, concealed, and folded in upon itself (GA 55: 87). As Heidegger puts it in

* Emmanuel Levinas, *Otherwise Than Being or Beyond Essence*, trans. Alphonso Lingis (The Hague: Martinus Nijhoff, 1981), 76.

> another context, "ousia names that within which, from the outset, earth and sky, sea and mountains, tree and animal, human being and God emerge and, as emerging, show themselves in such a way that, in view of this, they can be named 'beings'. *

When time began or if it never had a beginning has been debated throughout the whole history of physics, philosophy, and religion. When time begins is inseparable from questions of origin. Some have said it always was, it was eternal. Others have told us the beginning came from nothing.

> The ancient Greeks debated the origin of time fiercely. Aristotle, taking the no-beginning side, invoked the principle that out of nothing, nothing comes. If the universe could never have gone from nothingness to somethingness, it must always have existed. For this and other reasons, time must stretch eternally into the past and future. Christian theologians tended to take the opposite point of view. Augustine contended that God exists outside of space and time, able to bring these constructs into existence as surely as he could forge other aspects of our world. When asked, What was God doing before he created the world? Augustine answered, Time itself being part of God's creation, there was simply no before! †

As we have seen, more recent physics suggests time is not fundamental. In relativity time is not really different from space in its dimensionality. We think from the cosmological theory of the Big Bang, our universe started from a singularity which means there was a spacetime when spacetime did not exist at all. A singularity would be pure chaos where spacetime and information would be meaningless. It would not be nothing and it would not be something, it would be indefinite and indeterminate. Chaos is a very ancient notion.

> Hesiod's Theogony (8th-7th century BC) was the first attempt to synthesize these traditions, which probably dated back to the Assyrian and Babylonian civilizations. In recounting the stages in the emergence of the gods from primordial chaos, Theogony offers an answer to the eternal questions of cosmogony: who created the world; what were the basic materials from which it was made; which came first, the gods, the stars or the elements? Not only did Theogony have a strong influence on Greek thought, it also anticipated in some ways today's theories of the origin of the world - particularly the idea of primordial chaos. Since the universe appears to have an ordered structure (albeit an imperfect one), it seems logical to regard the state which preceded the Creation as one of disorder and confusion. This notion has provoked greater controversy than almost any other in the history of cosmogony.‡

I want to note here that in the statement, '*it seems logical to regard the state which preceded the*

---

* Martin Heidegger, *Basic Concepts of Ancient Philosophy*, trans. Richard Rojcewicz (Bloomington: Indiana University Press, 2008), 87.
See Being at the Beginning: Heidegger's Interpretation of Heraclitus (.pdf)

† Veneziano, Gabriele. Febuary 1, 2006. "The Myth Of The Beginning Of Time". Scientific American (online)
Brian Greene, *The Fabric of the Cosmos: Space, Time, and the Texture of Reality* (New York: Alfred A. Knopf, 2004), 272.

‡ John Gribbin, *In the Beginning: The Birth of the Living Universe* (London: Penguin Books, 1993), 23.
See Creation Chaos Time from Myth to Modern Cosmology (pdf)

Creation as one of disorder and confusion' that of course disorder and confusion are oppositional to 'logical' by definition, so the 'controversy' is a faux argument. 'Order' necessitates '*disorder and confusion*' but must relegate '*disorder and confusion*' definitionally, epistemologically, not ontologically, as if it were purely a matter of existence or nothing. Nothing is not chaos. Nothing is a logical negation of something. Nothing is a determinate idea from the idea of something. Nothing only gains an apparent identity based on its opposite - something. Something can be God, creation, existence, Being, Concept, etc. Basically, everything which is not nothing can be something. Both something (what we define it as) and nothing are from the start symbiotic ideas. They have an essential dependence on each other. The logic of nothing and something cannot have any meaning for chaos. The ancient notion of chaos is not logical in any sense. Chaos in the ancients came long before logic. However, it could be argued that this pre-history of chaos gave birth to logic, order, and sense. In the temporal transitions of the ancients and in the spookiness of quantum physics, chaos does not exclude such a birth as origin. It does not logically negate origin but only finds irrelevance to origin. In classicism, origin becomes but in quantum physics uncertainty has hazy relations with observations upon observations, updating information branches and flows which do not negate previous observations as much as move confluences of parallel information flows or more easily thought of as universes away from each other. This is called the quantum wave function.

The observer is not a subject or many subjects, but the wave function is caught up together with the (for lack of better word) ~~universal~~* wave function. Physics refers to this as information. Information is not some version or metaphysic of 'reality'. In the wave function cause and effect are an information update which cannot yet find an order. It cannot be universal or absolute. Information cannot yet be the same or other. Even the notion of '*is*' *is* information which recoils upon itself. The idea of '*is*' and '*is not*' are logical negations of the first order. To equate '*is*' and '*is not*' as contradictory is already to assume identity and its opposite. We must use '*is*' to use language. Language affords us no other alternative. Information as causal flow 'is' infinitely regressive. Information recycles upon itself from the inescapable idea of, on one hand, information as both [causal, ordered] information and, on the other hand, as exformation which '*is*' and '*is not*' information. Current physics has exhausted and proven as much as anything can be proven in physics, that uncertainty and indeterminacy give rise to certainty, determinacy, cause and effect, logic, order, 'reality' in some spooky relationship with exformation. The intertwining of information and exformation can have no absolute or even relative [as relativity]. It defies the '*is*' to the point where externality cannot yet '*be*' ascertained as opposed to internality. In the use of the term externality, we point not to nothing and not to something. Rather we point to the inability to be able.

Phenomenology in 20th century philosophy addresses the relevance of classical notions of 'reality' by asking us to simply observe not just what presents itself but **how** it presents itself. Phenomenological observation means simply looking at what presents itself in our lived experience without imprisoning phenomena in historic abstractions. I have discussed this in a previous post on Orwell and Levinas † and many other posts on my blog as well. As embodied humans we stand on the feeble precipice of knowledge. Physics as historically separated from philosophy is now coming full circle back

* Jacques Derrida is the primary postmodern philosopher associated with "erasure" or overstriking (sous rature) words like "universal," "truth," or "being." This technique signals that a word is necessary for language yet inadequate or deceptive, challenging the stability of traditional philosophical concepts.

† See George Orwell and Emmanuel Levinas Introspective: Socialism and the Other | Musings (mixermuse.com)

to a more archaic and lost richness, A richness which now holds the promise of not returning to chaos but perhaps gaining a hint of a bygone epoch of what may have once been indicated by the terms 'yawning gap'. The yawning gap at this point in the history of philosophy has the possibility of not being abducted by historic neutrality but allowing for the he or she which faces us in a very different way. If contemporary situatedness can allow us to behold the new richness of this precipice we might find a way into Ethics, responsibility for the other. Ethics which has rarely been given its saying and only covered over by what has been said.

We have seen how, without us being aware of it, modernity imposes an abstraction on our practicalities which reflexively defines 'reality' as subject and object existing as things in absolute time and space. We do this at the expense of our lived experience. Lived experience is dismissed as merely subjective. Lived time is ignored in favor of clock time. Lived space is thought in terms of linearity not lived regions we bring close and withdraw from. For example, how we inhabit the space of a cup of coffee we are drinking rather than the glasses on our face which are linearly closer to our eyes.

We have seen how exformation has a kind of externality which places a requirement on us. This externality cannot be captured by ready-made categories. What is more, this externality requires something from us. It requires us to continually check our reflexive understanding at the gap of an excess to our commonsense notions of 'reality'. What comes 'naturally' to us falls into a convenience of history which exformation no longer affords us. I have illustrated this unease as chaos, a yawning gap, which finds no rest in the absolutes of modernity. Now, in this section, I ask you to take this dynamic of an externality which we cannot reconcile or bridge over into any familiarities into Levinas' discussion of the other.

To be clear, I am going to use the word 'sensual' as the lived embodied experiences which we 'feel' practically on a daily basis, not just the restrictive form of sexual sensuality. Levinas asks us to take a phenomenological look at our sensual lived experience of the other to see if perhaps our historic practicality has enabled us to overlook something important about how we experience the other. Levinas grounds this experience in very practical ways in which we come face to face with the other. As a human, I am embodied. I do not just have modernity's senses as a thinking machine which takes in sense data. I have sensuality. My embodied sensibility comes before my thoughts which come in the form of reflection after the sensual encounter with the other. Here is where the experience of the other diverges from Exformation. Exformation is not a sensual experience of the other. However, I have used the discussion of exformation to illustrate something which I hope is more accessible to the reader. We think of physics as something 'real'. And now physics has arrived at a junction in history in which the 'real' has become problematic—essentially problematic. Now, I want to turn our attention to how we interact and reflect on our interactions with the other to see if something akin to the exteriority of exformation might be found in how we sensually experience the encounter with the other and what that may imply.

From the earliest Greek notion of *phusis* through its split into physics and philosophy into the Latin reduction to nature one notion has been dominant and authoritative all the way through Western history - rule, authority, origin and birth are all thought together with Being. All the way back to my discussion about the inability of 'thinking' to think the informational thought of Entangled particles should tell us something, something very important. The awkward inability to be able, to have words or thoughts for thinking exformation, the ***is*** of exformation which cannot be in spacetime, in cause and effect, but yet shouts at us to pay attention. We can say what it isn't because language affords us the 'not', the negation. However, we cannot own it. We cannot find positive footing from our history which discloses, brings into presence, into light. We can describe entangled, informational behavior, as how two observers will see its spookiness and how we can build quantum technologies for incredibly new and promising ways of thinking but the '***is***' of our inheritance cannot rest easily in 'spooky-action-at-a-distance'. Why is that?

How can we speak about anything without naming subjectivity? This is a truism. Any conception must arise from a thinking subject. Kant has shown that thought *a priori* rises from Categories of Understanding. Only the 'thing-in-itself' is left to ponder subjectivity without objective Categories of Understanding. For Kant, phenomenality, existence, my lived subjectivity, arises from the object of thought, the Categories of Understanding. But what is identity for Kant?

Kant posits that a person possesses a unity of apperception, which means the ability to recognize oneself as a continuous, self-aware entity over time. This unity allows us to perceive our own identity through different moments and experiences. Kant acknowledges that we understand our identity through empirical criteria. These include sensory experiences, memories, and the continuity of consciousness. When we recognize ourselves as the same person who existed in the past, we rely on these empirical markers. Interestingly, Kant also acknowledges the rationalist notion of identity. According to this perspective, identity is not solely dependent on empirical evidence. Instead, it is a necessary concept for practical purposes. While empirical criteria help us recognize our identity, the rationalist notion provides a foundational understanding that transcends specific experiences. Kant maintains that the rationalist notion of identity is "necessary and sufficient for practical use." In other words, even though empirical evidence informs our daily understanding of identity, the rational concept of identity serves as a practical framework for moral responsibility, decision-making, and social interactions. Kant's exploration of identity encompasses both empirical and rational dimensions, emphasizing the interplay between our lived experiences and our conceptual understanding.

> "For through the I, as a simple representation, nothing manifold is given; it can only be given in the intuition, which is distinct from it, and thought through combination in a consciousness. An understanding, in which through self-consciousness all of the manifold would at the same time be given, would intuit; ours can only think and must seek the intuition in the senses. I am therefore conscious of the identical self in regard to the manifold of the representations that are given to me in an intuition because I call them all together my representations, which constitute one. But that is as much to say that I am conscious a priori of their necessary synthesis, which is called the original synthetic unity of apperception, under which all representations given to me stand, but under which they must also be brought by means of a synthesis" *

Kant's apperception is the immediate "I think therefore I am" of René Descartes. But Kant tells us that apperception is not enough to explain identity. Therefore, the requirement for identity is self-awareness, self-consciousness. The self must have representations of itself as an awareness of awareness. Here again we find the solitary substance of one without another. The self knows itself from itself. The

* Immanuel Kant, *Critique of Pure Reason*, trans. Paul Guyer and Allen W. Wood (Cambridge: Cambridge University Press, 1998), 248.

other is a representation of self to itself. Identity once again effaces the relevance of the radical alterity of the other. *

The notion of identity can only reverberate, replay itself endlessly in the mode of A=A, to own my being, my representation of myself to myself. My subjectivity must be swallowed up by the *a priori*. Forget about that messy little thing-in-itself behind the curtain in the Kingdom of Oz, of Being. Levinas tells us,

> Without resorting to the truism that all reality that is in any way recognized is subjective, a truism that goes with the one that says that everything that is in any way recognized presupposes the comprehension of being, Kant, by distinguishing in the course of the solution of the Antinomies the temporal series of experience from the in-temporal (or synchronic?) series conceived by the understanding of the other, has shown in the very objectivity of an object its phenomenality: a reference to the fundamental incompletion of the succession, and hence to the subjectivity of the subject. †

Kant opposed the reduction of Hume to sensation and empiricism. Kant wanted to understand what was behind metaphysics. He tells us,

> "That the human mind will ever give up metaphysical researches is as little to be expected as that we should prefer to give up breathing altogether, to avoid inhaling impure air. There will therefore always be metaphysics in the world; nay, every one, especially every man of reflection, will have it, and for want of a recognized standard, will shape it for himself after his own pattern. What has hitherto been called metaphysics, cannot satisfy any critical mind, but to forego it entirely is impossible; therefore a Critique of Pure Reason itself must now be attempted or, if one exists, investigated, and brought to the full test, because there is no other means of supplying this pressing want, which is something more than mere thirst for knowledge." ‡

Kant in his Critique of Pure Reason used reason itself to determine its own limits as in the Categories of Human Understanding. Sensation was the raw data which was necessarily always filtered *a priori* by the Categories. We could never understand 'reality' itself, the thing-in-itself except from the Categories.

> Much of the difficulty in Hegel's work stems from his purpose: he sought to dismantle a monumental work of philosophy, Immanuel Kant's Critique of Pure Reason (1781), where Kant used reason to determine its own limits. Our senses are bombarded with stimuli – raw

---

* See Immanuel Kant (IEP)
And Kant on the identity of persons
And Kant's Theory of Self-Consciousness
And Identity of the Self in Kant

† Emmanuel Levinas, *Otherwise Than Being or Beyond Essence*, trans. Alphonso Lingis (The Hague: Martinus Nijhoff, 1981), 75.

‡ Immanuel Kant, *Prolegomena to Any Future Metaphysics*, trans. Paul Carus (Chicago: Open Court, 1902), 148–49.
See Immanuel Kant: Prolegomena to Any Future Metaphysics (.pdf)

> data which our minds receive and shape and organize, creating our perception of reality. There is an objective reality out there, which Kant called the 'thing-in-itself'. As we can't access that reality directly, but only filtered through our perceptions, the 'thing-in-itself' is beyond the scope of science and even reason, so it will forever be a mystery. Despite this, Kant knew that the mind thirsts for ultimate reality: "That the human mind will ever give up metaphysical research is as little to be expected as that we should give up breathing," he wrote. Hegel set out to prove Kant wrong. Like his contemporaries Fichte and Schelling, he felt that philosophy could find and understand Kant's thing-in-itself. *

Hegel thought that Kant stopped short at the thing-in-itself. The thing-in-itself was yet another metaphysics. It was a metaphysics of negation. For Hegel, metaphysics was the desire for subjectivity to possess, to own, its object, to find how it was ultimately the subject of all predicates. Therefore, according to Hegel, subject and object define each other. They are the same thing and are therefore, lifted up as self-consciousness in Hegel's work entitled, "The Phenomenology of Spirit". However, in Hegel's Jena Lectures (1805-1806),

> What is posited thereby is the concept of freely interrelated self-consciousnesses – but only the concept itself. Since it is only a concept, it is still to be realized; i.e., it is to transcend (aufzuheben) itself in the form of a concept and approach reality, in actuality, it itself occurs unconsciously in the dissolution of the problem and in the problem itself – unconsciously, i.e., so that the concept does not intrude into the [realm of the] object. †

Hegel tells us that Kant's work failed to arrive at the object, the universality of Concept. Some Hegelian scholars tell us that the "Phenomenology of Spirit" could not account for the universality of absolute Spirit, that it could not arise to Concept merely from itself. Individual self-consciousness could not give us absolute consciousness. Consciousness could not admit an exteriority required for absolute Idea or Concept.

> To anchor idealism immanently--so that we need not appeal to an outside and thus undermine our idealism--we need resistant objects, which, as we have seen, theoretical reason gives us. Secondly, we need totality--the absolute. There can be nothing outside of consciousness--no outside to consciousness. Individual consciousness cannot give us such totality--an absolute consciousness will be required. ‡

However, the "Phenomenology of Spirit" absolute consciousness was found to be lacking. Subjectivity is based on self-consciousness. Even in absolute Spirit's negation of itself it cannot arise into

---

* "Hegel," in *The Philosophy Book: Big Ideas Simply Explained* (New York: DK Publishing, 2011), 178.
See The Trouble with Hegel.

† Friedrich Wilhelm Joseph von Schelling, *The Grounding of Positive Philosophy: The Berlin Lectures*, trans. Bruce Matthews (Albany: State University of New York Press, 2007), 164.
See The Philosophy of Spirit (Jena Lectures 1805-6): PART I. Spirit according to its Concept

‡ Thomas Nenon, *Objectivity and System: A Study of Kant's Theory of Experience* (The Hague: Martinus Nijhoff, 1986), 133.
See Hegel, Reason, and Idealism

universality. Self-consciousness cannot arise to an identity of absolute consciousness except in abstraction where essence as time would have to turn back on itself. The multiplicity of self-conscious subjects would have to combine to achieve a universal self-consciousness of the mind. But negating a multiplicity of individual self-conscious subjects to achieve universal self-consciousness would be like "*bits of dust collected by its movement or drops of sweat glistening on its forehead because of the labor of the negative it will have accomplished. They would be forgettable moments of which what counts is only their identities due to their positions in the system, which are reabsorbed into the whole of the system*". Hegel finds this unacceptable as he always wants to move from abstractions to more concrete dialectics. Levinas writes,

> The reduction of subjectivity to consciousness dominates philosophical thought, which since Hegel has been trying to overcome the duality of being and thought, by identifying, under different figures, substance and subject. This also amounts to undoing the substantivity of substance, but in relationship with self-consciousness. The successive and progressive disclosure of being to itself would be produced in philosophy. Knowing, the dis-covering, would not be added on to the being of entities, to essence. Being's essence carries on like a vigilance exercised without respite on this very vigilance, like a self-possession. Philosophy which states essence as an ontology, concludes this essence, this lucidity of lucidity, by this logos. Consciousness fulfills the being of entities. For Sartre as for Hegel, the oneself is posited on the basis of the for-itself. The identity of the I would thus be reducible to the turning back of essence upon itself. The I, or the oneself that would seem to be its subject or condition, the oneself taking on the figure of an entity among entities, would in truth be reducible to an abstraction taken from the concrete process of self-consciousness, or from the exposition of being in history or in the stretching out of time, in which, across breaks and recoveries, being shows itself to itself. Time, essence, essence as time, would be the absolute itself in the return to self. The multiplicity of unique subjects, entities immediately, empirically, encountered, would proceed from this universal self-consciousness of the Mind: bits of dust collected by its movement or drops of sweat glistening on its forehead because of the labor of the negative it will have accomplished. They would be forgettable moments of which what counts is only their identities due to their positions in the system, which are reabsorbed into the whole of the system. *

But Hegel wants to take away the substantive from absolute. Substance all the way from the Latin transformation and deformation of *ousia* build on the notion of abstract stuff, matter, body in binary opposition to mind. Therefore, Hegelians have argued that the use of the word 'absolute' does not designate a non-relative essence but rather simply a relative position within Hegel's progression. Cambridge published a book entitled, Hegel's "Philosophy of Spirit", where the author Angelica Nuzzo challenges what could possibly be meant by such a subtlety of nuance in thinking of the absolute.

---

* Emmanuel Levinas, *Otherwise Than Being or Beyond Essence*, trans. Alphonso Lingis (The Hague: Martinus Nijhoff, 1981), 200–201.

> I have argued that contrary to what many interpreters seem to assume, there is simply no original, substantive "Absolute" in Hegel's philosophy, but that the adjective "absolute" (along with the adverb) is instead a systematically crucial, topological predicate that indicates the "place" or position of a certain determination or concept (and its reality) within the overall structure of philosophical thinking. Moreover, this position is not a static point or marker within a given whole but is rather a dynamical stage in the process through which the whole of philosophy is first constituted in the form of a complete system. This is, to be sure, the first step in a broader discussion that leads to the further question of what warrants the designation of "absolute" for a certain moment within such a process. In other words, what is it that makes a certain moment at stake at a specific stage of the systematic constitution of the whole of Hegel's philosophy an "absolute" moment? And, furthermore, is the "absoluteness" of all "absolute" structures and concepts the same? *

In naming 'universal', Is Hegel hierarchical or temporal? If not, what is meant by the absoluteness of absolute? Is universal just a shortened name for a particular movement otherwise called 'Concept'? I think we need to look at how absolute Spirit, self-consciousness' concrete fulfillment of itself juxtaposed to the other. to otherness, universalizes. Absolute Spirit must be itself by its other, absolute otherness. Absolute spirit cannot pull itself up by its own bootstraps. It needs what is not itself to become something other than itself. Exteriority to self-consciousness raises them both up to Concept.

> According to Hegel, the relationship between self and otherness is the fundamental defining characteristic of human awareness and activity, being rooted as it is in the emotion of desire for objects as well as in the estrangement from those objects, which is part of the primordial human experience of the world. The otherness that consciousness experiences as a barrier to its goal is the external reality of the natural and social world, which prevents individual consciousness from becoming free and independent. However, that otherness cannot be abolished or destroyed, without destroying oneself, and so ideally there must be reconciliation between self and other such that consciousness can "universalize" itself through the other. †

So universal Idea is the subject from which all predicates proceed not a metaphysical monad. But in order for self and other, interiority and exteriority, to rise above themselves, exteriority has to lose its radicality. In this case, exteriority can only be what it is by opposing subjectivity. This is a binary duality which can have no excess but only a resolve, a completion, a universal essence which we are led to believe still hold the interiority of the subject and the exteriority of the object, the other, together in their difference to each other. Concept cannot be Concept without other. Nor can Concept be

* Angelica Nuzzo, "The 'Absoluteness' of Hegel's Absolute Spirit," in *Hegel's Philosophy of Spirit: A Critical Guide*, ed. Marina F. Bykova (Cambridge: Cambridge University Press, 2019), 189–90.

† According to Hegel, the relationship between self and otherness is the fundamental defining characteristic of human awareness and activity, being rooted as it is in the emotion of desire for objects as well as in the estrangement from those objects, which is part of the primordial human experience of the world. The otherness that consciousness experiences as a barrier to its goal is the external reality of the natural and social world, which prevents individual consciousness from becoming free and independent. However, that otherness cannot be abolished or destroyed, without destroying oneself, and so ideally there must be reconciliation between self and other such that consciousness can "universalize" itself through the other.

Hegel: Social and Political Thought

Concept with the completion of self-consciousness in absolute Spirit. Can we detect a shift in what is meant by exteriority? In this case exteriority has no excess to the same of self-consciousness. Exteriority is made logically commensurate with its dialectical opposite. The other must oppose self-consciousness as what self-consciousness cannot know, must always reside as the binary duality of knower and known. There is no excess here for exteriority only a 'not' which must fill the place, mark the *telos* and completion of exteriority. But is that the sum of everything meant by exteriority? Well, at least quantum physics would differ on that point as we have seen. In dominant strains of classical narratives of philosophy and science the self, knowing, Concept, time and space, cause and effect, locality as mechanism must always cover over the ancient intuitions of chaos, lack of origin, and exteriority. Exteriority must always bow to the rule of origin to be recaptured, recuperated, reduced into the 'not' of knowledge, essence, and truth.

> But is subjectivity thus conceived in what is irreducibly its own? Hegel and Heidegger try to empty the distinction between the subject and being of its meaning. In reintroducing time into being they denounce the idea of a subjectivity irreducible to essence, and, starting with the object inseparable from the subject, go on to reduce their correlation, and the anthropological order understood in these terms, to a modality of being. In the Introduction to the Phenomenology of Mind, in treating as a "pure pre-supposition" the thesis that knowing is an instrument to take hold of the Absolute (a technological metaphor) or a medium through which the light of truth penetrates the knower (a dioptic metaphor), Hegel denies that there is a radical break between subjectivity and the knowable. It is in the midst of the Absolute that the beyond takes on meaning; essence, understood as the immanence of a knowing, is taken to account for subjectivity, which is reduced to a moment of the concept, of thought or of absolute essence. Heidegger says, in a remark at the end of his Nietzsche (Vol. II, p. 451), that the "current term subjectivity immediately and too obstinately burdens thought with deceptive opinions that take as a destruction of objective being any reference from Being to man and especially to his egoity." Heidegger tries to conceive subjectivity in function of Being, of which it expresses an "epoque": subjectivity, consciousness, the ego presuppose Dasein, which belongs to essence as the mode in which essence manifests itself. *

In classical narrative the 'not' of life, death, has become the marker for necessity to overcome, to conquer, to raise its pitch to the heights of Zarathustra's tragic mount. For Heidegger, authenticity begins when we resolutely accept that we are always on a journey toward our own death. Our own death is something that no one else can experience for us. Heidegger refers to this as being-towards-death. It is in this acceptance that authentic Being is disclosed. Facing death as mine opens up the question of Being itself. Authenticity resolutely arises when we accept and endure our own mortality rather than concealing it in the they-self, *das man*. †

* Emmanuel Levinas, *Otherwise Than Being or Beyond Essence*, trans. Alphonso Lingis (The Hague: Martinus Nijhoff, 1981), 200.

† See 5 - Being-towards-death
And Heidegger on "Being-Toward-Death"
And Being and Time part 6: Death
And What is the character of Heidegger's notion of authenticity in being-towards-death?

In our preliminary existential sketch, Being-towards-the-end has been defined as Being towards one's ownmost potentiality-for-Being, which is non-relational and is not to be outstripped. Being towards this possibility, as a Being which exists, is brought face to face with the absolute impossibility of existence. Beyond this seemingly empty characterization of Being-towards-death, there has been revealed the concretion of this Being in the mode of everydayness. In accordance with the tendency to falling, which is essential to everydayness, Being-towards-death has turned out to be an evasion in the face of death—an evasion which conceals. While our investigation has hitherto passed from a formal sketch of the ontological structure of death to the concrete analysis of everyday Being-towards-the-end, the direction is now to be reversed, and we shall arrive at the full existential conception of death by rounding out our Interpretation of everyday Being-towards-the-end...

The full existential-ontological conception of death may now be defined as follows: death, as the end of Dasein, is Dasein's ownmost possibility—non-relational, certain and as such indefinite, not to be outstripped...

Death is Dasein's ownmost possibility. Being towards this possibility discloses to Dasein its ownmost potentiality-for-Being, in which its very Being is the issue. Here it can become manifest to Dasein that in this distinctive possibility of its own self, it has been wrenched away from the "they". This means that in anticipation any Dasein can have wrenched itself away from the "they" already. But when one understands that this is something which Dasein 'can' have done, this only reveals its factical lostness in the everydayness of the they-self.

The ownmost possibility is non-relational. Anticipation allows Dasein to understand that that potentiality-for-being in which its ownmost Being is an issue, must be taken over by Dasein alone. Death does not just 'belong' to one's own Dasein in an undifferentiated way; death lays claim to it as an individual Dasein. The non-relational character of death, as understood in anticipation, individualizes Dasein down to itself. This individualizing is a way in which the 'there' is disclosed for existence. It makes manifest that all Being-alongside the things with which we concern ourselves, and all Being-with Others, will fail us when our ownmost potentiality-for-Being is the issue. Dasein can be authentically itself only if it makes this possible for itself of its own accord...

The ownmost, non-relational possibility, which is not to be outstripped, is certain. The way to be certain of it is determined by the kind of truth which corresponds to it (disclosedness). The certain possibility of death, however, discloses Dasein as a possibility, but does so only in such a way that, in anticipating this possibility, Dasein makes this possibility possible for itself as its ownmost potentiality-for-Being. The possibility is disclosed because it is made possible in anticipation. To maintain oneself in this truth—that is, to be certain of what has been disclosed—demands all the more that one should anticipate. *

According to Alexander Kojeve, a thinker deeply influenced by Hegel, death occupies a central place in Hegel's philosophy. For Hegel, acknowledging our own mortality is essential for authentic existence. Without death, we wouldn't be true individuals, free agents, or historical beings. Hegel saw

* Martin Heidegger, *Being and Time*, trans. John Macquarrie and Edward Robinson (Oxford: Blackwell, 1962), 303, 307, 308–9.

death as a manifestation of negativity. He believed that man's death is essentially voluntary, arising from risks willingly taken. In this view, death is not merely a biological event but a conscious choice. Hegel described humanity as "that night, that empty Nothingness," containing an infinite number of representations. The night of the world, the intimacy of Nature, and the pure personal-Ego all converge in this existential darkness. In his Philosophy of Right, Hegel argued against the right to suicide. He considered it a contradiction, asserting that there is no inherent right over one's own life. Interestingly, it would seem to be a contradiction to argue there is no inherent right to suicide and at the same time argue that death is as much a conscious choice as a biological event. Hegel's engagement with death extends beyond mere mortality—it touches on the essence of Being, authenticity, and the human condition. *

History must paint over death with a brush of redemption or unredeemable, the moment which leads to meaning or the moment which must be meaningless. There is a covering over of alterity in all these canopies upon which something must be drawn to fill the yawning gap. Exteriority must be settled, find fulfillment even if only by the marker of a tombstone. But every day, every moment of existence quantum physics tells us we fall from the exteriority of the spooky wave function which has an exteriority which cannot find meaning or meaninglessness, origin or the ''not' of origin, truth or falsity, authenticity or inauthenticity, all the permutations resulting from and in Concept and its absolute which must always overcome its 'not' without prejudice – from sheer logic. Seneca tells us, "*We suffer more in our imagination than in reality.*" The permutation of Oscar Wilde and George Bernard Shaw tells us, "There Are Only Two Tragedies. One Is Not Getting What One Wants, and the Other Is Getting It". † William Shakespeare, the master of human psychology and intricate character portrayal, explored the interplay between desire, drama, and reality in his timeless works. In "A Midsummer Night's Dream", Shakespeare deftly handles the theme of romantic desire that is prohibited by parents. When parents stand athwart desire, the forbidden nature of that desire makes it even more enticing. He illustrates this psychological truth through the play-within-the-play of "Pyramus and Thisbe". The myth of Pyramus and Thisbe, where young lovers defy parental prohibition to meet secretly, serves as a cautionary tale. Yet, ironically, it becomes the basis for the famous drama "Romeo and Juliet". True love remains attractive, even when it meets a tragic end. Perhaps the very defiance against parental restrictions adds to its allure. René Girard, influenced by Shakespeare, observed a pattern in human desire. Rather than labeling it as "narcissistic," Girard's "mimetic theory" suggests that all desire is learned from others. The cases often diagnosed as "narcissistic" are better understood as "pseudo narcissism", where mimetic interactions play a significant role. Desire, in its essence, is shaped by imitation and shared patterns of longing. Shakespeare's characters, driven by desire, seek eternity. Even when love meets tragedy, as in "Romeo and Juliet", the names of the ill-fated lovers are immortalized. Their early demise becomes part of the appeal—a defiance against mundane calculations and a testament to the power of love. Drama, with its portrayal of desire, captures the essence of human existence, transcending mere reality. Shakespeare recognized that our yearning for drama often surpasses the bound-

---

* See Death, Hegel, and Kojève (.pdf)
And Hegel, Death and Sacrifice
And GEORG WILHELM FRIEDRICH HEGEL from Philosophy of Right
And 4 Hegel's Conception of Tragedy

† See There Are Only Two Tragedies. One Is Not Getting What One Wants, and the Other Is Getting It

aries of everyday life. Through his plays, Shakespeare explores desire, authenticity, and the eternal human spirit on the grand stage of existence. *

At this point I want to make a brief remark about il y a, ego, and the self. None of these topics should be thought of as somehow different 'states of being' or in some unspecified way different from each other. It is better to think of these capacities as relationships like onion skins, as all playing together in the 'who' of me.

The echo of Narcissus

The echo of Narcissus is not a solitary voice. Echo resounds from the innumerable canyons of a play named History. All the actors get reduced to 'me here now' in commonsense, in relationships which embody all the ambivalences love, desire, alienation, the bad intent and threat of the other, the irrelevant, alien, remote, the played and the player, danger, evil, God, nothingness all captured in the service of the 'my' of self. It is the voice of historic branching which must raise existence to its highest crescendo where the gift it brings is drama. Drama exacerbates and annihilates exteriority. It assumes too much, dares too much, resounds as known and the 'un' of known. The branching of fictions all devised to block off, alter, what faces us, what exceeds us, what is unable to be able. This is not gnosis or a-gnosis (agnostic), knowledge or no knowledge (the 'not' of knowledge). Gnosis is a feminine Greek noun which means "knowledge" or "awareness". † Think of the quantum wave function which cannot easily find a 'place to be'. It is an excess which has no accounting. And 'it' is not an 'it'. The 'it' of history is the other who faces us which finds no place in echo and drama. The self must flee into origin, being, history to evade, the indeterminate in immersion of drama, sensuousness. The ego, the alter of ipseity, must inhabit itself, be at home with itself. The 'willing it thus' must arise in drama and sensuality. It must sooth itself in the refuge of historic voice and echo. The other is reciprocity to me as transactional. "*To present the knot of ipseity in the straight thread of essence according to the model of the intentionality of the for-itself, or as the openness of reflection upon oneself, is to posit a new ipseity behind the ipseity one would like to reduce.*" (Levinas, Otherwise Than Being or Beyond Essence, 1981, p. 203)

> The unjustifiable identity of ipseity is expressed in terms such as ego, I, oneself, and, this work aims to show throughout, starting with the soul, sensibility, vulnerability, maternity and materiality, which describe responsibility for others. The "fulcrum" in which this turning of being back upon itself which we call knowing or mind is produced thus designates the singularity par excellence. It can indeed appear in an indirect language, under a proper name, as an entity, and thus put itself on the edge of the generality characteristic of all said, and there refer to essence. ‡

The oneself does not rest in peace under its identity, and yet its restlessness is not a dialectical scission, nor a process equalizing difference. Its unity is not just added on to some content of ipseity, like

* See Shakespeare: Three Plays About Human Desire
And Shakespeare: Three Plays About Human Desire 1
And William Shakespeare: Desire

† See Gnosis

‡ Emmanuel Levinas, *Otherwise Than Being or Beyond Essence*, trans. Alphonso Lingis (The Hague: Martinus Nijhoff, 1981), 205.

the indefinite article which substantifies even verbs, "nominalizing" and thematizing them. Here the unity precedes every article and every process; it is somehow itself the content. Recurrence is but an "outdoing" of unity. As a unity in its form and in its content, the oneself is a singularity, prior to the distinction between the particular and the universal. It is, if one likes, a relationship, but one where there is no disjunction between the terms held in relationship, a relationship that is not reducible to an intentional openness upon oneself, does not purely and simply repeat consciousness in which being is gathered up, as the sea gathers up the waves that wash the shore.

The ego is not in itself like matter which, perfectly espoused by its form, is what it is; it is in itself like one is in one's skin, that is, already tight, ill at ease in one's own skin. It is as though the identity of matter resting in itself concealed a dimension in which a retreat to the hither side of immediate coincidence were possible, concealed a materiality more material than all matter —

> a materiality such that irritability, susceptibility or exposedness to wounds and outrage characterizes its passivity, more passive still than the passivity of effects. *

In the retreating ego it imagines itself in the drama of origin, self as origin, which must rule and reign lest the eternal night overtake it. This night, the il y a, must be flooded with certainty pronounced in praxis †, perchance to dream as 'daily life'. Physics has brought us before a gap, a void, a chasm which brings us towards a spooky alterity which has no home or place to lay its head in history. Levinas brings us before a radical alterity which finds no home in all our echoes. The inability of the self to be able, to be at home in itself, to root itself, branching rhizomic chaos pre-intentional sensuous lapse falling together with saying which must be forgotten in order to viscerally re-member it as the said, historic narrative taken up in echoing 'me'. The retreat to echo, to forget the lapse of interruption of saying, of exteriority before 'I' can gather into egoity foregoes ipseity, the naked bare self fleeing chaos, awareness not yet constituted with awareness of awareness. The lapse of time between the pre-intentional sensuous moment and its intentionalization denotes the transcendence that Levinas also equates with "the Saying".

His discussion of the Saying correlates with his treatment of sincerity, introduced already in Existence and Existents. Otherwise than Being radicalizes his notion of sincerity, insisting that the structure of sensibility is always as if punctuated by sensuous lapses. It is thanks to such time lapses that we are open and able to communicate because, as we have seen, proximity is an affective mode that motivates dialogue. While all sensuous lapses are not necessarily openings to intersubjective communication, proximity and vulnerability are the loci of transcendence-in-immanence and the birth of signification (whether words are actually uttered or not). For Levinas, there is more in living affectivity than Heidegger's conception of being speaking through language captured. This is clear the moment we under-

* Emmanuel Levinas, *Otherwise Than Being or Beyond Essence*, trans. Alphonso Lingis (The Hague: Martinus Nijhoff, 1981), 197.

† Emmanuel Levinas, *Existence and Existents*, trans. Alphonso Lingis (The Hague: Martinus Nijhoff, 1978), 58–60.

The word praxis is from Ancient Greek: πρᾶξις, romanized: praxis. In Ancient Greek the word praxis (πρᾶξις) referred to activity engaged in by free people. The philosopher Aristotle held that there were three basic activities of humans: theoria (thinking), poiesis (making), and praxis (doing). Corresponding to these activities were three types of knowledge: theoretical, the end goal being truth; poetical, the end goal being production; and practical, the end goal being action. See Praxis (Origins)

stand signification originally as an affective proto-intentionality and not as some thought, already formulated, that the I thereupon chooses to communicate to another. *

Sensibility is punctuated by sensuous lapses. Sensibility rises from sensuous lapses. We first encounter the world sensually and then bring them into historic narrative as if that endows sensuousness with origin, the Idea of right exercised by free will. We authenticate our lived experience with right [*Recht*] in the field of freedom. It is as if embodiment is an after-glow of idea. In this case Idea must ever after cover over its in-adequation as the never-ending toil of Sisyphus who must eternally push a stone uphill only for it to once again fall back to the earth. Embodiment is diachronous to idea – meaning, not temporally synchronized to idea. An unaccounted-for excess interrupts us sensually in our embodiment before Idea claims its right in the freedom of a self. As such, the other as unaccounted-for externality is anarchic. It resists synchronization to my lived time. To the classical idea of universal time. In anarchic externality, the saying of the other speaks from a lapse to my sensuality, from a radical alterity of externality. As embodiment, my lived sensibility, the saying of the other is from a time not my time which can only be made adequate to my idea by force as branching from classical narrative as the field of freedom in which my idea finds origin.

However, saying like exformation is not caused by or a result of origin. As diachronous, the other is not embodied in my lived time and space but speaks from an externality which is unequivocal to me. Saying is the "lapse of time" between "the pre-intentional sensuous moment" and Husserl's intentionalization into voices of consciousness. Embodiment branches from the lapse of radical externality which must retreat affectively before alter-ipseity can take hold as affective ego and become self from which idea, commonsense, understanding and knowledge can rule. The radical externality of the other addressing me is pre-intentional, pre-originary to my sensual lived time and lived space.

Saying as pre-intentional sensuousness falls together as historic retreat from the radical alterity of the face-to-face encounter with the other. Classical narratives as time and space come with historic, cultural, linguistic abstractions such as linear space and clock time. Even physics since Einstein has clearly shown that these notions come from history and classic physics before the advent of relativity. Absolute time and space are an abstraction not only to my sensuous experience of time and space but is also alien to the common-sense notions of absolute time and space which we have internalized from history and language. This is not philosophy telling us about classical narrative of time and space. This is physics which has unorthodoxly pulled these historic notions out from under our feet. We can remain in ever growing vacuous abstractions of idea, in history, or we can seriously consider the notions even implied by relativity. It might be interesting to note that agrarian time has been translated to clock time as linear 'now' moments where technology has moved our sense of time from years and days toward seconds and milliseconds. Entanglement is perhaps analogously similar to Levinas' notion of the alterity of the other with one very fundamental exception. Instead of exformation, embodiment would require our analogy to utilize the concept of singularity, the analogously 'ex' of exformation. We will refer to this as Levinas does as radical alterity or the Other as designated by a capital 'O'.

> For Levinas, the oneself may indeed appear as a faceless neutrality, "on the edge of the generality characteristic of all said," but this appearance is already a mask covering its "nameless

* See 2.4.3 The Saying and the Said at Levinas (Stanford)
Also, (Levinas, Otherwise Than Being or Beyond Essence, 1981), page 43

> singularity" (OB 106), its recurrence as a point of identity bearing alterity. This nameless singularity is irreducible to being, but it also inscribes itself as a trace in the midst of being; it borrows a name from being in order to show itself in the said, in order to matter in the world, and not just in the pure elsewhere of pre-originality—wherever and whatever that may be. However, this name of being is only a mask which is constantly unmasked, or unsaid, in its singular exposure to the Other. To repeat the formulation with which I began this paper: Anyone is responsible for any Other who happens to come along. The indeterminacy of the interval in the instant of hypostasis provides a logic for articulating selfhood as the Other-in-the-same, both separate and responsible, different and exactly equivalent to Others; it allows us to describe how the self remains itself (or, more exactly, becomes itself) while substituting for the irreplaceable, non-exchangeable Other. [*]

For Levinas, singularity as absolute uniqueness does not happen in a vacuum. Its actual uniqueness consists in its oscillation and confluence of the saying and the said. The lapse discussed earlier in the encounter with the other which Levinas refers to here as saying is not in my lived time and space. Phenomenology, postmodernism, and even analytic philosophy in the 20th century along with quantum physics are all exposing insufficiencies with classical narratives. The exteriority of the wave function and chaos find synergies with how we live physics, space and time, historic narratives of language, and the linguistic turn in analytic philosophy. For Levinas the encounter with the other is a rupture of externality in the saying, a lapse, not yet made commensurate by history and narrative nor even to the sensual, feeling, 'me'. But in the interruptive 'branch' I first live the after effect of the externality of the other as embodiment. In the clash of my lived time and space and unaccounted-for externality of the other cannot be made congruent with me except in a mask, a façade of the same, the echo of classicism on Being and Idea. For physics, a singularity is exformation caught up in the exteriority of the wave function which can no longer be thought of as spacetime but rather information which passes into chaos, without organization, without classical science and historic narrative. In Levinas' notion of singularity, the "*indeterminacy of the interval*" is a lapse of exteriority which cannot be made classical or commensurate with me in 'my reality'.

> The phenomenology of internal time consciousness explained the past entirely by this retentional effort; the past is the represented, by recall and first by primary retention; it is held by the force of the present. In principle everything is retained, and still recoverable, representable. Yet the past passes, and passes of itself, irretrievably. There is loss, falling away irrevocably, lapse of time. The bond with the past is a bond with a dimension of oneself which one cannot regain possession of once more, which prevents complete self-possession — and which yet holds on to one, holds one like a bond. Ageing is this temporalization — by virtue of the temporalization of one's time, one is being carried beyond one's powers. [†]

---

* Lisa Guenther, "Nameless Singularity: Levinas on Individuation and Ethical Singularity," *Epoché: A Journal for the History of Philosophy* 14, no. 1 (2009): 184–85.

Thomas Berns and Asja Szafraniec, "Levinas and the Question of the 'Political'," in *The Oxford Handbook of Levinas*, ed. Michael L. Morgan (Oxford: Oxford University Press, 2019), 565.

† Emmanuel Levinas, *Otherwise Than Being or Beyond Essence*, trans. Alphonso Lingis (The Hague: Martinus Nijhoff, 1981), 22–23.

The interruption of temporality by externality cannot be temporalized. It cannot be measured or spanned.

> It is then the temporalization of time, in the way it signifies being and nothingness, life and death, that must also signify the beyond being and not being; it must signify a difference with respect to the couple being and nothingness. Time is essence and monstration of essence. In the temporalization of time the light comes about by the instant falling out of phase with itself — which is the temporal flow, the differing of the identical. *

But time and distance as life must arise again into being and existence as if there were no lapses, as if non-existence could only be classical and metaphysical. In the branch from exteriority, we first live the other, being with the other as embodiment which even then, cannot yet be reduced to the idea of the other, to the monstration of essence. From embodiment yet once again, there is difference which must be rescued and accounted for by difference.

> When stated in propositions, the unsayable (or the an-archical) espouses the forms of formal logic; the beyond being is posited in doxic theses, and glimmers in the amphibology of being and beings — in which beings dissimulate being. The otherwise than being is stated in a saying that must also be unsaid in order to thus extract the otherwise than being from the said in which it already comes to signify but a being otherwise. Does the beyond being which philosophy states, and states by reason of the very transcendence of the beyond, fall unavoidably into the forms of the ancillary statement? †

As embodiment, we branch into the said. We are spoken to by radical exteriority. We are called by the singularity of the other, in speaking to us. Always before we can retreat from him or her, we are taken captive from a radical exterior which can only find retreat from difference as the same. The saying of the other as singularity which our sensualness finds no lived place or time for reflexively pushes us away from the other into Being, logos, and historic self-interest. Difference raises itself into Being as Idea. Here the echo resounds, the same yet different. Once again, for the first time, we step into the light of reason, the most common of sense, the moment of the Big Bang which can only be reckoned afterwards, where time zero is never found only the spirit of gravity in Idea must fall again from many-worlds. "I think; therefore I am." This idiom exposes the bare certainty of origin in modernism. But the gravity of idea is lived as experience, as affect-weighted attraction in which presence and idea bobble on the surface of lived experiences. This is where we saw earlier where embodiment, idea, and neutrality as modes of lived presence and absence find their way into phenomenology.

The way I think about phenomenology is as Einstein's relativity. For the most part, relativity sees no breaks in spacetime. Sure, the curvature of spacetime is deformed by mass but for the most part transitions are smooth and continuous. A singularity in general relativity is a 'metaphysical' break in spacetime, a rupture which most relativistic physicists understand a point in the theory which does not

* Emmanuel Levinas, *Otherwise Than Being or Beyond Essence*, trans. Alphonso Lingis (The Hague: Martinus Nijhoff, 1981), 64.

† Emmanuel Levinas, *Otherwise Than Being or Beyond Essence*, trans. Alphonso Lingis (The Hague: Martinus Nijhoff, 1981), 61.

make mathematical sense. It is relegated to 'something we do not understand yet'. But for quantum physics breaks in spacetime are the rule rather than the exception. So, a singularity is simply the bare exposure of spacetime as exformation which is not exceptional to spacetime but spacetime is probably emergent from exformation as the quantum wave function. Likewise, phenomenology does not see breaks in lived phenomenon. We live transitions in phenomena experientially analogous to smooth transitions of spacetime distortions by mass. But ruptures in phenomenology like the lapse of radical exteriority for Levinas cannot exist except as abstraction, metaphysics. But the observer effect in the branching of the wave function has also met with such abrasive skepticism early on. Even the inventor of relativity could not abide the spooky of quanta and spent decades trying to find a classical answer in locality and unified field theory. Postmodernism has also found breaks in linguistics and language which have incurred the wrath of unmitigated obscurity and merely language games. All of this as species threats of extinction in climate change and nuclear weapons with authoritarianism on the rise indicate a kind of sociological existential crisis. Even Levinas has not been unscathed by criticisms of the radical alterity of the other. This comes in the form of what we think of as the 'bad faith other'.

The popular notion of 'othering' someone as making them not like me and therefore less than me is a misnomer or the wrong name applied to a certain behavior. The name which should be used in this case is sameness, making someone the same as my idea of them. This is exactly what Levinas counters. Levinas calls 'othering' someone totalizing them, making them the same as my idea of them, e.g., lazy, inferior, stupid, sinner, evil, etc. We need look no further than the extreme right's catch all word for everything they do not like in the absurdly misunderstood term, 'wokeness'. There is no sense of otherness in their use of 'wokeness' since there is no other in their use of the term. 'Othering' in this sense is only a monologue with oneself. It has nothing to do with an 'other' apart from me and my idea. Levinas would call this violence towards the other or more precisely the murder of the other. This is what he means by totalizing the other - leveling them off to my idea of them.

African Americans have been using the term 'woke' for generations in the United States. Women in the United States just got the right to vote one hundred years ago. Feminism made explicit what many women encounter in a world dominated by men. Ethnic minorities have encountered obstacles of discrimination and inequality. Lesbian, gay, bisexual, transgender and queer are harassed and are victims of violence. In a phenomenological sense, this demonstrates the violence of suspending our individual phenomenal experiences of the world in favor of despotism rising from a frantic, fanatical feverishness marking the exasperated pleas as a bygone dominant universal epoch of righteous claims of rule and origin at any cost including horrific violence and extermination of evil others. In mental psychosis evil others and voices pop up magically. We are seeing this on a worldwide sociological level now. It is as if sociologically, unconsciously we cannot tolerate the possibilities given by a dying history taken as the 'not' of order, disorder, the bad faith other of order as maddening chaos. All of this is group psychosis in which externality must be quashed at any cost.

The reactionary movement we are currently experiencing in the political right in our country is a desperate and futile attempt to force and give artificial life-support to universal narrative of a fading age, an epoch, which can no longer sustain civilization. While relativity was rattling the cages of classic physics in the early 20th century, phenomenology was asking about the concrete ways we live 'reality'. Phenomenology was asking why we downplay our personal experiences over the dominant cultural narrative of commonsense neutrality and transactionalism. For Husserl, the point of phenomenology was to seek what was the truth of personal or 'local' experience. The word 'veridical' means truthful, genuine, not illusory.

> A perfect hallucination is one that is indistinguishable from a veridical perceptual experience. In principle, for every possible veridical perceptual experience there is a corresponding hallucination that cannot be distinguished from it. Nothing about the experience gives it away as non-standard in any respect. *

Husserl stood at the precipice of the oncoming abyss in the early 20th century. He understood the approaching storm which would undermine centuries of historic narrative. His strategy was cooler heads prevail. Let's stop, *epoché*, bracket and reevaluate.

> The point of the local epoché can perhaps best be brought out if we follow Husserl in applying it to the case of perceptual experience. The phenomenologist is supposed to perform his or her descriptions from a first-person point of view, so as to ensure that the respective item is described exactly as it is experienced.†

Phenomenology in a relativistic sense asks the truth of how the individual experiences phenomena. It also brings into question the universal validity of our current epoch. The inability of Being and essence to be 'in itself' is already an announcement of the end of our current classical epoch. There is no going back to reinstate a dying epoch. Philosophy and physics have become the voice of the modern-day prophet. The road ahead can only lead to extinction in the war of all against all. In the 19th century, Nietzsche prophetically saw the approaching storm which required a reevaluation of all values. However, he failed to gain clarity as to how that would happen and what it would mean. His idea was a restart of the greatness of the ancient Greeks. He saw this as the Übermensch, the superman, the ascending life of the 'new *man*'.

> Elevated is then your body, and raised up; with its delight, enraptureth it the spirit; so that it becometh creator, and valuer, and lover, and everything's benefactor.
>
> When your heart overfloweth broad and full like the river, a blessing and a danger to the lowlanders: there is the origin of your virtue.
>
> When ye are exalted above praise and blame, and your will would command all things, as a loving one's will: there is the origin of your virtue.
>
> When ye despise pleasant things, and the effeminate couch, and cannot couch far enough from the effeminate: there is the origin of your virtue.
>
> When ye are willers of one will, and when that change of every need is needful to you: there is the origin of your virtue.

---

* William Fish, *Philosophy of Perception: A General Introduction* (New York: Routledge, 2010), 82.
See Husserl on Hallucination: A Conjunctive Reading.

† Dan Zahavi, *Husserl's Legacy: Phenomenology, Metaphysics, and Transcendental Idealism* (Oxford: Oxford University Press, 2017), 54.
See Epoché, perceptual noema, hýle, time-consciousness and phenomenological reduction

> Verily, a new good and evil is it! Verily, a new deep murmuring, and the voice of a new fountain!
>
> Power is it, this new virtue; a ruling thought is it, and around it a subtle soul: a golden sun, with the serpent of knowledge around it. *

This is a clear statement of Nietzsche's echo back into himself. Nietzsche is the pinnacle of "I am that I am". This is not to imply a religious criticism. This is a criticism and iconoclasms of classicism's inability to recognize the higher man. This is the self as self-made, the self as god-like from Nietzsche's ancient Greek exaltations. But this is not a novel as Nietzsche pretends. This is the ancient path of Narcissus.

Substitution

This section will eventually get into some more difficult examinations of Levinas' writings on substitution. If you find it too difficult, I will return to it later with a simpler explanation in the last section entitled "Ethics and Love". I want to take a step back at this point to try to get some perspective in preparation for this section. I spent a lot of time on the quantum wave function trying to illustrate a couple main points. The first point is that most folks generally have a lot of confidence in science for good reasons. And in the 20th century science really jumped off a cliff no matter how some would like to restate that point. The geniuses who discovered quantum physics and their predecessors have amply written and publicly spoken about the devastating consequences to classical science and philosophy from their discoveries. Certainly, there is a way quantum physics seeks after and finds how classical science can be accounted for in limited cases of quantum physics, but this is not business as usual. There is an exteriority to exformation and the quantum wave function which cannot be tamed by classicism. And this demands the same focus on exteriority from philosophy. The 20th century is a case and point for how philosophy has grappled with the erosion of its certainties. The second point is since we can have confidence that there is much more under the surface of 'reality' than we ever thought there was from classical history. We cannot partition off our macro world as if it was self-evidently distinct from what quantum physics is telling us. Quantum physics has already made inroads into the macro world benefiting humanity in many ways. If *Homo sapiens* can survive long enough there is every reason to believe that we, as self-conscious, may have some spooky connection to the chaos and uncertainty of quantum physics. In any case, whatever comes out of all this in that regard, most of us will not be around when those discoveries come into fruition. So, some of us that are not in the business of physics would like to try to intuitively make some sense out of what phenomena are telling us. We know that a very different philosophy is implicated by 20th century quantum physis. What seems most certain is that alterity lies beneath the covers of 'reality' and it affects us – who we think we are, what 'are' *is,* and unavoidable responsibility implicated from the vantage of our standing on this historic precipice.

With that in mind, Levinas tells us that radical exteriority faces us every day. Specifically, Levinas tells us the other is not the echo of our ideas of the other but very different. We also know the philo-

* Friedrich Nietzsche, *Thus Spake Zarathustra*, trans. Thomas Common (New York: Modern Library, 1917), 75–76. See Thus Spake Zarathustra: First Part, XXII. The Bestowing Virtue

sophical tradition of essence, whether determined in classical or modernist narratives, has serious deficiencies in what quantum physics is showing us. I have tried to be very careful to use the word 'analogy' when discussing issues between philosophy and quantum physics which may impact each other and especially in regard to Levinas' idea of the alterity of the other. To imply that my associations in this regard are true or certain would be ludicrous. Might there be a science/philosophy connection between quantum spookiness and the he or she who faces us? Certainly, but one key point quantum physics plainly tells us is that 'certainty' is not what it pretends and what we simply assume it *is*. However, I think there is an intuitive feeling that there must be something to all this philosophically. There are little direct written connections between postmodern philosophy, Levinas, and quantum physics. However, just as the tree of life myth grew up pluralistically in very ancient cultures so too have odd associations grown between 20th science and philosophy. Mostly, Levinasian scholars base their research on the history of philosophy which led up to Levinas. So, to imply that what Levinas is telling us that quantum physics is telling us something about quantum physics would be wrong. There are many differences which cannot be made commensurate. In the next few sections I plan to bring out some of these differences in Levinas.

To start the discussion of Levinas' idea of substitution I want to briefly give some context to this. Let's recall the various ways the radically exterior of the he or she who faces us gets leveled off and effaced from historic narrative which has found its way into practical everyday common sense assumptions such as:

- 'We are all the same and/or different' where 'different' here implies the bad faith echo of the 'other' which is not the other at all only my idea of the other.
- 'It's me for or against them' which I referred to as transactionalism coming from the self of echo.
- We are all merely machines whether that is modernism, reductionary evolution, merely biological, God's children, or all the other ways in which the immeasurable difference gets reduced and leveled off.

We are finding out that what we thought physics was might have a lot more to do with *phusis* as the yawning gap of Hesiod than what we have been led to believe from our historic narratives of science. Likewise, if there is radical exteriority between myself and the other, we will never find resolve in the history of logic, Being and Concept, common sense, etc. which has led us towards binary dualities. We can continue to tell ourselves the historic fables which amount to nothing other than timeless echoes of ourselves or we can recognize our debt to the other which is 'not me' nor the 'not' of the 'not me' but an excess which can never *be* recoverable by me. To doggedly hold on to echoing retreat as ego is to find oneself analogically in the 'wave function-like' state of *il y a* or analogously to whatever horrific awareness corresponds to in the chaos of the quantum wave function. The horrific state of sheer awareness chaotically - without Being, Concept, history, self, temporality/spatiality, etc. and even without the solace of nothingness. Ipseity, the bare self, then would be the 'luxury' of retreat, of sleep for insomnia. Ipseity as a retreat from il y a fashions ego from the order of existence as being, thinking, functioning. I would remind you of the National Institute of Health studies I cited earlier which understand mental health as ranging from psychotic disorders to healthy behavior having to do with how well the sense of self is developed. To repeat, these extremes range from,

> 1. "Minimal" self, also referred to as "basic" or "core" self or as "**ipseity**." This is a pre reflective, tacit level of selfhood. It refers to the implicit first-person quality of consciousness, ie, the implicit awareness that all experience articulates itself in first person perspective as "my" experience. In other words, all conscious acts are intrinsically self-conscious, a feature sometimes designated as "self-affection." "Minimal" or "core" self constitutes the foundational level of selfhood on which other levels of selfhood are built.
>
> 2. "Narrative" or social self. This refers to characteristics such as social identity, personality, habits, style, personal history, etc. Psychological concepts such as "self-esteem" or "self-image" refer to this level of selfhood. This level is widely understood to presuppose the sense of existing as a subject of experience ("minimal self") and often involves reflective, metacognitive processes, in which one's self is largely an object of awareness. *

The ramifications of this are quite startling if you think about it. The more a society/history caters to the transactional self, the echo of self, the self, which is more isolated within itself, the unhealthier all its selves become. The more a society/history caters to the social self, a sense of the otherness of others, the healthier all its selves become. It seems to me that in the United States we have become more and more insulated within ourselves. Technology seems to have greatly accelerated this tendency in some ways. If this is the case, we are breeding a whole society of unhealthy folks which leads us towards sociological psychosis. This kind of disorder or traditional ways of thinking of chaos as the 'not' if order moves in the direction of the lack of social cohesion as the war of all against all. The more the self is trapped in the mode of transnationalism, the more the place of the other gets replaced with the narcissistic self. Narcissism moves in the direction of psychosis and ipseity, the bare isolated self. A society which breeds such environments will bring forth more extremists, zealots, fanatics, terrorists, etc. in which cohesion of socially accepted facts become fantasies of one's own choosing. Levinas seems to me to tell us that there is more to this than an unhealthy sociological function. This moves an individual towards the bare self which arises in its retreat from exteriority as the 'there is' of il y a.

This reinforces what Levinas is telling us, albeit in the most extreme terms, about il y a and ipseity. So, contrary to popular rhetoric, the more sociality reenforces sameness, the more it reinforces the isolated self of ipseity. This counters diversity and acceptance of difference which is the basis of sociality. Furthermore, Levinas tells us that the difference between me, the self, is not bridgeable. This does not mean that we are all living in different 'realities'. Levinas goes even further than beyond differences in 'realities'. He calls this difference 'proximity' and means something very specific by this. Levinas tells us that there is a difference in proximity. But that difference in proximity is not difference itself but the impossibility of being indifferent to the other. For Levinas proximity is the inability to be indifferent to the other. The converse of that would also be true in that if we are indifferent to the other there is no proximity which Levinas refers to as ego. One historic narrative of lived spatiotemporality is the mechanistic modernist contention that we exist in linear extension of space and 'now' moments of time. This assumption is not proximity at all but yet one more abstraction in which narcissism thrives, the inability to recognize anything other than oneself as a mechanism in a machine of other mechanisms.

---

* Josef Parnas and Mads Gram Henriksen, "Disordered Self in the Schizophrenia Spectrum," *World Psychiatry* 13, no. 3 (October 2014): 222.

Disturbance of Minimal Self (Ipseity) in Schizophrenia: Clarification and Current Status.

However, if there is proximity to the other there is non-indifference which is due to responsibility to the other. This non-indifference he calls *"the-one-for-the-other"*.

> This difference in proximity between the one and the other, between me and a neighbor, turns into non-indifference, precisely into my responsibility. Non-indifference, humanity, the-one-for-the-other is the very signifyingness of signification, the intelligibility of the intelligible, and thus reason. The non-indifference of responsibility to the point of substitution for the neighbor is the source of all compassion. It is responsibility for the very outrage that the other, who qua other excludes me, inflicts on me, for the persecution with which, before any intention, he persecutes me. Proximity thus signifies a reason before the thematization of signification by a thinking subject, before the assembling of terms in a present, a pre-original reason that does not proceed from any initiative of the subject, an anarchic reason. It is a reason before the beginning, before any present, for my responsibility for the other commands me before any decision, any deliberation. Proximity is communication, agreement, understanding, or peace. Peace is incumbent on me in proximity, the neighbor cannot relieve me of it. *

In linguistics and structuralism Saussure tells us that language is nothing more than a system of signs. The signs are signifiers and signified. Perhaps we could also think of this as syntax and semantics. People think of the signified as a kind of coherent meaning which we all simply understand. But that is not what Saussure thinks. He thinks the signified is really a whole new set of signifiers. So, for Saussure there is no metaphysical kind of meaning to the signified that we all just assume we privately know what it is. What we really have is a learned pattern of signifiers in language which only and ever point to other signifiers.

For Levinas, signifiers and signified are a result of the said *visa vie* the third other which I will get into later. For now, the said accounts for the entire historical fascinations and narrations of essence and Being. Levinas writes that the veracity, the truth or facts, of the subject. The subject as given by historical narrative is a representation of itself to itself. It is an assemblage of elements of thought into a structure. In historic narrative the subject is an arrangement of structures. The subject is a signifier in a signified in which the signified is another assemblage or arrangement. All of this is strained through the unity of a presence which is read for re-presencing throughout all the narratives of history. And what shall we make of the signifyingness of the signifier and the signified? In this case signifyingness would then be intelligibility. This would be the essence of Being, the *ipsum esse*.

> As Ipsum Esse Subsistens, God is Ipsum Esse (i.e., Existence or Act of Existence Itself, subsistent of Itself or subsisting by Itself) (Aquinas, Summa Theologiae, I, q. 4, a. 2). As such, Ipsum Esse Subsistens contains within Itself or Himself the whole perfection of esse. †

This assemblage of subjectivity is what Levinas calls the said and "*The veracity of the subject would*

* Emmanuel Levinas, *Otherwise Than Being or Beyond Essence*, trans. Alphonso Lingis (The Hague: Martinus Nijhoff, 1981), 166–67.

† Thomas Aquinas, *Summa Theologiae*, I, q. 4, a. 2.
See 3.13 Developing Aquinas' Missing Metaphysical Concept of God Today: Ipsa Essentia Subsistens as Infinite Self-Fulfillment Itself.

have no other signification than this effacing before presence, this representation." (Levinas, Otherwise Than Being or Beyond Essence, 1981, p. 246). Then Levinas asks what is to be made of communication or the subject as a speaking that is absorbed into the said? In writing of this in "Otherwise Than Being or Beyond Essence", a section entitled, "c: The Subject as a Speaking that is Absorbed in the Said" he writes,

> Nor would it [the subject] have another signification if one attends to the communication of essence manifested to the other, if one takes the saying as a pure communication of a said. The manifestation to the other and the interhuman, intersubjective understanding concerning the being that manifests itself can in turn play its part in this manifestation and this being. The veracity of the subject would be the virtue of a saying in which the emission of signs, insignificant in their own figures, would be subordinate to the signified, the said, which in turn would be conformed to the being that shows itself. The subject would not be the source of any signification independently of the truth of the essence which it serves. A lie would be only the price that being's finitude costs it. A science would be able to totalize being at all levels of its esse by fixing the ontological structures that articulate being. Subjectivity, the ego and the others would be the signifiers and signified in which the subjective representation of being is realized. [*]

Levinas refers to these narratives of the said as thematizations. In historical narratives and even common sense thinking itself is based on the said. We believe that knowledge in the widest meanings of the word comes first and then anything else. Levinas turns this totally around. Levinas tells us that radical alterity (as I think at least a good case has been made by quantum physics) comes first in the saying and then all our confidences and narratives come later in the said. In this case the whole structuralist and postmodern concerns of signifier and signified may well be true. But before that there is the question of signifyingness itself. Parts fitting together might be contrived to try to explain what makes the glue of signifier and signified hold together but the question of signifyingness cannot easily fit into yet another assemblage. This is a kind of kicking the can down the road and hoping no one notices.

Levinas tells us that the "*signifyingness of signification*" and what makes intelligibility, reason, logic, *logos*, is not some metaphysical meaning or historic narrative as a network of signifiers we think of as signified but our non-indifference in proximity to the other. Non-indifference in proximity to the other is substitution of *"the-one-for-the-other"*. Signifyingness is not networks of additional signifiers but substitution as *"the-one-for-the-other"*. Here 'the one' is me of not only responsibility to the other but substitution as expiation of the other. The topic of expiation is another difficult topic in Levinas. I will also return to it later with a simpler explanation in the last section entitled "Ethics and Love". I find it interesting the Levinas, a devout Jew, would use the term expiation. Most commonly, the term is used in Christianity as what Jesus accomplished by dying on the cross. As the dogma goes, Jesus took on the sins of the world, like a sacrificial lamb. Jesus atoned for us by becoming sin for us and thereby relieving us of the punishment of sin. Certainly, some may draw allusions to this in Levinas' use of the term expiation but there is much more to this than that.

---

* Emmanuel Levinas, *Otherwise Than Being or Beyond Essence*, trans. Alphonso Lingis (The Hague: Martinus Nijhoff, 1981), 132.

Expiation

From our fading modernist notions, we naturally think that my conscience arises within me. But let's bring to mind what the National Institute of Health studies show and also what just makes sense. When someone isolates themselves and everything is about 'me' to the point of extreme self-consciousness, this is called the pre-reflective self. In other words, they have diminished capacity to reflect on themselves. Another way to think this is that their conscience does not bother too much and in the worst case not at all. They speak in the first person a lot and it's always about their experiences. This is the minimal self of ipseity. It is also the basis for transactionalism. Everything by hook or by crook is to benefit them. There is a part of all of us which has that capacity. Most of us are aware that if you sear your conscience long enough it will simply vanish. We call this narcissism. Levinas refers to this as ipseity. It is the minimal bare ego which is an essential part of all of us. While all of us can have that mental health pathology most of us have a better developed sense of ego.

According to Levinas and mental health studies layered on this is a more developed ego which I think of as the ego Freud tells us about. This notion of ego is sandwiched between id (or ipseity) and superego (or perhaps not ego at all but might we think the radical alterity of the other for Levinas). In this case, the ego has fashioned itself in varying degrees between narcissistic transactionalism and a sense of sociability and conscience. This is the ego many of us inhabit throughout our lives. In this higher ego we have a conscience. But at this level narratives of history work through common sense. This is where notions of essence and Being, cause and effect, absolutes, reductionisms, binary dualities all get sifted down into simple common sense. This is also where dominate histories like modernism set the stage for what we think of as reality. In addition to all these influences many wonderful simple folks already have found a way towards a more intuitive and developed sense of self. The more developed self has a feeling, an intuition, about the externality of the other. The self is where we feel other folks' pain. This is called empathy. This is truly a generous person not because they feel like they have to be but because they have a highly developed sense of sociality. This is at the top of the National Institute of Health spectrum of a fully developed self.

Now, let's get back to conscience. If someone has a conscience they are not on the bare self, pathological end of the spectrum. They are on the other end of sociability. They are reflective and conscientious about themselves. They have a well-developed sense of others and feel other pain and suffering in a fashion that can be named expiation. The point is that conscience does not arise in ourselves. It comes from without, from the other. Now we can start to see why Levinas' idea of the saying, proximity, diachrony, the radical alterity of the other culminates in Ethics. As a philosopher Levinas is telling us that the other cannot be subsumed by 'me'. When the other is an object of my ambitions, domain, control, power, or even assumed as the same as 'me' there is a reduction which makes itself amenable to, susceptible to, useful to, assumed to be known and understood by, and sets the stage for the less developed end of the mental health spectrum where ipseity and transactionalism reign pathologically supreme.

We have seen from the previous discussion on quantum physics that we have not understood it if it does not blow our socks off or as Bohr puts it, they have "not understood it". The yawning gap of the quantum wave function will never be bridged by the observer effect of 'me' as if we were masters of the quantum universe. All these shenanigans come from branching into reality, historic narratives condensed down into common sense from the lived 'observer effect'. The externality of the quantum wave function cannot have no god or 'me', no absolute, no essence, or being. Neither is it 'beyond' all that. It is an excess which we cannot wrap our brains around, but we can be aware that externality has no common ground to us but also in some spooky way we spring from this excess. But not as origin, as somehow an instant in which we poof-ed. The point is that accusation turning into persecution is not from within. This is from saying where we face the radical externality of the other. Facing this other as externality arises as 'me' from il y a to ipseity to socio-historic ego to self. Facing the he or she (whatever pronoun of lack of pronoun) is the root of the root and the bud of the bud and the sky of the sky of a tree called life * – my life.

Levinas is telling us that expiation is the opposite of the echo of narcissism in the assemblage and arrangements of systems of essence and Being. Levinas asks - but what about the 'me' which is not an ego which cannot be generalized, the me that I am?

> Here the unicity of the ego first acquires a meaning — where it is no longer a question of the ego, but of me. The subject which is not an ego, but which I am, cannot be generalized, is not a subject in general; we have moved from the ego to me who am me and no one else. Here the identity of the subject comes from the impossibility of escaping responsibility, from the taking charge of the other. Signification, saying — my expressivity, my own signifyingness qua sign, my own verbality qua verb — cannot be understood as a modality of being; the disinterestedness suspends essence. As a substitution of one for another, as me, a man, I am not a transubstantiation, a changing from one substance into another, I do not shut myself up in another identity, I do not rest in a new avatar. As signification, proximity, saying, separation, I do not fuse with anything. Have we to give a name to this relationship of signification grasped as subjectivity? Must we pronounce the word expiation, and conceive the subjectivity of the subject, the otherwise than being, as an expiation? †

> Man is not to be conceived in function of being and not-being, taken as ultimate references. Humanity, subjectivity — the excluded middle, excluded from everywhere, null-site — signify the breakup of this alternative, the one-in-the-place-of-another, substitution, signification in its signifyingness *qua* sign, prior to essence, before identity. Signification, prior to being, breaks up the assembling, the recollection or the present of essence. On the hither side of or beyond essence, signification is the breathlessness of the spirit expiring without inspiring, disinterestedness and gratuity or gratitude; the breakup of essence is ethics. ‡

Signification as saying cannot come into presence in the fashion of Being or essence. Proximity is not mediated by spacetime. Analogously, it might be thought more like all classicism's historic narratives explained in multitudes of the said as collapse or branching play no role in observer effect from the quantum wave function. But this is not mere physics as things that go bump in the night. This is spooky physics which is no longer spooky but radical alterity which faces us, faces me, as he or she. Proximity is *signifyingness* without classical mediation. Proximity as *signifyingness* evokes my responsi-

* See E. E. Cummings

† Emmanuel Levinas, *Otherwise Than Being or Beyond Essence*, trans. Alphonso Lingis (The Hague: Martinus Nijhoff, 1981), 13–14.

‡ Levinas, Otherwise Than Being or Beyond Essence, 1981, p. 72

bility to the other. Here "*the subject comes from the impossibility of escaping responsibility*". This is the one for the other or substitution. Substitution is "*not a transubstantiation, a changing from one substance into another, I do not shut myself up in another identity, I do not rest in a new avatar. As signification, proximity, saying, separation, I do not fuse with anything*". Neither is this God or Jesus or yet another religious account. This is what Levinas calls expiation which involves me intimately. Levinas writes, "*The uniqueness of the self is the very fact of bearing the fault of another.*" * This is not ego as ipseity, the pre reflective, tacit level of selfhood implicit in the first-person quality of consciousness, the implicit awareness that all experience articulates itself in first person perspective as "my" experience, the conscious which acts always intrinsically self-conscious, self-affection or auto-affection. This is self as socially one for the other not from the core self as totally caught up within itself but facing an alterity which makes all such notions of me, ego, self, etc. indebted before any such thing as 'my freedom', 'my consciousness', 'my presence'. This is the subject which does not return back to itself to establish itself but incurs an unpayable debt. It is not even consciousness which assembles itself, but which draws itself from indebtedness to radical alterity which it cannot own, negotiate with, even finding its origin in. This is pure signifyingness which has not yet found signifier and signified.

In this sense self arises as responsibility thrown back on itself in *signifyingness* from saying, the interruption or lapse of radical exteriority. So, self does not even arise from imagined decisions in the freedom of ego or the assumptions of substance, Being, or essence. Self arises from responsibility to the other which cannot be denied. We have mildly referred to this as sociality. Levinas writes, "*Obsessed with responsibilities which did not arise in decisions taken by a subject "contemplating freely," consequently accused in its innocence, subjectivity in itself is being thrown back on oneself. This means concretely: accused of what the others do or suffer, or responsible for what they do or suffer.*" † To turn away from the suffering of the other is to turn towards fictions of ego, echo, Narcissus. The self always stands accused in a passivity beyond all passivity because all the whys, wherefores, cogitations of whodunit cannot come into the said. This passivity of the self is taken as persecution. In transactionalism the ego squelches self, sociability. But persecution leads to action as expiation, the basis of "the-one-for-the-other".

> This accusation can be reduced to the passivity of the self only as a persecution, but a persecution that turns into an expiation. Without persecution the ego raises its head and covers over the self. ‡

The self without accusation and responsibility is the ego of Narcissus who can only hear his own echo. Remember Echo was the nymph who loved Narcissus. "*As Narcissus lay dying Echo mourned him and echoed his words.*" All Echo could do was echo Narcissus' words. The ego without accusation, without expiation, without "the-one-for-the-other" covers over the self. To be self is to be responsible for the other showing itself in expiation of the other.

* (Levinas, Otherwise Than Being or Beyond Essence, 1981), page 213

† (Levinas, Otherwise Than Being or Beyond Essence, 1981), page 213

‡ Emmanuel Levinas, *Otherwise Than Being or Beyond Essence*, trans. Alphonso Lingis (The Hague: Martinus Nijhoff, 1981), 112–13.

> Peace then is under my responsibility. I am a hostage, for I am alone to wage it, running a fine risk, dangerously. This danger will appear to knowing as an uncertainty, but it is transcendence itself, before certainty and uncertainty, which arise only in knowledge. To require that a communication be sure of being heard is to confuse communication and knowledge, to efface the difference, to fail to recognize the signifyingness of the-one-for-the-other in me. I am extracted from the concept of the ego, and am not measured by being and death, that is, escape the totality and structures. I am reduced to myself in responsibility, outside of the fundamental historicity Merleau-Ponty speaks of. Reason is the one-for-the-other! One is immediately inclined to call such a signification lived, as though the bipolarity of the lived and the thematized, to which Husserl's phenomenology has habituated us, did not already express a certain way of interpreting all meaning in function of being and consciousness. As though the responsibility of the-one-for-the-other could express only the naivety of lived experience that is unreflected but promised to thematization. As though the-one-for-the-other of responsibility, the signification of fraternity, could not "float above the waters" of ontology in its irreducible diachrony. As though the interval or the difference that separates the one from the other, which the non-indifference of the-one-for-the-other did not annul, could only be gathered up in a theme or be compressed into a state of soul. To intelligibility as an impersonal logos is opposed intelligibility as proximity. But does the reason characteristic of justice, the State, thematization, synchronization, re-presentation, the logos and being succeed in absorbing into its coherence the intelligibility of proximity in which it unfolds? Does not the latter have to be subordinated to the former, since the very discussion which we are pursuing at this moment counts by its said, since in thematizing we are synchronizing the terms, forming a system among them, using the verb to be, placing in being all signification that allegedly signifies beyond being? Or must we reinvoke alternation [the repeated occurrence of two things in turn: "the regular alternation of stressed and unstressed syllables"] and diachrony as the time of philosophy?
>
> If the preoriginal reason of difference, non-indifference, responsibility, a fine risk, conserves its signification, the couple skepticism and refutation of skepticism has to make its appearance alongside of the reason in representation, knowing, and deduction, served by logic and synchronizing the successive. *

Levinas wants to reverse the commonly understood order where impersonal *logos* give rise to personal, to self. Levinas tells us that *logos* itself arises from the self's responsibility to the other as expiation. Perhaps a quirk but as I discussed much earlier for Heraclitus *logos* would mean more like saying than logic as we understand today. Anyway, without this reversal Levinas writes *logos* would be and has been the object of skepticism (e.g., Nietzsche). Even Positivists distrusted logic in favor of empirical and stochastic observation. *Logos* would arise as the negative, as the power embedded in negation, the 'not', most clearly exhibited by Hegel. Levinas tells us thematization is synchronizing the terms in a system, the System, starting with verbs of being and existence and moving beyond Being to pure, absolute signification itself (e.g., state of soul, Concept, etc.). The coupling of "*skepticism and refutation of*

* Emmanuel Levinas, *Otherwise Than Being or Beyond Essence*, trans. Alphonso Lingis (The Hague: Martinus Nijhoff, 1981), 167–68.

skepticism has to make its appearance alongside of the reason in representation, knowing, and deduction, served by logic and synchronizing the successive" meaning effectively, rationality must float above and beyond indebtedness to the other. But if saying is preoriginal, prior to reason, a nonnegotiable, non-indifference of the one for the other, proximity to the other as pure *signifyingness* which cannot found itself then even reason and logic itself must answer to responsibility and Ethics.

Levinas tells us that the chasm between myself and other cannot come into my 'reality' or any narrative of 'reality'. My 'reality' left to itself, to the ego of historic narrative, does not include the other as radical alterity. The 'my' which takes 'reality' to itself excludes the other to the point of retreat which must always be the negative of what it cannot endure. The other is not the other if that absolutely means my idea of the other. The radical alterity of the other cannot have anything to do with my 'reality'. My analogous way of thinking about this is as the hypothetical awareness in the quantum wave function which has no branching into classicism as the observer effect. The wave function could be said to be passivity more passive than all passivity. Might this be an untraceable lapse which the observer effect branching cannot give an account for? The observer effect from the other can be thought of as a kind of spooky branch into the 'me here now' which cannot find its source in a return to itself in the 'me here now'. In the passivity of the wave function a void, a yawning gap, which knows nothing of origin "*undoes thematization, and escapes any principle, origin, will, or ἀρχή, which are put forth in every ray of consciousness. This movement is, in the original sense of the term, an-archical.*" (Levinas, Otherwise Than Being or Beyond Essence, 1981, p. 197)

Therefore, I am indebted to the other not because I have an altruistic ethic of the other, but because the other in an observer-like saying awakens me as existent. But I cannot find or recover an origin in the call of the other from the externality of the other. The calling of the other as Levinas' saying cannot found, originate, cause and effect, rule and authority, etc. There is no metaphysic which binds me to the other as in the Christian mythos which makes us all children of the parent God with a hereditary and eternal connection which we will find when we die and go to heaven as long as we do x, y, and z. This is reciprocity and transactionalism of the *meta*. The externality of the other in the retreat of ego covers over self and its non-indifference of one for the other as signifyingness. Responsibility to the other is not because they exist like me, and I 'should' be nice to them (as long as they buy what I am selling). There is not a case of "you do this, and I will do that" as transactionalism. This goes much further. Expiation is constituent of selfhood.

If Levinas is correct and I think he is, the self which is not hanging on the horrific brink of ipseity and il y a, the psychotic state of affairs, but *is* a debt which I cannot repay. Even more,

> The more I answer the more I am responsible; the more I approach the neighbor with which I am encharged the further away I am. This debit which increases is infinity as an infinition of the infinite, as glory. *

I am not merely called by the other into self. The very selfness of self arises from the undeniable call of the other. Healthy sociality is not simply an option, it is a necessity for me to even range from ipseity to self. Even further, Levinas tells us that responsibility not only rests on debt to the other but on expi-

* Emmanuel Levinas, *Otherwise Than Being or Beyond Essence*, trans. Alphonso Lingis (The Hague: Martinus Nijhoff, 1981), 93.

ation. Expiation is what even Levinas refers to as the "nonsensical idea" that I am responsible for the other even as far as taking on their burden, their shame as if it were mine.

Essence on the verge of abysmal meaninglessness can no longer arouse itself in an attempt to once more complete itself. Essence at our abyss "*stretching on indefinitely, without any possible halt or interruption, the equality of essence not justifying, in all equity, any instant's halt, without respite, without any possible suspension, is the horrifying* ***there is***". (Levinas, Otherwise Than Being or Beyond Essence, 1981, p. 289) In essence's frenetic gasp to delete itself, it more and more insists on itself. Violence as the only recourse unconsciously wants annihilation, believing that the metaphysics of nothing yet has a promise for the dying or for Nietzsche's heroic affirmation of eternally recurrent tragedy. But Levinas tells us that even nothing cannot die. Instead, *there is*. In naming il y a, could it be that we also might think of awareness which cannot branch from the quantum wave function? Awareness which cannot cover itself with the fig leaves of Being and essence? Awareness which can only persist as the wave function persists not as nothing but as awareness of no-thing, as insomniacs' awareness. Might we think unassembled exformation without spacetime or cause and effect, chaos which cannot gather itself from absolute indeterminateness which would not afford an observer effect fleeing refuge from the no exit of sheer awareness, no retreat from pure immediacy? Levinas writes,

> Essence stretching on indefinitely, without any possible halt or interruption, the equality of essence not justifying, in all equity, any instant's halt, without respite, without any possible suspension, is the horrifying there is behind all finality proper to the thematizing ego, which cannot sink into the essence it thematizes. *

Essence absolutely depleted finds no *logos*, no word or theme. Narcissus' echo eternally obliviated where only the worm of bare awareness cannot live or die, cannot fall from the wave function in the observer effect, cannot live or die, cannot sleep or wake, and neither can it even echo itself as self-conscious or awareness of awareness but must only stare interminably without even underworld or echo of itself back to itself. This is the vacuum of one without another. Perhaps we again think analogically of branching from the wave function as *il y a*. Pure signification not yet called (observed) from radical externality, not yet the saying of the other, the observer. The other's grace calling-saying-observing-falling from unimaginable, horrific *il y a* having no retreat. I will get into this more a little further down, but I see it as a kind of volitional grace for ego to even have the possibility of retreat from ipseity facing *il y a* and to even be able to abandon its freedom as essence, as its own echo and return to the persecuted self, "*the-one-for-the-other*" which is unable to be able. As we will see, ego is a retreat which cannot escape from radical exteriority and the indebtedness of selfhood. But ego is a kind of emotive, historically based escape route from the self of responsibility for the other. Ego is a volitional devolution from the self, the-one-for-the-other, an egoistic echo from the persecuted self (to be discussed further down), to dramatize itself transactionally.

> It is inasmuch as the signification of the-one-for-the-other is thematized and assembled, and through the simultaneity of essence, that the one is posited as an ego, that is, as a present or as a

---

* Emmanuel Levinas, *Otherwise Than Being or Beyond Essence*, trans. Alphonso Lingis (The Hague: Martinus Nijhoff, 1981), 163.

> beginning or as free, as a subject facing an object. But it is also posited as belonging to essence, which when assembled cannot leave anything outside, has no outside, cannot be worn away. This way for the subject to find itself again in essence, whereas essence, as assembled, should have made possible the present and freedom, is not a harmonious and inoffensive participation. *

Yet, at the end of the epoch of Being, essence has worn itself down to "*incessant buzzing*" which finds it harder and harder to ignore the meaninglessness of repetition. Sociologically, ego wears thin with only a bare rumbling, its ceaseless buzz, which is losing it egoist imaginings increasingly once again towards *there is*.

> It is the incessant buzzing that fills each silence, where the subject detaches itself from essence and posits itself as a subject in face of its objectivity. A rumbling intolerable to a subject that faces itself as a subject, and assembles essence before itself as an object. But its own subtraction is unjustifiable in an equal woven fabric, of absolute equity. The rumbling of the there is is the non-sense in which essence turns, and in which thus turns the justice issued out of signification. There is ambiguity of sense and non-sense in being, sense turning into non-sense. It cannot be taken lightly. †

How much longer must egoist repetition exhaust itself?

With that right does the idealist extract the ego from being and confer upon it a transcendental status, when the subject returns to being in the very stability of its status? (Levinas, Otherwise Than Being or Beyond Essence, 1981)

But Levinas tells us that all the while imagined freedom as choice awaits the dying echo of *there is,* the egoity of Hamlet which cries, "To die, to sleep – to sleep, perchance to dream – ay, there's the rub, for in this sleep of death what dreams may come..." But what if the 'dream' is nightmare beyond all imaginings, one without any other? The inability to be able to die to dreams as perhaps we could analogously imagine as the unescapable experience of stark and bare awareness, of chaotic unassembled exformation in which the possibility of retreat is impossible, one without another, where even 'oneself' has hollowed itself out. But essence only plays well in the vacuousness of ego. Levinas calls signification the *evacuation of Being's essence for the other*.

> But the forgetting of ambiguity would be as little philosophical. It is in its ex-ception and expulsion as a responsible one that a subject outside of being can be conceived. In signification, in the-one-for-the-other, the self is not a being provisionally transcendental and awaiting a place in the being it constitutes. Nor is it the absolute being of which phenomena would express

---

* Emmanuel Levinas, *Otherwise Than Being or Beyond Essence*, trans. Alphonso Lingis (The Hague: Martinus Nijhoff, 1981), 163.

† Emmanuel Levinas, *Otherwise Than Being or Beyond Essence*, trans. Alphonso Lingis (The Hague: Martinus Nijhoff, 1981), 163.

> only the coherent dream. The one in the-one-for-the-other is not a being outside of being, but signification, evacuation of Being's essence for the other. *

The vacuous repetition of il y a can only dream as Narcissus' stark and bare awareness which cannot branch from itself, from the saying (observation) of the other, nor even yet *be* the ego of essence and its own echoing freedom. The saying of the other (the observer) speaks ego and ipseity from dark simultaneity admitting no temporality, from annihilation as chaotic unassembled exformation in which no possibility for retreat is possible. What remains after all *evacuation of Being's essence*, and every freedom of the ego has been exhausted? Perhaps the self as *substitution for the other*. Could it be that the absurdity of the *there is* may constitute a signification, a signification not yet signifier or signified of *the-one-for-the-other*? Could it be that beneath essence there is a surplus over classically esteemed sensibility the lowly, the insignificant, nonsense, and absurdity?

> The self is a substitution for the other, subjectivity as a subjection to everything, as a supporting everything and supporting the whole. The incessant murmur of the there is strikes with absurdity the active transcendental ego, beginning and present. But the absurdity of the there is, as a modality of the-one-for-the-other, signifies. The insignificance of its objective insistence, recommencing behind every negation, overwhelms me like the fate of a subjection to all the other to which I am subject, is the surplus of nonsense over sense, through which for the self expiation is possible, an expiation which the oneself indeed signifies. **The there is is all the weight that alterity weighs supported by a subjectivity that does not found it.** But one must not say that the there is results from a "subjective impression." In this overflowing of sense by nonsense, the sensibility, the self, is first brought out, in its bottomless passivity, as pure sensible point, a dis-interestedness, or subversion of essence. †

Levinas tells us "*Behind the anonymous rustling of the there is*" is a "*passivity more passive than all passivity*" and otherwise than being, signification meaning the-one-for-the-other makes possible the subjectivity of the self as substitution, expiation.

> Behind the anonymous rustling of the there is subjectivity reaches passivity without any assumption. Assumption would already put in a correlation with an act this passivity of the otherwise than being, this substitution prior to the opposition of the active and the passive, the subjective and the objective, being and becoming. In the subjectivity of the self, substitution is the ultimate retraction of passivity, the opposite of the assumption in which the receptivity which the finitude of a transcendental I think describes is completed, or which it presupposes. The identity of the chosen one, that is, the assigned one, which signifies before being, would get a foothold and be affirmed in essence, which negativity itself determines. To support

---

* Emmanuel Levinas, *Otherwise Than Being or Beyond Essence*, trans. Alphonso Lingis (The Hague: Martinus Nijhoff, 1981), 163.

† Emmanuel Levinas, *Otherwise Than Being or Beyond Essence*, trans. Alphonso Lingis (Pittsburgh: Duquesne University Press, 1998), 163–64.

> without compensation, the excessive or disheartening hubbub and encumberment of the there is is needed. *

Expiation as substitution of the one-for-the-other is the who resulting from "*persecuted subjectivity*". "*Persecuted subjectivity*" as a "*passivity more passive than all passivity*" has no origin (i.e., essence, Being, etc.). Levinas calls this the anarchy of passivity. Let's try to get some perspective on the self and the ego. The self facing the externality of the other, the radical alterity of the other has no way to understand, put into context, make sense of the lapse of radical externality. This reflects a "*passivity more passive than all passivity*" in the self's inability to be able. Il y a is really an orientation of signifyingness, of this absolute 'inability to be able'. The inability to be able results in a retreat into historic answers and commonsense assumptions. The reason il y a is so horrific is because of the inability of self to be able as the-one-for-the-other. This is persecution of the self. Due to persecution, the self in egoity wants to withdraw to the bare self of ipseity. The bare self, already obligated but unable to be able to meet its obligation, is persecuted in the face of the lapse of radical externality, of the other it cannot understand, or come to grips with. Therefore, the self's retreat is its bare self, is il y a, the horror it cannot bear. Historic narratives have been fashioned as distraction, as allusion, as essence, Being, God. etc. The ego then arises from ipseity, the bare self facing the horror of *il y a*.

The retreat from the inability to be able is the ego which imagines itself to be free and, as Narcissus, rooted absolutely in itself as essence, Being, 'reality', etc. This is the synchronic play of language and philosophy in the sensible arena of signifier and signified. This is where identity is formed. This arises from the lapse of the saying into the echo of the said. The said recuperates and collapses into consciousness as time. It is a change imposed upon duration. The unitary operator in the quantum wave function has no temporality except from semiclassical branching as an independent variable. Synchronic temporality is already classicism, already the said of history and narration. Radical exteriority "*does not exhaust the signifyingness of the sensible and of immediacy, is its play, logical and ontological, as consciousness*". From radical externality, from diachronic other as saying, as observer, signifyingness not yet sensible, classical, clothed with signifiers and signified. In the wave function we find all possibilities not yet observed into actuality. For Levinas the saying of the other arrives diachronously, anarchically as if branching (the observer effect) as 'me', as conscious. Signifyingness is saying collapsing into said is as *logos*.

> In the sensible as lived, identity shows itself, becomes a phenomenon, for in the sensible as lived is heard and "resounds" essence, the lapse of time and the memory that recuperates it, consciousness; the time of consciousness is the resonance and understanding of time. But this ambiguity and this gnoseological function of sensibility, this ambiguity of the understanding and intuition that does not exhaust the signifyingness of the sensible and of immediacy, is its play, logical and ontological, as consciousness. This play does not begin by caprice; we will have to show the horizon in which it occurs. But this play does not undo the responsibilities that arouse it. In analyzing the sensible in the ambiguity of duration and identity, which is already the ambiguity of the verb and the noun that scintillates in the said, we have found it already

* Emmanuel Levinas, *Otherwise Than Being or Beyond Essence*, trans. Alphonso Lingis (Pittsburgh: Duquesne University Press, 1998), 164.

> said. Language has been in operation, and the saying that bore this said, but goes further, was absorbed and died in the said, was inscribed. Or, if one likes, our analysis concerned the time that marks historiography, that is, the recuperable time, the recoverable time, the lost time that can be found again. As the time narrated becomes, in the narrative and in writing, a reversible time, every phenomenon is said, characterized by the simultaneity of the successive in a theme. In the remission or detente of time, the same modified retains itself on the verge of losing itself, is inscribed in memory and is identified, is said. The lived, a "state of consciousness," a being, designated by a substantive, is distended, in the time of lived experience, into life, into essence, into a verb but across the opening that the diastasis of identity works, across time, the same finds again the same modified. Such is consciousness. These rediscoveries are an identification — of this as this or as that. Identification is ascription of meaning. Entities show themselves in their meanings to be identical entities. They are not first given and thematized, and then receive a meaning; they are given by the meaning they have. But these rediscoveries by identification occur in an already said. *

Signifyingness makes meaning possible as essence and existence. Consciousness or awareness rises from signifyingness. But signifyingness bears the trace of il y a, the impersonal rustling of the sheer terror of awareness with signifier and signified, being, essence, temporality, devoid of any object but unable to find self-consciousness. The other speaking or observing commands me as unpayable debt, responsibility which cannot remain il y a. Responsibility is Ethics. Responsibility stifles the primordial horror of il y a temporalizes me into the said. The ethical relation emerges from signifyingness, the anonymity of il y a, giving rise to self-consciousness made ready for the said and the burden of justice in the third other to be discussed next.

So, the extremity of the retreat of ipseity facing il y a as signifyingness makes the signifier or signified possible as the said. The said stands for something very different from the horror that ipseity in the throes of il y a portrays. The said is a placeholder, the echo, for the self arising from signifyingness, unable to be able. The self is already beholden in facing the radical alterity of the other. This is what Levinas refers to as persecution. In the quote below Levinas is addressing the obvious question of whether or not we have duped ourselves and simply reiterated the history of essence in perhaps a more subtle fashion.

> In this exposition of the in itself of the persecuted subjectivity, have we been faithful enough to the anarchy of passivity? In speaking of the recurrence of the ego to the self, have we been sufficiently free from the postulates of ontological thought, where the eternal presence to oneself subtends even its absences in the form of a quest, where eternal being, whose possibles are also powers, always takes up what it undergoes, and whatever be its submission, always arises anew as the principle of what happens to it? It is perhaps here, in this reference to a depth of anarchical passivity, that the thought that names creation differs from ontological thought. It is not here a question of justifying the theological context of ontological thought, for the word creation designates a signification older than the context woven about this name. In this

* Emmanuel Levinas, *Otherwise Than Being or Beyond Essence*, trans. Alphonso Lingis (The Hague: Martinus Nijhoff, 1981), 102–3.

context, this said, is already effaced the absolute diachrony of creation, refractory to assembling into a present and a representation. But in creation, what is called to being answers to a call that could not have reached it since, brought out of nothingness, it obeyed before hearing the order. Thus in the concept of creation ex nihilo, if it is not a pure nonsense, there is the concept of a passivity that does not revert into an assumption. The self as a creature is conceived in a passivity more passive still than the passivity of matter, that is, prior to the virtual coinciding of a term with itself. The oneself has to be conceived outside of all substantial coinciding of self with self. Contrary to Western thought which unites subjectivity and substantiality, here coinciding is not the norm that already commands all non-coinciding, in the quest it provokes. Then the recurrence to oneself cannot stop at oneself, but goes to the hither side of oneself; in the recurrence to oneself there is a going to the hither side of oneself. A does not, as in identity, return to A, but retreats to the hither side of its point of departure. Is not the signification of responsibility for another, which cannot be assumed by any freedom, stated in this trope? Far from being recognized in the freedom of consciousness, which loses itself and finds itself again, which, as a freedom, relaxes the order of being so as to reintegrate it in a free responsibility, the responsibility for the other, the responsibility in obsession, suggests an absolute passivity of a self that has never been able to diverge from itself, to then enter into its limits, and identify itself by recognizing itself in its past. Its recurrence is the contracting of an ego, going to the hither side of identity, gnawing away at this very identity — identity gnawing away at itself — in a remorse. Responsibility for another is not an accident that happens to a subject, but precedes essence in it, has not awaited freedom, in which a commitment to another would have been made. I have not done anything and I have always been under accusation — persecuted. The ipseity, in the passivity without arche characteristic of identity, is a hostage. The word I means here I am, answering for everything and for everyone. Responsibility for the others has not been a return to oneself, but an exasperated contracting, which the limits of identity cannot retain. Recurrence becomes identity in breaking up the limits of identity, breaking up the principle of being in me, the intolerable rest in itself characteristic of definition. The self is on the hither side of rest; it is the impossibility to come back from all things and concern oneself only with oneself. It is to hold on to oneself while gnawing away at oneself. Responsibility in obsession is a responsibility of the ego for what the ego has not wished, that is, for the others. This anarchy in the recurrence to oneself is beyond the normal play of action and passion in which the identity of a being is maintained, in which it is. It is on the hither side of the limits of identity. This passivity undergone in proximity by the force of an alterity in me is the passivity of a recurrence to oneself which is not the alienation of an identity betrayed. What can it be but a substitution of me for the others? It is, however not an alienation, because the other in the same is my substitution for the other through responsibility, for which, I am summoned as someone irreplaceable. I exist through the other and for the other, but without this being alienation: I am inspired. This inspiration is the psyche. The psyche can signify this alterity in the same without alienation in the form of incarnation, as being-in-one's-skin, having-the-other-in-one's-skin.

In this substitution, in which identity is inverted, this passivity more passive still than the passivity conjoined with action, beyond the inert passivity of the designated, the self is absolved of itself. Is this freedom? It is a different freedom from that of an initiative. Through substitu-

> tion for others, the oneself escapes relations. At the limit of passivity, the oneself escapes passivity or the inevitable limitation that the terms within relation undergo. In the incomparable relationship of responsibility, the other no longer limits the same, it is supported by what it limits. Here the overdetermination of the ontological categories is visible, which transforms them into ethical terms. In this most passive passivity, the self liberates itself ethically from every other and from itself. Its responsibility for the other, the proximity of the neighbor, does not signify a submission to the non-ego; it means an openness in which being's essence is surpassed in inspiration. It is an openness of which respiration is a modality or a foretaste, or, more exactly, of which it retains the aftertaste. Outside of any mysticism, in this respiration, the possibility of every sacrifice for the other, activity and passivity coincide. *

In the passage above Levinas asks how can we be sure that we are not yet once again weaving a tale of origin, ontology, and essence? How can we be sure that the saying is not already the said? The said already effaces diachrony as creation. I think what he means is that the said always papers over something which cannot be made commensurate with my time or my space. By 'anarchic creation' "*what is called to being answers to a call that could not have reached it since, brought out of nothingness, it obeyed before hearing the order.*" "*The self as a creature is conceived in a passivity more passive still than the passivity of matter, that is, prior to the virtual coinciding of a term with itself.*" "*A does not, as in identity, return to A, but retreats to the hither side of its point of departure.*" Ok, you might be thinking this sounds really mystic. Let's look for an analogy (if Levinasian philosophers have made it this far, which would be a miracle, this will surely finish them off from my quantum nonsense). But doesn't the wave function also have a ring of mystic to it relative to classicism? Reread the quotes I just cited in terms of what physicists might be telling us about the quantum wave function.

- "*called into being*" => from the observer effect branching into classicism?
- "*answers to a call that could not have reached it since, brought out of nothingness, it obeyed before hearing the order*" => branching of the wave function perhaps intuitively thought of as 'nothingness' in classicism's terms? The metaphysical notion of nothingness would not be a reciprocity as if the decohered branch could be reversed back into the wave function. But the wave function prior to spacetime in its simultaneity already includes in superposition what we think of as 'my branch' plus all the other branches (e.g., many-worlds or perhaps better overlapping confluences, etc.) In obeying before hearing the order seems a lot like the fall into classicism in which reciprocity is not only irreversible but also cannot even be made commensurate.
- "*The self as a creature is conceived in a passivity more passive still than the passivity of matter, that is, prior to the virtual coinciding of a term with itself*" => the self, conceived as branching from the wave function, the self, *more passive still than the passivity of matter*, the self, as not ontological, not epistemological not essence or classicism, and the self "*prior to the virtual coinciding of a term with itself*". The self, branching from the wave function is not a term of identity which coincides with itself.

* Emmanuel Levinas, *Otherwise Than Being or Beyond Essence*, trans. Alphonso Lingis (The Hague: Martinus Nijhoff, 1981), 215–18.

- "*A does not, as in identity, return to A, but retreats to the hither side of its point of departure*" => The observer as subject as one possible branch of innumerable branches is not identical to itself but retreats to the hither side of its point of departure? Could this insinuate that the branched 'self' observed/spoken from the wave function having already included in superposition what we think of as 'my branch' plus all the other branches – could this refer to the '*hither side of its point of departure*" simultaneously "*brought out of no-thingness*"
- "*anarchy of passivity*" => the chaos of the ~~universal~~[*] wave function

I suppose if there is anything to this analogy it could be thought of as more of an intuition on the part of Levinas. Again, Levinas arrives at his philosophy from his scholarly knowledge of the history of philosophy without resorting in any way to quantum physics. But since, in this volume, I am looking for ways to communicate Levinas ideas with hopefully *easier* analogies I think this *could* be helpful. Is this a *real* thing? I guess that all hinges on what we mean by 'real' to give a *cliché* philosophical answer.

The Third Other

Levinas does not simply criticize the classical history of *logos*, essence, Being, etc. He thinks it legitimately arises from the encounter of the third other. So far, I have tried to elucidate how Levinas understand the encounter with the other as radical alterity. But we do not just encounter an other. We encounter others. The ego is not a solidary phenomenon. It cannot be taken up completely into a generalized concept, but it faces multiple others, third parties, which necessitates decisions. Here, the third party can also be extended as communities, politics, any and all injustices which responsibility cannot deny.

> To say that the ego is a substitution is then not to state the universality of a principle, the quiddity of an ego, but, quite the contrary, it is to restore to the soul its egoity which supports no generalization. The way by which, from this situation, the logos arises to the concept of the ego passes through the third party. The subject as an ego is not an entity provided with egoity as an eidetic structure, which should make it possible to form a concept of it, and make the singular entity be its realization. [†]

Responsibility as previously described is not just to an other but others. From facing others, the ego has to concern itself with justice.

> To be sure — but this is another theme — my responsibility for all can and has to manifest itself also in limiting itself. The ego can, in the name of this unlimited responsibility, be called upon to concern itself also with itself. The fact that the other, my neighbor, is also a third party with respect to another, who is also a neighbor, is the birth of thought, consciousness, justice

---

* Jacques Derrida is the primary postmodern philosopher associated with "erasure" or overstriking (sous rature) words like "universal," "truth," or "being." This technique signals that a word is necessary for language yet inadequate or deceptive, challenging the stability of traditional philosophical concepts.

† Emmanuel Levinas, *Otherwise Than Being or Beyond Essence*, trans. Alphonso Lingis (The Hague: Martinus Nijhoff, 1981), 237.

> and philosophy. The unlimited initial responsibility, which justifies this concern for justice, for oneself, and for philosophy can be forgotten. In this forgetting consciousness is a pure egoism. *

Responsibility not only places a demand on me but in the encounter of others self-consciousness is born. The third party places a demand on me to decide justice. Justice itself arises in the impossibility of contradiction in the saying between two radical alterities who face me. Justice is demanded of responsibility. From this inescapable demand *logos* arising as historic narrative in the necessity of justice as the comparable if incomparables. An equality of inequalities must assemble order, thematization, language, knowledge, intelligibility in the court of justice. Even the history of essence and Being arises from impossible justice.

> If proximity ordered to me only the other alone, there would have not been any problem, in even the most general sense of the term. A question would not have been born, nor consciousness, nor self-consciousness. The responsibility for the other is an immediacy antecedent to questions, it is proximity. It is troubled and becomes a problem when a third party enters.
>
> The third party is other than the neighbor, but also another neighbor, and also a neighbor of the other, and not simply his fellow. What then are the other and the third party for one another? What have they done to one another? Which passes before the other? The other stands in a relationship with the third party, for whom I cannot entirely answer, even if I alone answer, before any question, for my neighbor. The other and the third party, my neighbors, contemporaries of one another, put distance between me and the other and the third party. "Peace, peace to the neighbor and the one far-off" (Isaiah 57: 19) — we now understand the point of this apparent rhetoric. The third party introduces a contradiction in the saying whose signification before the other until then went in one direction. It is of itself the limit of responsibility and the birth of the question: What do I have to do with justice? A question of consciousness. Justice is necessary, that is, comparison, coexistence, contemporaneousness, assembling, order, thematization, the visibility of faces, and thus intentionality and the intellect, and in intentionality and the intellect, the intelligibility of a system, and thence also a copresence on an equal footing as before a court of justice. Essence as synchrony is togetherness in a place. Proximity takes on a new meaning in the space of contiguity. But pure contiguity is not a "simple nature." It already presupposes both thematizing thought and a locus and the cutting up of the continuity of space into discrete terms and the whole — out of justice.
>
> Thus one would understand, in proximity, in the saying without problems, in responsibility, the reason for the intelligibility of systems. The entry of a third party is the very fact of consciousness, assembling into being, and at the same time, in a being, the hour of the suspension of being in possibility, the finitude of essence accessible to the abstraction of concepts, to the memory that assembles in presence, the reduction of a being to the possible and the reckoning of possibles, the comparison of incomparables. It is the thematization of the same on

* Emmanuel Levinas, *Otherwise Than Being or Beyond Essence*, trans. Alphonso Lingis (The Hague: Martinus Nijhoff, 1981), 238.

> the basis of the relationship with the other, starting with proximity and the immediacy of saying prior to problems, whereas the identification of knowing by itself absorbs every other. *

The 'said' arises from the imposition of the third party. Even consciousness must be called upon, brought forth, into the present demand. Justice cannot arise from empiricism or even from logic. All of that is after the fact and irrelevant to the demand of justice. Proximity faces multiple diachronies from which justice is required as the origin of appearing, as the very origin of origin itself.

> Consciousness is born as the presence of a third party. It is in the measure that it proceeds from it that it is still disinterestedness. It is the entry of the third party, a permanent entry, into the intimacy of the face to face. The concern for justice, for the thematizing, the kerygmatic discourse bearing on the said, from the bottom of the saying without the said, the saying as contact, is the spirit in society. And it is because the third party does not come empirically to trouble proximity, but the face is both the neighbor and the face of faces, visage and visible, that, between the order of being and of proximity the bond is unexceptionable. Order, appearing, phenomenality, being are produced in signification, in proximity, starting with the third party. The apparition of a third party is the very origin of appearing, that is, the very origin of an origin. †

The un-resolvability of proximity with the necessity of justice for unreconcilable radical alterities cannot be ignored but must remain ambivalent and an enigma. There will never be a final justice as there are never finalities to radical alterity and proximity. But neither can there be an existential requirement to disregard justice. There is no justice in the death of a loved one, the intolerability of war, the destitution of poverty and neither is there any escape from having to decide and suffering the devastating pain in which decision fails no matter what justice and wisdom demands from us. Illeity is an external embodiment of the self. ‡ Justice draws us out of ourselves in order to decide, to be held accountable for responsibility. All of histories externalizing as Being, Concept, God, truth, etc. is the placeholder for what I must do as disinterested from my personal desires, acquisitions, narcissistic echoes of myself. The demand for justice which is not 'my justice' is not abnegate-able. Even in spurning the demand we have bad conscience, painful feeling we get when you say something hurtful, or do something to a friend that we regret, for example or when we feel like we haven't done a good enough job. Bad conscience mostly exists as a social function, a sin against conscience. In the impossibility of this demand, we cannot rest in ourselves as sufficient, as able to be able, and yet having to act externalizing us, not at ease in our insufficiency. Those who do not feel bad conscience have acquiesced to narcissism, illeity as externalizing themselves into shells, vacancies of responsibility. In either case, responding to the demand for justice or neglecting the demand for justice, the impossibility of justice is maintained.

---

* Emmanuel Levinas, *Otherwise Than Being or Beyond Essence*, trans. Alphonso Lingis (The Hague: Martinus Nijhoff, 1981), 281–2.

† Emmanuel Levinas, *Otherwise Than Being or Beyond Essence*, trans. Alphonso Lingis (The Hague: Martinus Nijhoff, 1981), 285–6.

‡ See Illeity

The way leads from responsibility to problems. A problem is posited by proximity itself, which, as the immediate itself, is without problems. The extraordinary commitment of the other to the third party calls for control, a search for justice, society and the State, comparison and possession, thought and science, commerce and philosophy, and outside of anarchy, the search for a principle. Philosophy is this measure brought to the infinity of the being-for-the-other of proximity, and is like the wisdom of love. But, come out of signification, a modality of proximity, justice, society and truth itself which they require, must not be taken for an anonymous law of the "human forces" governing an impersonal totality.

It is through its ambivalence which always remains an enigma that infinity or the transcendent does not let itself be assembled. Removing itself from every memorable present, a past that was never present, it leaves a trace of its impossible incarnation and its inordinateness in my proximity with the neighbor, where I state, in the autonomy of the voice of conscience, a responsibility, which could not have begun in me, for freedom, which is not my freedom. The fleeting trace effacing itself and reappearing is like a question mark put before the scintillation of the ambiguity: an infinite responsibility of the one for the other, or the signification of the Infinite in responsibility. There is an ambiguity of the order that orders to me the neighbor who obsesses me, for whom and before whom I answer by my ego, in which being is inverted into a substitution, into the very possibility of gift — and of an infinite illeity, glorious in the human plot hatched in proximity, the subversion of essence into substitution. In it I could not arise soon enough to be there on time, nor approach without the extraordinary distance to be crossed augmenting before every effort to assemble it into an itinerary. Illeity overflows both cognition and the enigma through which the Infinite leaves a trace in cognition. Its distance from a theme, its reclusion, its holiness, is not its way to effect its being (since its past is anachronous and anarchic, leaving a trace which is not the trace of any presence), but is its glory, quite different from being and knowing. It makes the word God be pronounced, without letting "divinity" be said. That would have been absurd, as though God were an essence (that is, as though he admitted the amphibology of being and entities), as though he were a process, or as though he admitted a plurality in the unity of a genus. Does God, a proper and unique noun not entering into any grammatical category, enter without difficulties into the vocative? It is non-thematizable, and even here is a theme only because in a said everything is conveyed before us, even the ineffable, at the price of a betrayal which philosophy is called upon to reduce. Philosophy is called upon to conceive ambivalence, to conceive it in several times. Even if it is called to thought by justice, it still synchronizes in the said the diachrony of the difference between the one and the other, and remains the servant of the saying that signifies the difference between the one and the other as the one for the other, as non-indifference to the other. Philosophy is the wisdom of love at the service of love. *

* Emmanuel Levinas, *Otherwise Than Being or Beyond Essence*, trans. Alphonso Lingis (The Hague: Martinus Nijhoff, 1981), 287–9.

6

ETHICS AND LOVE

Levinas prefers to think of philosophy as the "wisdom of love" rather than the "love of wisdom" (the usual translation of the Greek term "φιλοσοφία"). In his view, responsibility toward the Other (Autrui) precedes any "objective searching after truth". As we have seen, for Levinas, the inter-human relation—referred to as the "face-to-face" relation—is a central philosophical issue. Love plays a foundational role in this relation, shaping our ethical and political interactions with others. Levinas recognizes the significance of erotic love for the subject and, ultimately, for inter-human relations. His description of erotic love and the caress indicates an interlacing of egoistic pleasure and selfless engagement with the other in the sexual relation. He also writes of love for one's neighbor, paternal love, and even the likens the love of God to the relation to the other. [*] Here are some quotes from Levinas on love: [†]

> "Faith is not a question of the existence or non-existence of God. It is believing that love without reward is valuable." (Emmanuel Levinas, *Difficult Liberty: Essays on Judaism*, trans. Sean Hand[‡]
>
> "To approach the Other in conversation is to welcome his expression, in which at each instant he overflows the idea a thought would carry away from it. It is therefore to receive from the Other beyond the capacity of the I, which means exactly: to have the idea of infinity. But this also means: to be taught. The relation with the Other, or Conversation, is a non-allergic relation, an ethical relation; but inasmuch as it is welcomed, this conversation is a teaching. Teaching is not reducible to maieutics [as a midwife]; it comes from the exterior and brings me more than I contain. In its non-violent transitivity, the very epiphany of the face is produced." [§]
>
> "I will say this quite plainly, what truly human is - and don't be afraid of this word - love. And I mean it even with everything that burdens love or, I could say it better, responsibility is actually love, as Pascal said: 'without concupiscence' [without lust]... love exists without worrying about being loved." [#]
>
> "For others, in spite of myself, from myself." [○]

---

* See Emmanuel Levinas (Wikipedia)
And 4 - Levinas: Love, Justice and Responsibility
And 4 Levinas: Love, Justice and Responsibility 1
And Levinas and the Wisdom of Love: The Question of Invisibility

† See Emmanuel Levinas Quotes
And Emmanuel Levinas Quotes 1
And Emmanuel Levinas Quotes 2

‡ Emmanuel Levinas, *Difficult Freedom: Essays on Judaism*, trans. Sean Hand (Baltimore: Johns Hopkins University Press, 1990), 143.

§ Emmanuel Levinas, *Totality and Infinity: An Essay on Exteriority*, trans. Alphonso Lingis (Pittsburgh: Duquesne University Press, 1969), 51.

Emmanuel Levinas, *Entre Nous: Thinking-of-the-Other*, trans. Michael B. Smith and Barbara Harshav (New York: Columbia University Press, 1998), 108.

○ Emmanuel Levinas, *Otherwise Than Being or Beyond Essence*, trans. Alphonso Lingis (The Hague: Martinus Nijhoff, 1981), 118.

"Au moment même où tout est perdu, tout est possible." (Translation: "At the very moment when everything is lost, everything is possible.") [*]

"Love remains a relation with the Other that turns into need, transcendent exteriority of the other, of the beloved. But love goes beyond the beloved... The possibility of the Other appearing as an object of a need while retaining his alterity, or again, the possibility of enjoying the Other... this simultaneity of need and desire, or concupiscence and transcendence, constitutes the originality of the erotic which, in this sense, is the equivocal par excellence." [†]

"If one could possess, grasp, and know the other, it would not be other." [‡]

"The very relationship with the other is the relationship with the future." [§]

"Politics is opposed to morality, as philosophy to naivete." [#]

"Memory as an inversion of historical time is the essence of interiority." [○]

"Of course we do not live in order to eat, but it is not really true to say that we eat in order to live; we eat because we are hungry. Desire has no further intentions behind it... it is a good will." [**]

"A miracle entails a degree of irrationality-not because it shocks reason, but because it makes no appeal to it." [††]

"What could an entirely rational being speak of with another entirely rational being?" [‡‡]

I have been young and now I suppose I am considered old at 69 years old. I have lived through and embodied love as a young man and as an older man. I understand and hold dearly the passions of love I had in my youth, the excitement, the joy, and the desire. But through hard knocks I have found out what derails love and also the kind of love that lasts waking up every morning after decades of marriage still filled with passion, excitement, and joy and also filled with contentment and commitment which is always fresh and deeply meaningful.

Mature love gives these ideas an affective and meaningful context. The first thing that must go to have mature love is thinking you are always right, not really listening to your lover, and being able to say the simple words humbly and sincerely, "I am sorry". When you do not have to be insecure or take offense at these things there is a kind of real peace in letting go. Even more, when you can sincerely empathize and understand your lover's pain, you take it on as your own, as what Levinas refers to as

---

* Emmanuel Levinas, *Existence and Existents*, trans. Alphonso Lingis (The Hague: Martinus Nijhoff, 1978), 104.

† Emmanuel Levinas, *Totality and Infinity: An Essay on Exteriority*, trans. Alphonso Lingis (Pittsburgh: Duquesne University Press, 1969), 254–55.

‡ Emmanuel Levinas, *Time and the Other*, trans. Richard A. Cohen (Pittsburgh: Duquesne University Press, 1987), 90.

§ Emmanuel Levinas, *Time and the Other*, trans. Richard A. Cohen (Pittsburgh: Duquesne University Press, 1987), 77.

\# Emmanuel Levinas, *Totality and Infinity: An Essay on Exteriority*, trans. Alphonso Lingis (Pittsburgh: Duquesne University Press, 1969), 21.

○ Emmanuel Levinas, *Totality and Infinity: An Essay on Exteriority*, trans. Alphonso Lingis (Pittsburgh: Duquesne University Press, 1969), 56.

** Emmanuel Levinas, *Totality and Infinity: An Essay on Exteriority*, trans. Alphonso Lingis (Pittsburgh: Duquesne University Press, 1969), 134.

†† Emmanuel Levinas, *Difficult Freedom: Essays on Judaism*, trans. Sean Hand (Baltimore: Johns Hopkins University Press, 1990), 143.

‡‡ Emmanuel Levinas, *Totality and Infinity: An Essay on Exteriority*, trans. Alphonso Lingis (Pittsburgh: Duquesne University Press, 1969), 203.

expiation. You suffer for them and with them. There is a comfort in sharing each other's joy and pain which, as Levinas might write, increases indebtedness "*as an infinition of the infinite, as glory*". *

The Four Loves

I want to discuss the four ancient Greek words for love. Love is a concept that has thoroughly confused philosophers for ages. For philosophers, Hegel is much easier to comprehend than love. The Greeks' reflection on love confused us from the beginning by compounding love into loves. The four words for love in ancient Greek were: *agápe*, *éros*, *philía*, and *storgē*.

Agápe (ἀγάπη) is unconditional love. It has been used to refer to the love of a parent for a child or for a spouse. It was used in Christianity for God's love.

Éros (ἔρως) is passionate love. It is sensual desire or longing. It does not have to be erotic love. According to Plato it can initially be a feeling, but it can also see beauty in a person. It can also go beyond that to the appreciation of beauty itself. Éros helps the soul recall the knowledge of beauty. It also helps us understand spiritual truth. Erotic-based love aspires to the non-corporeal, spiritual plane of existence; that is, finding its truth, just like finding any truth, leads to transcendence. Lovers and a very few philosophers seek truth through éros. For Plato éros is the ideal form of youthful beauty. For the Greeks, éros leads to the uncreated, the lack of origin and genealogy.

Philía (φιλία) is friendship. It can also be affection. Aristotle thought philía was dispassionate virtuous love. It demonstrates loyalty.

Storgē (στοργή) means affection. It is the natural affection in family. †

All four loves co-arise, budding branches of the same tree. The roots of this tree are sunk below in excess exteriority. *Agápe* is unconditional love. This is love which goes out towards another and never returns unto itself.

For me, what is compelling about Orpheus' gaze is unconditional love. The gaze of Orpheus has been interpreted by many philosophers:

- Another interpretation or usage of the gaze of Orpheus is by Geoffrey Sirc (1953-). Sirc uses Orpheus's moment of violation as argument for creative form in writing versus the standard polished text. Urging the adolescent writer to break free of formal notions of form, Sirc views the journal as the media through which Orpheus yearns for Eurydice. Sirc suggests that if "the Work is freed of concern, the gaze is transgressive, then we're clearly not talking about the polished text, especially one oriented dutifully around the tiny truths available through an analysis of middle-brow media." Sirc's primary reference to this is in Kurt Cobain's Journals. ‡
- Jacques Lacan's (1901-1981) perspective on the gaze of Orpheus is more a matter of desires

* "The more I answer the more I am responsible; the more I approach the neighbor with which I am encharged the further away I am. This debit which increases is infinity as an infinition of the infinite, as glory." (Levinas, Otherwise Than Being or Beyond Essence, 1981), page 187-188

† Aristotle, *Nicomachean Ethics*, trans. Terence Irwin (Indianapolis: Hackett Publishing, 1999), 119–20, 128.
"Greek Words for Love," Wikipedia, last modified February 24, 2026, https://en.wikipedia.org/wiki/Greek_words_for_love.
See Greek words for love (Wikipedia)
And Greek Mythos

‡ See The Gaze of Orpheus (Wikipedia)

and yearning. On one hand we have Orpheus gazing towards the underworld, which serves to dissolve the connection between Orpheus and his desire, Eurydice. On the other hand, Orpheus' role in the upper world is to use his creativity and artistic talent to transform his desires into a recreated form. Lacan uses the topography of the myth to construct his mirror stage. Mark Linder suggests: "The mirror stage is not an isolated event or situation that results in a particular configuration of vision: it is both a loss (of primordial polymorphous, autoerotic wholeness) and an 'achieved anxiety' (a precocious anticipation of an impossible maturity or return to wholeness." *

- Maurice Blanchot (1907-2003) uses the myth of Orpheus to explore the non-dialectical movement of art, particularly literature's self-realization. Unlike Hegelian dialectics, where synthesis leads to a higher truth, Blanchot's Orpheus sacrifices Eurydice but does not attain the work itself. Instead, he affirms the impossibility that underpins the work's origin. Orpheus's gaze upon Eurydice becomes a symbol of the necessity of obliqueness and indirection when approaching existence. Blanchot draws inspiration from Martin Heidegger's concept of "The Origin of the Work of Art," emphasizing that the work is primarily hidden, shrouded in darkness, and only reveals itself through concealment. Blanchot understands Eurydice's disappearance as symbolizing a loss, which is compensated for by Orpheus's song—a creative act that transcends mere representation. Blanchot believes that the myth exemplifies the need for indirect approaches to truth and being. †

Orpheus' gaze of Eurydice must go under never to return to life, to Orpheus. But Orpheus loves her unconditionally even without her presence. Eurydice will go under into Orpheus' heartbreaking pain for the rest of his life, but it will not diminish his love for her. To the contrary, in absence his heart grows fonder. Any parent that has had their child become a young adult will understand the heartbreak of losing 'my baby' during their adolescence. And for those of us who never got to see our child grow up - no one can imagine that pain. Unconditional love marks the place of love which cannot return except as a fullness which grows in absence. And what does *Eros* tell us? Plato writes about Socrates' speech in the "Symposium",

> "What then is Love?" I asked; "Is he mortal?" "No." "What then?" "As in the former instance, he is neither mortal nor immortal, but in a mean between the two." "What is he, Diotima?" "He is a great spirit (daimon), and like all spirits he is intermediate between the divine and the mortal." "And what," I said, "is his power?" "He interprets," she replied, "between gods and men, conveying and taking across to the gods the prayers and sacrifices of men, and to men the commands and replies of the gods; he is the mediator who spans the chasm which divides them, and therefore in him all is bound together, and through him the arts of the prophet and the priest, their sacrifices and mysteries and charms, and all, prophecy and incantation, find their way. For God mingles not with man; but through Love. all the intercourse, and converse

* See The Gaze of Orpheus (Wikipedia)

† "The Gaze of Orpheus," Wikipedia, last modified February 12, 2026, wikipedia.org. See The Gaze of Orpheus (Wikipedia)

> of god with man, whether awake or asleep, is carried on. The wisdom which understands this is spiritual; all other wisdom, such as that of arts and handicrafts, is mean and vulgar.
>
> He is by nature neither mortal nor immortal, but alive and flourishing at one moment when he is in plenty, and dead at another moment, and again alive by reason of his father's nature. But that which is always flowing in is always flowing out, and so he is never in want and never in wealth; and, further, he is in a mean between ignorance and knowledge. The truth of the matter is this: No god is a philosopher. or seeker after wisdom, for he is wise already; nor does any man who is wise seek after wisdom. Neither do the ignorant seek after Wisdom. For herein is the evil of ignorance, that he who is neither good nor wise is nevertheless satisfied with himself: he has no desire for that of which he feels no want." "But-who then, Diotima," I said, "are the lovers of wisdom, if they are neither the wise nor the foolish?" "A child may answer that question," she replied; "they are those who are in a mean between the two; Love is one of them. For wisdom is a most beautiful thing, and Love is of the beautiful; and therefore Love is also a philosopher: or lover of wisdom, and being a lover of wisdom is in a mean between the wise and the ignorant. *

Eros is erotic love and excess to erotic love. *Eros* is "*neither mortal nor immortal, but in a mean between the two*". *Eros* is misty, not fully inhabiting binary dualisms but playing with them. *Eros* is something other. As such, *Eros* interprets and spans the chasm which divides. *Eros* binds together all the oppositions of history's narratives in jest and fun. The gods do not mingle directly with humans but through only through their intercourse of love, one moment awake and the same next moment asleep. Dare we think of the play? Socrates tells us wisdom is spiritual. The Greek word for "spiritual" in the phrase "*The wisdom which understands this is spiritual*" is "*δαίμων*" (pronounced daimon or daemon). A daimon is a lesser divinity or spirit that exists in the space between mortals and gods. It embodies a liminal state, sharing characteristics with both humans and deities. These daimones are often personifications of abstract concepts or forces of *phusis*. They can also be akin to spirit guides, chthonic heroes, or even the deities themselves. Plato's dialogues frequently mention the daimon, emphasizing its role as an intermediary between the divine and the human. The Symposium itself provides a succinct description of the daimon's role:

> "The daimonic ... interprets and carries over to the gods things from men, and to men things from the gods, from the one prayers and sacrifices, and from the other orders and rewards for sacrifice. It fills the space between both and thus binds the all to itself." †

Essentially, the daimon acts as an interpreter, bridging the gap between human and divine realms – binary dualisms. It facilitates communication, guiding humans toward the divine order that remains hidden from their direct perception. It cannot come to presence or its 'not' of absence but exceeds them. But here is "*the evil of ignorance*", *Eros* is "*neither mortal nor immortal*", "*alive and flourishing*

---

* Plato, *Symposium*, trans. Benjamin Jowett (Oxford: Clarendon Press, 1871), 520–22.
See Socrates' Speech from Plato, Symposium

† Josiah Dvorak, "Diotima's Daimon: The Role of Eros in the Symposium," *Philomathes: For the Undergraduate in the Humanities* 3 (Spring 2019): 44.
See The Philosopher Within: The daimōn in Plato (.pdf)

at one moment when he is in plenty, and dead at another moment". *Eros "is always flowing in is always flowing out, and so he is never in want and never in wealth; and, further, he is in a mean between ignorance and knowledge*". *Eros "is neither good nor wise*", prior to narrative and also "*Love is of the beautiful; and therefore Love is also a philosopher: or lover of wisdom, and being a lover of wisdom is in a mean between the wise and the ignorant*". The excess of love is both wise and ignorant, not yet collapsed, not yet branching from the observer effect of classicism but on the verge.

Storgē is the love of family. It is the bond which first becomes responsibility for the Other (Autrui) of radical alterity. *Storgē* can also be the love between parents and their children. *Storgē* is a deep appreciation of inner beauty. It can also be used to describe the love for one's country or even a sports team. *Storgē* is the beginning of community. *

Philía as understood in the context of ancient Greek philosophy, goes beyond mere friendship. It represents a sincere and platonic bond that develops through shared experiences, mutual respect, and appreciation. Unlike the physical attraction epitomized by *Eros*, *Philía* encompasses the deeper connections formed between free individuals. These connections can include intimacy and are not limited to conventional friendships. Plato, in his dialogues, explored the nature of *philía*, recognizing that it could extend beyond what we commonly think of as friendship. *Philía* is friendship but also mutual respect and appreciation. This is the possibility for significance in community where community can have an intimacy in which friends have commitment and responsibility to each other partially based on shared experiences. † But all these loves are about different ways of love which roam, coming together or falling apart. Love reflects a split which cannot be bridged, taken as projective echo, or disregarded from its necessity. Love desires and sinks below desire. Love wants and needs but remains love in absence, not becoming, not complete, torn and strewn over its many longings.

In Plato's Symposium, Aristophanes makes a speech about the original human. He said that the original human was androgynous, a combination of male and female. In the Symposium Aristophanes states, "*the primeval man was round, his back and sides forming a circle; and he had four hands and the same number of feet, one head with two faces, looking opposite ways, set on a round neck and precisely alike; also, four ears, two privy members, and the remainder to correspond. He could walk upright as men now do, backwards or forwards as he pleased, and he could also roll over and over at a great pace, turning on his four hands and four feet, eight in all, like tumblers going over and over with their legs in the air; this was when he wanted to run fast*". However, these androgynous beings became a threat to the gods. Aristophanes says of them that "*Terrible was their might and strength, and the thoughts of their hearts were great, and they made an attack upon the gods; of them is told the tale of Otys and Ephialtes who, as Homer says, attempted to scale heaven, and would have laid hands upon the gods*". Rather than kill humans, Aristophanes tells us that Zeus decided to split these being into two, male and female. Zeus thought if this did not work, he might split them again so they would "hop about on a single leg". Aristophanes notes that,

---

* See Storge
And The Ancient Greeks Were Right About Love

† See Lysis (dialogue) - Wikipedia
And Philia Love: A Deep Dive into the World of Platonic Affection | A Simplified Psychology Guide
And Plato on True Love | Psychology Today

> "After the division the two parts of man, each desiring his other half, came together, and throwing their arms about one another, entwined in mutual embraces, longing to grow into one, they began to die from hunger and self-neglect, because they did not like to do anything apart; and when one of the halves died and the other survived, the survivor sought another mate, man or woman as we call them,–being the sections of entire men or women,–and clung to that." *

After this, Zeus made some more modifications by putting the genitals on the front of the human being instead of the rear – design mishap according to some. He also made it possible for humans to regenerate with the new arrangement of genitalia. Aristophanes tells us that this created an ancient desire for us to complete ourselves in the opposite gender.

> "So ancient is the desire of one another which is implanted in us, reuniting our original nature, seeking to make one of two, and to heal the state of man." †

For Aristophanes, the conclusion is,

> And such a nature is prone to love and ready to return love, always embracing that which is akin to him. And when one of them meets with his other half, the actual half of himself, whether he be a lover of youth or a lover of another sort, the pair are lost in an amazement of love and friendship and intimacy, and one will not be out of the other's sight, as I may say, even for a moment: these are the people who pass their whole lives together, and yet they could not explain what they desire of one another. For the intense yearning which each of them has towards the other does not appear to be the desire of lover's intercourse, but of something else which the soul of either evidently desires and cannot tell, and of which she has only a dark and doubtful presentiment.
>
> Suppose Hephaestus, with his instruments, to come to the pair who are lying side by side and to say to them, 'What do you mortals want of one another?'
>
> They would be unable to explain. And suppose further, that when he saw their perplexity he said: 'Do you desire to be wholly one; always day and night in one another's company? for if this is what you desire, I am ready to melt and fuse you together, so that being two you shall become one, and while you live live a common life as if you were a single man, and after your death in the world below still be one departed soul, instead of two–I ask whether this is what you lovingly desire and whether you are satisfied to attain this?'–
>
> There is not a man of them who when he heard the proposal would deny or would not acknowledge that this meeting and melting into one another, this becoming one instead of two, was the very expression of his ancient need.
>
> And the reason is that human nature was originally one and we were a whole, and the desire and pursuit of the whole is called love. There was a time, I say, when we were one, but now because of the wickedness of mankind God has dispersed us, as the Arcadians were

---

* Plato, *Symposium*, trans. Benjamin Jowett (New York: Charles Scribner's Sons, 1888), 44.

† Plato, *Symposium*, trans. Benjamin Jowett (New York: Charles Scribner's Sons, 1888), 191d.

dispersed into villages by the Lacedaemonians. And if we are not obedient to the gods, there is a danger that we shall be split up again and go about in basso-relievo, like the profile figures showing only one half the nose which are sculptured on monuments, and that we shall be like tallies. Wherefore let us exhort all men to piety in all things, that we may avoid evil and obtain the good, taking Love for our leader and commander.

Let no one oppose him–he is the enemy of the gods who opposes him. For if we are friends of God and at peace with him we shall find our own true loves, which rarely happens in this world at present. I am serious, and therefore I must beg Eryximachus not to make fun or to find any allusion in what I am saying to Pausanias and Agathon, who, as I suspect, are both of the manly nature, and belong to the class which I have been describing. But my words have a wider application–they include men and women everywhere; and I believe that if our loves were perfectly accomplished, and each one returning to his primeval nature had his original true love, then our race would be happy. And if this would be best of all, the best in the next degree must in present circumstances be the nearest approach to such a union; and that will be the attainment of a congenial love.

Wherefore, if we would praise him who has given to us the benefit, we must praise the god Love, who is our greatest benefactor, both leading us in this life back to our own nature, and giving us high hopes for the future, for he promises that if we are pious, he will restore us to our original state, and heal us and make us happy and blessed.

This, Eryximachus, is my discourse of love, which, although different to yours, I must beg you to leave unassailed by the shafts of your ridicule, in order that each may have his turn; each, or rather either, for Agathon and Socrates are the only ones left. *

Androgyny is "*prone to love and ready to return love*". "*And when one of them meets with his other half, the actual half of himself, whether he be a lover of youth or a lover of another sort, the pair are lost in an amazement of love and friendship and intimacy, and one will not be out of the other's sight, as I may say, even for a moment: these are the people who pass their whole lives together, and yet they could not explain what they desire of one another. For the intense yearning which each of them has towards the other does not appear to be the desire of lover's intercourse, but of something else which the soul of either evidently desires and cannot tell, and of which she has only a dark and doubtful presentiment.*" Love defies explanation though many have tried. Love laughs at knowledge. We love all our lives but never have a clue about the 'is' of love. Love is beyond physicality. Love tells us something or maybe nothing about the soul which it tramples it to death. But what the something or nothing *is* cannot come to narrative, and history. Its exteriority as alterity, chaos which cannot be bridged, historically demands a bridge. Something or perhaps nothing lurks infinitely deep below the surface which never floats into words. The beauty of ancient Greek androgyny tells us something very significant about language and how history entraps it as gender and sexual identity long before that became 'Woke'. Even more, Zeus screwed up and had to make a revision to put the genitals on the front of the human being instead of the rear although many ancient Greeks might disagree with that epistemology. Wow, I guess the ancient Greek really discovered 'Woke" not wild-eyed radical liberals!

---

* Plato, *Symposium*, trans. Benjamin Jowett (New York: Charles Scribner's Sons, 1888), 44–47.
See Symposium By Plato (Translated by Benjamin Jowett)

Thus, éros is love that is "*the desire and pursuit of the whole*". According to Hesiod, *Eros* is: "*...the fairest of the deathless gods; he unstrings the limbs [makes the limbs go limp] and subdues both mind and sensible thought in the breasts of all gods and all men.*" * Hesiod tells us that Eros was one of the oldest deities, born from Chaos alongside Gaia (the Earth) and Tartarus (the Underworld). *Eros*, the non-generative, *without arche* – an-archy, without origin (remind us of something?), parentless god from Aristophanes is neither divine nor mortal. "*At the beginning there was only Chaos, Night (Nyx), Darkness (Erebus), and the Abyss (Tartarus). Earth, the Air and Heaven had no existence. Firstly, black winged Night laid a germless egg in the bosom of the infinite deeps of Darkness, and from this, after the revolution of long ages, sprang the graceful Love (Eros) with his glittering golden wings, swift as the whirlwinds of the tempest. Eros rises from deep, infinite, darkness, darkness which exceeds itself, exceeds light, not born of sperm and egg (virgin birth?). Eros swift, instantaneous as whirlwinds of a tempest, branching from chaos!*" † If I didn't know better, I would say that the Greeks understood quantum physics but to make us feel better let's call it intuition.

"*Eros mated in the deep Abyss with dark Chaos, winged like himself, and thus hatched forth our race, which was the first to see the light.*" ‡ Later Eros is spoken of as the child of night (Nyx). He is also spoken of as the son of Aphrodite,

> **[Hera addresses Athene :] We must have a word with Aphrodite. Let us go together and ask her to persuade her boy [Eros], if that is possible, to loose an arrow at Aeetes' daughter, Medea of the many spells, and make her fall in love with Iason . . .**
>
> **He [Eros] smites maids' breasts with unknown heat, and bids the very gods leave heaven and dwell on earth in borrowed forms.**
>
> **Once, when Venus'son [Cupid, aka Eros] was kissing her, his quiver dangling down, a jutting arrow, unbeknown, had grazed her breast. She pushed the boy away. In fact the wound was deeper than it seemed, though unperceived at first. [And she became] enraptured by the beauty of a man [Adonis].**
>
> **Eros drove Dionysos mad for the girl [Aura] with the delicious wound of his arrow, then curving his wings flew lightly to Olympos. And the god roamed over the hills scourged with a greater fire."** §

Wow! Was this pornography according to the Greeks? Let's not think too much about this. I am not so sure women got the better end of this deal judging from the history of patriarchy.

---

* Hesiod. *Theogony*. Translated by Hugh G. Evelyn-White. (Cambridge, MA: Harvard University Press, 1914), line 120.

† Aristophanes. *The Birds*. Translated by Eugene O'Neill Jr. (New York: Random House, 1938), lines 693-697.

‡ Aristophanes. *The Birds*. Translated by Eugene O'Neill Jr. (New York: Random House, 1938), lines 693-697.

§ Apollonius Rhodius, *The Voyage of Argo*, trans. E.V. Rieu (London: Penguin Classics, 1959), 3.25–29.
Seneca:
Seneca, *Phaedra*, trans. Frank Justus Miller (London: William Heinemann, 1917), 186.
Ovid:
Ovid, *Metamorphoses*, trans. Frank Justus Miller (London: William Heinemann, 1916), 10.525.
Nonnus:
Nonnus, *Dionysiaca*, trans. W. H. D. Rouse (Cambridge, MA: Harvard University Press, 1940), 48.470.

Love and Pain

All existence and essence rings hollow in the face of love. While I am not a Christian, I certainly find beautiful passages in the Bible. I think of 'the peace that passes all understanding' (Philippians 4:7) which must come at the cost of pain. One of the most beautiful expositions of love is 1 Corinthians 13:

> If I speak in the tongues of men and of angels, but have not love, I am only a resounding gong or a clanging cymbal.
>
> If I have the gift of prophecy and can fathom all mysteries and all knowledge, and if I have a faith that can move mountains, but have not love, I am nothing.
>
> If I give all I possess to the poor and surrender my body to the flames, but have not love, I gain nothing.
>
> Love is patient, love is kind. It does not envy, it does not boast, it is not proud.
>
> It is not rude, it is not self-seeking, it is not easily angered, it keeps no record of wrongs.
>
> Love does not delight in evil but rejoices with the truth.
>
> It always protects, always trusts, always hopes, always perseveres.
>
> Love never fails. But where there are prophecies, they will cease; where there are tongues, they will be stilled; where there is knowledge, it will pass away.
>
> For we know in part and we prophesy in part,
>
> but when perfection* comes, the imperfect disappears.
>
> When I was a child, I talked like a child, I thought like a child, I reasoned like a child. When I became a man, I put childish ways behind me.
>
> Now we see but a poor reflection as in a mirror; then we shall see face to face. Now I know in part; then I shall know fully, even as I am fully known.
>
> And now these three remain: faith, hope and love. But the greatest of these is love. *

Here is my pain from love for the son I lost at the age of 20 years old:

To Chris,

This song is dedicated to my dear son Chris. I still carry his heart...

Michael Hedges made E.E. Cumming's poem a song with help from David Crosby and Graham Nash.

[i carry your heart with me(i carry it in]

BY E. E. CUMMINGS

i carry your heart with me(i carry it in
my heart)i am never without it(anywhere
i go you go,my dear;and whatever is done
by only me is your doing,my darling)
i fear
no fate(for you are my fate,my sweet)i want
no world(for beautiful you are my world,my true)

---

* MIT https://web.mit.edu/jywang/www/cef/Bible/NIV/NIV_Bible/1COR+13.html
1 Corinthians 13:1–13 (NIV).

and it's you are whatever a moon has always meant
and whatever a sun will always sing is you
here is the deepest secret nobody knows
(here is the root of the root and the bud of the bud
and the sky of the sky of a tree called life;which grows
higher than soul can hope or mind can hide)
and this is the wonder that's keeping the stars apart *
i carry your heart(i carry it in my heart)

To Hear the Song: I Carry Your Heart (YouTube) at https://www.youtube.com/watch?v=GwczXm64eRk

To Hear a Reading: I Carry Your Heart With Me by E.E.Cummings - Poetry Reading at https://www.youtube.com/watch?v=MF5H7l9jEkY

†

This is a song I composed, arranged, and recorded about Chris - Distance at https://www.mixermuse.com/Mark/Alterity/Distance.mp3

And my daughter, who also had to carry that burden from her youth bravely and courageously without even the benefit of the painful wisdom which can come from age. I also have pain for her having to go through this tragedy. She is the third other in which justice is an impossibility, but neither is it optional.

This is the place where Levinas wants to take us. This is life which gurgles up from the depths which I have discussed throughout this project. Substitution, expiation, the third other are difficult notions in Levinas. In a way his ideas are much more difficult than quantum physics' radical externality

* See [i carry your heart with me(i carry it in]

† Mark Randall Dreher, *Untitled Photograph of Two Children and a Dog*, photograph, 2026.

and spooky chaos. But his ideas are rooted in experience, Ethics, and love. Not just for the ones we love but for ones we have not loved, not known, neighbors, acquaintances, the homeless we see on the street, the stranger or sojourner, the widow, the dejected by societal norms which have to suffer a silent violence. Whatever we are facing now in the face of radical externality - loving, bereaving, desiring, from depths unseen crying "Ethics" and in the very saying there is hope.

Persecution and Accusation: The Secret Life of Bad Conscience

As I mentioned earlier, I want to try to explain Levinas' idea of substitution, expiation, and justice in language which many of us can relate to more easily. How can we relate to what Levinas is telling us about accusation? I want to relate something about myself which I think will apply to many of us. From the time that memory became available to me at a very early age, I have had something happen to me which continued well into my sixties before I really had a handle on what was going on. A recurrent memory and emotion would simply pop up in my mind and emotions. Ranging from the most seemingly insignificant situations to highly significant situations from many decades ago until recently with no relation to why or what I was doing at the time. An embarrassing or even stupid thing I did would just flash into my mind. Immediately, I would get mad at myself all over again for the dumb, insensitive, or hurtful thing I did. Over and over through the years these same situations would pop into my mind. And it is not like I have a great memory for these kinds of things. These situations can be summarized as bad conscience. Perhaps others would just write it off as the other person's fault. I never did that. I always blamed it on me alone for being stupid, wrong, bad etc. Many times, I would even silently and verbally say something like stupid me! But no matter what I was doing these same incidents would almost magically appear over and over again. And again, I would inwardly go through throwing a fit about them *ad infinitum*. In my inability to deal directly with them I developed a practice of denial.

One day not too long ago it occurred to me (and I should have known better) that what was really going on was that I still had hurt and pain from these situations. I soon developed another practice when these situations would poof out of nothingness. Instead of getting mad I would ask myself, where is the hurt and pain you are feeling about this incident? Guess what, one by one these incidents stopped popping up in my brain and emotions. What I figured out was that my anger and disappointment could not resolve these recurrent situations. It only pushed them temporarily back into my subconscious where they were poised and ready to spring back on me when least expected. The point is I had a kind of secret embarrassment about myself that I thought was unique to me. And my knee jerk reaction to them was to be angry and disappointed with myself. So, anger and disappointment may have pushed them back temporarily, but that was only a short-term fix. What I really needed to do was face the real enemy – my hurt, pain, and disappointment with myself. In a nutshell, this is the surface phenomenon of what Levinas is trying to get at with his discussion of accusation. My conscience accused me, and I kept getting mad at myself for what I knew was true – I screwed up.

Accusation comes from without. Accusation always begins from my interaction with the other. After accusation confronts me then comes my interpretation of accusation as persecution. At the point of persecution there is a choice between two paths. I can blame the other or I can blame myself. Certainly, I could choose to justify myself, blame the other, hold myself innocent of all charges. I can try the accuser in the court of self and not so surprisingly find myself guilty or not guilty. My choice was to persecute myself. This is probably because I tend to be more of an introvert rather than an extrovert. Had I been more extroverted I could have easily accused the other for my persecution. It could have thought to myself it was all their fault I f***** up. But in either case, all that is really like taking a pill or a drink to postpone the inevitable eternal recurrence of the same - bad conscience. So, accusation brings persecution. This is where we need to begin to understand something much more profound with what Levinas' is trying to tell us. And guess what, when I practice the habit of letting myself feel the pain something else can come to the fore – expiation. In a way I took on the guilt and punishment, digging down into the emotions of hurt and pain of the other. In so doing I practiced expiation. Even more, Levinas is trying to tell us something much more fundamental than guilt and remorse as a psychological affect. Levinas is telling us that me, the self, is not built up from my own reductionary bootstraps. We branch from radical alterity of the other, the saying of the other which cannot be recuperated into spacetime, historic narrations of the other – my echo of the other in transactionalism.

Practicality: How can we relate to substitution, accusation, persecution, and expiation?

Having had two wonderful, loving children I can say there is no way a lover can ever break your heart like your child. Having a child is an other-worldly kind of love. The ancient Greeks had a special word for it - agápe (ἀγάπη) or unconditional love. If you have children there is no escape – you will have your heart broken, you will suffer pain for them. The most straight forward example of this is when they are on their way to young adulthood, and you have to let go of them as 'my babies'. Ouch, that bites the big one. You want to treat them as your babies their entire lives, but you know that is not the best thing for them. Treating them like babies is for 'me' not for them. For them, I must put aside 'me' and focus on an impossible justice. The impossible wisdom of love as justice imposes necessity on me to feel pain which I know I will never really recover from. I have to let them go for their sake. This hurts badly and for the rest of my life it will be there more or less. This is a case where the other is no longer about me, about my needs and wants but about them even if it hurts me like hell. This is a case of substitution based on justice. I substitute their need and wants for my needs and wants. My devotion to them comes first. Expiation for your children is literally laying your life down for them. Being the first in line to take incoming fire for them and without question. In Levinas the philosophical ideas of substitution, accusation, persecution, and expiation may be a little hard to understand.

I think the practical effect of what Levinas is telling us may seem somewhat distant in our youth. This brings me to the legacy of pain and heartache. As I have gotten older, I have discovered that certain kinds of emotional events concerning others which I had when I was younger like anger, blame, resentment, bitterness, etc., were not what they seemed through those years. All of us have experienced failures dealing with others which we regret. We wish we could have done or said something differently. We wish we had behaved differently and achieved a different outcome. This is a guilty conscience. We do not want to admit guilt and shame. I have news for younger folks, unless you sear your ability to have a conscience, these experiences will weigh on you more and more as you get older. Denial is like throwing gas on this fire. Regrets will keep getting worse and you will have to put more and more effort into squelching them as you get older. They will pop up when you least expect it. The strategies of denial and frustration with yourself will not vanquish these inopportune moments. And it only gets worse as we age. Even these kinds of experiences are opportunities to learn something about ourselves, to grow, to gain awareness of ourselves and others. There is a richness in being honest with ourself that money cannot buy.

When we are young, we avoid heartache like the plague. In youth, we learn quickly that it is better not to love than to love and have to live through the death of love. These emotional scars are a form of naive fleeing which simply sends pain underground for a while. We really have no choice but to love again because most of us feel like love is not optional. With wisdom and age, we may learn that there is something about the death of love that never dies. When you sign up for love, in a way, you sign your own death certificate. One clear example of this is those that we dearly loved who die before we do, our parents, our siblings, relatives and sometimes our own children.

There is also another kind of denial which is more insidious – unintentionally hurting others. We are all so busy trying to manage life that we can so easily slip into an insensitivity to what others are going through. We may 'see' others through the lens of our life's echoes. We may barely realize that something we did or said affected another person negatively. Maybe they felt put down, dismissed, pushed into a place that made them uncomfortable. These oversights are also something that our subconscious will replay over and over again with increasing intensity as we get older. Unless we have some kind of pathological narcissistic/anti-social disorder there is something built into us which is indebted to the other person and their well-being. Violating this, even unintentionally, leaves a mark in the subconscious which does not dissipate with time but remains, festers, and requires more and more effort to find temporary relief. I believe this is why many older folks get highly negative or turn to booze or drugs to weld this increasingly heavy load.

But there is a kind of peace, 'making friends with the horror', that can be made with these issues. All these knee-jerk reactions are an attempt to change pain into something which shifts blame to someone or something else. But this does not promote healing. Denial is a habitual practice. To break a habit requires another kind of habitual practice. The first thing to do is to mentally recognize the feeling as pain and heartache not as anger, bitterness, blame. The next thing to do is to let yourself feel the pain. This is not a cure, pain will still be there, but it allows us to find a kind of center in the midst of the pain, a kind of fullness for having the capacity to still feel pain. The fear of pain is much worse than the feeling of pain. The fear of pain hollows out the soul. And pain is there for a reason, to learn wisdom about the other.

But what shall we do? The ol' master said it truly, "*Tis better to have loved and lost than never to have loved at all*". To get old is to come face to face with this heartache. But there is a funny thing about gaining some wisdom along the way, the pain is worth having. The other, the one you loved is no longer 'there' but in a way they still are in the love you hold for them – you carry their heart in your pain of losing them. There is a richness to this that all the transactionalism, power, fame, glory, money can never give you. There is worthwhileness about living and loving that brings deep pain but also deep joy in the 'having been' which in some odd way still 'is'. There is a contentment or at least a sense of having loved someone and losing them which has a singularity unique to that person. There is an odd kind of fulfillment, even loss one would not trade that experience for anything. It is as if that would betray one's precious love in their absence. This is a debt that cannot be repaid. This is a longing in which,

The more I answer the more I am responsible; the more I approach the neighbor with which I am encharged the further away I am. This debit which increases is infinity as an infinition of the infinite, as glory. *

In my opinion Levinas brings to the fore the far-reaching implications of these kinds of pain. Levinas gives us an account which gives a very different account of the self than narratives of classicism, habituated into commonsense, would have us believe. This goes much deeper than the psychology of how all this plays out in our day-to-day practicalities. Levinas finds in these embodied experiences a kind of pre-sensuality which penetrates below all the historic narratives we tell ourselves. We tend to think of ourselves as free agents, rooted in selfhood from biology, from God, from all the narratives of history which give us a feeling of self-creation, of independent beings, free from any necessary relationship with other independent beings. But what if the deep-rooted feeling is totally wrong?

The Impossibility of Love for Self-Consciousness

The Greek myth of androgyny goes much further than simply biological gender. It goes to the impossibility of love. In Greek mythology, Psyche was the deification of the human soul. She was portrayed in ancient mosaics as a goddess with butterfly wings (because psyche was also the Ancient Greek word for 'butterfly'). The Greek word psyche literally means "soul, spirit, breath, life or animating force". In the second century a story is told of Eros and Psyche,

> The story tells of the struggle for love and trust between Eros and Psyche. Aphrodite was jealous of the beauty of mortal princess Psyche, as men were leaving her altars barren to worship a mere human woman instead, and so she commanded her son Eros, the god of love, to cause Psyche to fall in love with the ugliest creature on earth. But instead, Eros falls in love with Psyche himself and spirits her away to his home. Their fragile peace is ruined by a visit from Psyche's jealous sisters, who cause Psyche to betray the trust of her husband. Wounded, Eros leaves his wife, and Psyche wanders the Earth, looking for her lost love. Eventually she approaches Aphrodite and asks for her help. Aphrodite imposes a series of difficult tasks on Psyche, which she is able to achieve by means of supernatural assistance. After successfully completing these tasks, Aphrodite relents and Psyche becomes immortal to live alongside her husband Eros. Together they had a daughter, Voluptas or Hedone (meaning physical pleasure, bliss). †

Self-consciousness (Psyche) haunted by Love (Eros), wounded from jealousy and betrayal, symbolizing accusation and persecution of the other, wanders the earth. Self-consciousness lost to itself, unable to find itself as itself, is entrapped by the god of love. But the merely human, self-consciousness, can never find love. Only by finding the work of immortality can Psyche achieve the progeny of love. The '*is*' of love, its mortality, cannot '*be*'. Only the immortal can find the progeny of love. Self-consciousness is condemned by mortality to '*be*' torn by the inability of love to find its essence in exis-

* Emmanuel Levinas, *Otherwise Than Being or Beyond Essence*, trans. Alphonso Lingis (The Hague: Martinus Nijhoff, 1981), 93.

† Apuleius, *The Golden Ass*, trans. P. G. Walsh (Oxford: Oxford University Press, 1994), 78–122.
See Eros and Psyche

tence. Immortality is the mythos of a radically other in which the self is forever unable to complete itself from itself. Androgyny is not about gender. The self has no essence, no immortal basis in its own echo. The self arises in accusation, persecution, substitution, and expiation from the radical alterity of the Other (Autrui). Levinas writes of the radical alterity of the Other (Autrui),

> Immemorial, unrepresentable, invisible, the past that bypasses the present, the pluperfect past, falls into a past that is a gratuitous lapse. It can not be recuperated by reminiscence not because of its remoteness, but because of its incommensurability with the present. The present is essence that begins and ends, beginning and end assembled in a thematizable conjunction; it is the finite in correlation with a freedom. Diachrony is the refusal of conjunction, the non-totalizable, and in this sense, infinite. But in the responsibility for the other, for another freedom, the negativity of this anarchy, this refusal of the present, of appearing, of the immemorial, commands me and ordains me to the other, to the first one on the scene, and makes me approach him, makes me his neighbor. It thus diverges from nothingness as well as from being. It provokes this responsibility against my will, that is, by substituting me for the other as a hostage. All my inwardness is invested in the form of a despite-me, for-another. Despite me, for-another, is signification par excellence. And it is the sense of the "oneself," that accusative that derives from no nominative; it is the very fact of finding oneself while losing oneself. *

The radical alterity of the Other (Autrui), "*Immemorial, unrepresentable, invisible*" cannot '*be*', as the 'my' of reality, mortality, existence - echo. All the pasts of my presence, my mortality, fall into "*gratuitous lapse*" from a past which was not my past. Doomed to wander the earth vainly attempting to *be* "*recuperated by reminiscence*" by its own echo. All the while in saying, I face the "*incommensurability*" other. Self-conscious' thematization wanders to and fro' in illusion, only ever reaching its own presence as essence and the mirages of its own freedom. Diachrony, the Other (Autrui) whose past is not my past, whose time is not my time, who cannot exist temporally or spatially in my idea, my presumed freedom, in branching as 'me'. Androgyny is not the split of gender, it is a diachrony of time which refuses conjunction. The Other (Autrui) is "*non-totalizable*" which Levinas tells arises as a sense of infinity. Love is embodied shadowing of the radical alterity of the Other (Autrui). Love infinitely yearning never completing itself with the other. The '*is*' can only recuperate and repeat itself like Echo but the excess of love is responsibility. Love is recognition of its inability to be able to '*be*' the other. And yet... -"*the immemorial, commands me and ordains me to the other*". Echo substitutes itself for the other in the allusion of completeness, essence, and Being.

The other is not the origin of history - my story. My idea of the other cannot find the other in me. The vanity of finding the other in me, as my idea of the other, is auto-affection. Jesus is quoted as saying in order to find yourself you must lose yourself and whoever loses themselves for the sake of the other will find themselves (my paraphrase from Matthew 10:39). *Echo* as the beautiful nymph, full of love for Narcissus, loves him to his end and hers. All the while Narcissus only hears his own echo. Narcissus cannot 'see' love, a radical externality to himself. Echo is diachronous to him. Echo loves Narcissus girding him, yearning for him, even to the point of Narcissus ignoring her love for him by

---

* Emmanuel Levinas, *Otherwise Than Being or Beyond Essence*, trans. Alphonso Lingis (The Hague: Martinus Nijhoff, 1981), 11–12.

turning her into his essence, into echoes of himself. She waits by the stream until he ages and dies and still, she waits and waits until she dies – but at the end, from Echo's corpse a flower grows on the tree of life marking the trace where Narcissus dies. Echo bespelled in love, abandoned as love, as unrequited love lives on in the flowering of love. The myth of Eros and Narcissus is a good analogy for the direction Levinas wants to take us towards the other. However, even beyond the reach of love, responsibility and justice faces us in the alterity of the Other (Autrui). Infinitely beneath love, unreachably beneath the self, 'me', cannot remain in itself. It has an indebtedness from which we cannot choose.

The radical alterity of the Other (Autrui) - the Other (Autrui) "*has the face of the poor, the stranger, the widow, and the orphan.*" Levinas emphasizes that Ethical experience begins with this encounter preceding any specific rules, principles, commonsense, essence, Concept, and Being. The stranger, the sojourner, the immigrant – here is responsibility where even love cannot *feel* its depths. This other to whom I face who calls forth responsibility. Not as altruism but as debt which cannot be repaid. Infinitely more,

> The more I answer the more I am responsible; the more I approach the neighbor with which I am encharged the further away I am. This debit which increases is infinity as an infinition of the infinite, as glory. *

The sojourner is not echo suited for political transaction or neglect. The sojourner exceeds me as an excess which cannot return to myself, my disdain, my rejection. The anti-immigrant fevered pitch in our country is no different from the birth of Jesus - cast out of Nazareth, run out of town by a mob, to be born in a stable in Bethlehem. Conventional ethical commonsense is borne on the whims of the isolated self's discretion, the echo of self to oneself. Levinas' Ethics arises from responsibility facing the exteriority of the Other (Autrui) in which the 'I' arises from indebtedness.

> The Face of the Other. Levinas argues that the Other discloses himself as a face. The face is the "way in which the other presents himself, exceeding the idea of the other in me". "The face approaches with a glance, a word, a gesture or a movement of the whole body. It addresses, expresses an appeal to, and makes demands of the Self". The Other, which reveals himself as a face, touches the self to the very core of its being. The look of the Other pierces the heart of the self so that the self cannot evade the Other's call for help, generosity and sacrifice. This look is the look of the poor, the orphan and the widow that begs for mercy and compassion. The presence of the Other that strikes the very being of the self Levinas would call passivity, affectivity, vulnerability, persecution and trauma. †

The other is not the 'said' of my idea. The other facing me exceeds the "*idea of the other in me*", might we think 'says me' as pure signifyingness before representation. Pure signifyingness implies the self, me, is not my own but signifies its absolute inability to be able (to be its own origin) - its complete

---

* Emmanuel Levinas, *Otherwise Than Being or Beyond Essence*, trans. Alphonso Lingis (The Hague: Martinus Nijhoff, 1981), 93.

† Emmanuel Levinas, *Totality and Infinity: An Essay on Exteriority*, trans. Alphonso Lingis (Pittsburgh: Duquesne University Press, 1969), 50–51, 198–199.

See Ethics as Optics of the Divine

indebtedness to the radical exteriority of the Other (Autrui). The *logos* of alterity, diachrony, cannot be made commensurate with my *logos*. I am answerable from "*Immemorial, unrepresentable, invisible, the past that bypasses the present*", "*incommensurability with the [my] present*".

Even more impossible than branching as self-origin, is the impossibility of justice and the third other. Proximity as radical alterities 'speaking' from diachronous third others makes me responsible for justice, for decision between diachronies, for *logos* not my *logos*. Just as *agápe* cannot return to itself beyond all reaches of notions of love, my obligation to others calls me to decide impossible justice between them. My requirement and impossibility are accusation which I take as persecution. Am I required to mythically die for the sins of the world, to expiate the world – only to get nailed to a cross of impossible responsibility?

Responsibility cannot grasp, cannot echo, cannot return to itself but owes its very self to the Other (Autrui) who faces me. Radical alterity of the Other (Autrui) presences as face[s] of the other coming from a dimension of height, a dimension of transcendence that presents the stranger. * The responsibility I owe to others comes back to me as the nymph Echo. The words I speak Echo beyond my loves, beyond my origin, coming back without her, without excess, anarchically as classicisms, as narratives, as 'words I say' exceeding themselves from exteriority - from her, him, they. The gift of the stranger, the sojourner, the immigrant, the orphan, the destitute – they are the gift of the tree of life.

Can we think - my branch? If so, I cannot observe myself. From saying the other *observes* me. This is the debt I owe which I cannot repay. My own echo does not arouse 'me' into awareness of awareness. Rather the other mythologized in the nymph Echo observes Narcissus who only sees himself. Without the saying/observation of the other, Narcissus is merely awareness in superposition of the wave function. The il y a of the quantum wave function is horrific awareness which is incapable of branching of its own. Bare awareness without any possibility of branching, of collapse as *seen* or observed – insomnia in the entire, chaotic environment of the quantum wave function can never sleep in perpetuity. The saying of the other is grace beyond all measure.

Awareness of Awareness: Who is the Observer?

As has been mentioned, it is far too early to draw any conclusions about consciousness' relationship to the quantum wave function. But ever since the creation of quantum mechanics the very founders of quantum mechanics have openly suspected, speculated, and wrote that there must be some kind of relationship to consciousness. Any conclusions are a long way off – well past any of our lifetimes. So, as my repeated disclaimers on this subject, I am going to make some huge speculations in this section. Namely, if I, self-consciousness, branch in some fashion from the quantum wave function, who is the observer? It cannot be me, since in this scenario, I am the one falling from superposition at birth, through life, to death, returning (but never left) back to the wave function (insert snide remarks about mysticism here) .

Quantum physics has brought us before a spooky exteriority which may have been intuited throughout history. Quantum physics is compelling us to reconsider our most familiar ideas and beliefs about reality. Quantum physics brings us to an unbridgeable abyss which cannot be bridged by history and commonsense. Similarly, Levinas brings us face to face with radical alterity in which the self does

* Totality and Infinity: An Essay on Exteriority, Alphonso Lingis (trans.), Pittsburgh: Duquesne University Press, 1969.

not exist in its own freedom. I am not self-made, self-created or composites from Concept, from soul, from essence, from Being, from substance, from matter and neutrality. Neither am I composites of the 'not' of all these culminating in the totality of nothingness as neutrality. I suppose religious people could tell us that God is the *logos*, the word, as observer of the wave function. Perhaps biocentrists might inform us that spontaneous collapse or branching functions as the observer. Biocentrists believe that intelligence and consciousness are not exclusive to humans but are inherent in various forms of life, including animals, plants, and even ecosystems. However, in either case quantum physics has made it clear that the singular observer, the scientist who conducts the actual experiment, is an indispensable observer. This singularity is an indispensable co-determinate factor in the probability of what is observed. There is something spooky and extremely intimate in the observer effect. This is not the objective and neutral 'nature' of history. And neither is it absolute subjectivity. It is radical excess which cannot rest in our historic and commonsense certainties.

There are no quantum discontinuities in classicism's narratives. There are no spooky disjunctive breaks from relativity's relatedness except perhaps in the nonsensical mathematics of a singularity. General relativity's singularity is simply the place marker for the 'not', what we do not understand and cannot understand. There is no structural undergirding in which we, being, can stand relative to classicisms narratives, exhaustions, and totalizing 'nots'. Throughout classicism, the self has an analogical relatedness relative to God, to essence, to Being, or to Nietzsche's affirmation of "I am that I am" in the face of eternal recurrence of the same. Nothingness, the catacombs of meaning, the absolute 'not' which seals the final deal are all built upon logical negations whose counter identity depends on identities, as tautological historic narratives. History provides us with metaphysical absolutes, whether it be essence, Being, Concept or neutrality. Relationships to these are then set accordingly in verticali**ty** and/or horizontali**ty** (where the '**ty**'s signify totality). But each in their own fashion finds a field of asymptotic correspondence to their *telos*, their culmination. Similarly, Einstein's relativity also and much more precisely establishes relationships to the absolute speed of light. In all these philosophical historic narratives the 'not' is always ready to totalize retaining themselves as a placeholder for radical excess - looming metaphysical horizons effacing any and all exteriorities. This is totality, reduction to the same, final certainty.

At the end of epochs what survives most from classicism is the absolute 'not' of neutrality and echoes of reactionary fanaticism. Nietzsche recognized this dynamic as the destitution of meaning, as nihilism, as the last man of mediocrity. The 'not' seals the tomb of totality in finality, the impossibility of anything further in saying and resigned to eternal recurrence of the said. This is hollowing out of meaning from the inside. Barely noticeable we find ourselves acclimated to the infinite recesses of neutrality, mere existence, mediocrity, where the ones with the most toys win. Accompanying this is the rise of pathologies, mental illnesses, suicides and the other absolute 'nots' of fanaticism, autocracy, 'wokeness'. When everything is the same everything is nothing. Classicism's narratives have finally brought us once again face-to-face with impending ipseity, the bare flight from *il y a*. Meaning dwindles as,

awareness

(pure signification, a "*passivity more passive still than the passivity*", as *il y a* devoid of responsibility) of

awareness

(ipseity's flight towards self-consciousness into ego) of

awareness

(the exhaustion of epoch as the totality of the 'not' - history's exhaustion of the unmoved mover, God, Essence, Being, Concept, Absolute – its inability to be able, to face the horror of 'not's nothing)

in lifelong circularity until even echo is undone by *il y a's* horrific, insomniac awareness - death into pure signifyingness without another.

At the beginning of classicism, Aristotle's' unmoved mover moves all, including humans, as actuality. Here noncontradiction reigns supreme. Logic is born from this essence. Aristotle, the unmoved mover moves all, including humans, as actuality. At the end of classicism, neutrality is the absolute which must forever go underground.

We are unaware how history seals us in a totality which makes it very difficult for us to reckon with radical alterity. Historic narratives give us the freedom to retreat into dreams, to flee from pain and sorrow into allusion. Could it be that all our evasions, self-righteousness, anger, bitterness, revenge, and resentments mask pain? Could it be that love with all its beauty and tragedies points away from Narcissus' echo - ever towards Orpheus' other which cannot *be* as mine, as my idea or history's idea? What if radical alterity facing me, holding me responsible, indebted before 'I am that I am', before my world and freedom, my pain and conscience, bring us to an unbridgeable chasm? Could we face accountability before we are even able to be able? If so, perhaps we can begin to understand what Levinas means by substitution, accusation, persecution, and expiation. All our loves and pain bring us irretrievably gaping before a void as our inability to be the master of our fate, to be the neutral substance of a thinking thing – devoid of exteriority.

Is it possible that we are not our own? Do we still want to maintain that classic spacetime, cause and effect, reality, Being, essence... all our favorite confidences in historic narratives, gelled into commonsense, are somehow magically underlying the findings of quantum physics? Do we really want to be an echo of ourselves, our certainties, our confidences... one without another? The established findings of 20th century quantum physics have made it impossible for us to go back to classicism. It seems undeniable that the spooky exteriority of the quantum wave function is here to stay. Could this affect our deep-rooted philosophical notions which bubble up into commonsense? Doesn't the observer effect inform us that the metaphysic of neutrality is simply a historic abstraction?

Could the observer effect tell us something about ourselves? How could 'I' *be* both the observer and the observed of the quantum wave function? Perhaps this myth depends on the 'how' of observation, my historic narrative of myself. If we are thinking about mind or self-consciousness, 'me here now', in terms of the wave function observer effect, I do not observe myself into being, into classicism. The observer effect is not some kind of absolute solipsism or narcissism of self-creation. This conclusion would be like the physicist who measures a classical particle, observes/branches together from/in the wave function discovering that he is the particle. The physicist is not a necromancer of himself. This would fall into the camp of everything is relative to me and my perception which would only be the echo of narcissism yet again. In the quantum wave function, information cannot observe itself and thereby 'branch itself' into probable values of a classical particle. This is way too simplistic.

Observation in the wave function is a confluence and a singularity. What do I mean by this? In actualizing, branching, a definite measurement is achieved. But this measurement does not come from a predetermined certainty. In this sense actuality is singular to the observer. I would refer to this singularity as responsibility not as echo. The observer has a responsibility for what is observed. How is the observer observing? What are the conditions under which observation can produce an answer? At some level the underlying *phusis* of quantum mechanics plays a role in self-consciousness and consciousness of other people. How the consciousness is directed, in what context it is taken, certainly has something to do with pre-judging how the answer may arise. If judgement is what I expect to see, isn't this my own echo in transactionalism then game theory is on, and strategy wins the day. However, if the observer understands how observation preconditions what is to be seen, this is called responsibility. Responsibility has to do with how we see not what we see.

Now, let's think about the other's role in my observation. The wave function does not collapse alone with a single conscious observer. The wave function branches with confluences of others. Observation not only brings about spacetime, cause and effect, the physical universe and nature. It also brings along with it historic narrative, classicism, embodiment, lived biological, emotional, experience – me here now already clothed with reality. This confluence also brings along the other, others, the stranger, lovers, children. But these others in my observation are not my actualization of the Other (Autrui) who faces me. The Other (Autrui) who faces me is a radically other observer than me. The Other (Autrui) who faces me is radically exterior to me the observer. Here again, what is imposed on us in observation is context - how observation preconditions what is to be seen. Here is where pre-judgement conditions what is actual, what is seen. If the physicist in a quantum experiment does not set up the conditions for what is to be seen correctly, nothing will be seen but chaos. Here by chaos what I mean is nonsense, the full environmental bath of the ~~universal~~[*] wave function. The physicist must exercise extreme rigor setting up the conditions of what is to be seen. This is responsibility. I can take the other to simply be my observation and nothing else. In this case the other has no otherness but only me echoing myself to myself. Or I can take on my responsibility for what becomes actual. Taking on responsibility for what becomes actual is what Levinas calls Ethics.

In the face of others, the third other, my responsibility requires me to make judgements about justice. The other and 'third other' are both radical externalities to me. As such my responsibility for judging, for justice, imposes a necessity for rendering a decision between two radical externalities. Since it is impossible for me to have a smooth, relativistic understanding of the other, adding another other doubles-down on those impossibilities. Adding many others reaches an asymptotic limit of impossibility in which responsibility cannot withdraw from daunting impossibilities. I come back to my metaphor of the crucifixion of Jesus taking on the sins of the whole world in expiation of the sins of the other. For me, a non-Christian, I can think of no better analogy, mythos, than how responsibility, radical exteriority, and sacrifice of the 'me' bearing the burden of impossible responsibility for justice in the face of the radical alterities can be mythologized. This is the God-Man (Koinē Greek: *θεάνθρωπος*, romanized: *theánthropos*; Latin: *deus homo*) or what Noah Harari reverses in his book "Homo Deus". Jesus calls himself the "son of man". Jesus is the mortal self encumbered with the instinct for survival just like all of us. But narcissism is not an option. Jesus must not only recognize the radical alterity of the other but also place himself at the pinnacle of impossible justice, expiation for others. The contrary of this expiation is entitlement, of right, of echo, of transactionalism. A lifetime of this kind of habitual irresponsible observation pre-conditions, locks the self ultimately into ipseity as perpetual retreat from *il y a*, from sheer awareness where death cannot die - return to *il y a*. *Il y a* has no capacity to recognize

[*] Jacques Derrida is the primary postmodern philosopher associated with "erasure" or overstriking (sous rature) words like "universal," "truth," or "being." This technique signals that a word is necessary for language yet inadequate or deceptive, challenging the stability of traditional philosophical concepts.

the other. *Il y a* has no capacity for self-consciousness. *Il y a* has no ability for self-origin or self-creation. *Il y a* is incapable of responsibility to radical externality, to the other. In life, awareness of awareness preconditions the 'how' of observation. In the wave function, without branch, there is only spooky quantum haze of awareness – there is no branching as awareness of awareness. In the wave function observer or observed do not fall into life. And awareness, however we imagine it, must have some node of pure signification in the quantum wave function.

What has been shown in the course of these volumes are that in the spooky, not physical and also not the 'not' of physical, there is no pre-metaphysical self, essence, Concept, neutrality which gives rise to the quantum wave function. The quantum wave function's u~~niversals~~* have some pre-original, anarchic, hazy a-temporal and a-spatial, simultaneous chaos which already anticipated all actualities. The quantum wave function's chaos brings into question every naming of 'is', 'are, 'will be', 'has been', essence, Being, Concept, life and death. However we want to reference the observer effect, the fall into classicism's narratives is not from some imagined self, essence, Being, Concept which would create itself, branch itself as if from some supposed metaphysical essence or reality outside the wave function. Branching as existence, as me, is from the externality of an observer, of the face of externality saying, observing – invocations from the tree of life.

As you can see, the quantum wave function presents a difficulty I and many others, way smarter than me, have struggled to understand, to put into words, to find a common *logos*, as if to reconcile it to some common principle or philosophy. The way I look at this is – this is radical alterity. This is why we are not cogs in a machine. However, this is not a threat. This is the joyous, gracious, everlasting beauty of the gaping abyss we face – who faces us in all the externalities of lovers, children, parents, friends, orphans, widows, prisoners, immigrants. They are the observers which evoke me from the gaping void. They are the ones with whom the self, me, has impossible responsibility for justice. They are the ones to which I am indebted to the point of expiation. This is the tree of life from which I grow (*rhizo-phusis to chaogenesis*) to the face (visage) of the other which has always already interrupted my genesis.

Hasn't quantum physics brought us to this abyss where the absolute '*is*' or its totalizing anti-identity, its 'not', cannot find refuge? Hasn't quantum physics brought us before this gaping void, this excess in which classicism's binary duality's and totalizations can no longer remain within itself before this radical exteriority, before the quantum wave function? When all that remains is classical history's inability to rouse itself from the 'not', from sealing itself in its narcissistic tomb of anonymity, all that awaits awareness - is *il y a*. The gaping void of the exhaustion of classicism can no longer allow exteriority in the tombstone of the 'not'. What remains of the day after the totality of classicism, exhausts itself - after classicism has been summed up, reduced, to the exhaustion of waking insomnia? *Il y a* pure signification cannot '*be*'. It is without ownness. It is without signifier and signified. It is the void which has no other, no life. It is unobserved pure signification.

Is it possible that we are unable to be able to create ourselves, find our origin, because we cannot observe ourselves from the quantum wave function's chaos – from radical exteriority which cannot be merely an 'it'? [1] In the quantum wave function, awareness of chaos without self-consciousness, *il y a*,

* Jacques Derrida is the primary postmodern philosopher associated with "erasure" or overstriking (sous rature) words like "universal," "truth," or "being." This technique signals that a word is necessary for language yet inadequate or deceptive, challenging the stability of traditional philosophical concepts.

cannot be born, collapse, or branch - *Il y a* as pure signification cannot observe itself. The entire quantum wave function is pure signification which has not been assigned a value, has not branched into classicisms. The face of the other, the androgynous mother combined with the father as love, as *agápe*, *éros*, *philía*, and *storgē*, as substitution, accusation, persecution, and expiation falling into being from love, pain, and responsibility. Born already indebted, before recognition is possible – responsible prior to any origin. And crawling towards death, back towards what we never left - the inability to be able. Once again, returning to pure signification, awaiting whom awakens as 'me'. Might this be the arising as growth of the tree of life?

CONCLUSION

In conclusion, life branches as love and justice. Justice seeks after truth. Justice is the pursuit of knowledge in the face of chaos. Justice weighs and balances. Justice requires blindness in the inescapable pursuit of sight. Justice looks for justification, for proof, identity, order, logic, and proportionality. The 'not' of justice is echo. It is the leveling off into the same. It is merciless. It shows itself as error, incongruence, disorder, contradiction, disproportionate, dogma, and totalitarianism. The 'not' of justice is its depletion, its silence, its error, its inability to be its 'isness'. But the impossibility of justice exceeds the 'not' of justice. Justice cannot find truth. In exceeding the 'not' of justice, impossibility demands justice account for itself, return back to itself, even if only as accusation and persecution, the pangs of what cannot find completion.

Love feels after wholeness. Love yearns for union, emergence, and community. Love requires singularity which cannot remain neutral. The 'not' of love is need and want. Need shows itself as hate, alienation, violence, estrangement, apathy, intolerance, and disdain. But the tragic pain of love exceeds the 'not' of love engulfing both binary polarities of love. In exceeding love, responsibility demands Ethics. But Ethics cannot find satisfaction. Love demands, accuses, and persecutes hollowing itself out in its mortal passivity, its inability as one-for-the-other.

Our androgynous split is not gender, it is unsettled chaos of uncertainty. Psyche falls into self-consciousness as obligation for what cannot come to presence or absence. In narrations of Being and essence, all too mortal wandering the earth in the impossibility but requirement for justice. Love, already accountable, responsible as substitution, accusation, persecution, and expiation. But without reason, without echo, only a gaping void as face, as observer. *Eros* woos from beyond, incapable of mortality, incapable of branching from chaotic exteriority. Nevertheless, impels from still, silent voice. We think, we feel, we live embodied.

Branching, Plato's Forms, Aristotle's unmoved mover have found their way through wandering narratives of history teleologically gaping towards insurmountable uncertainty. Life rending itself as justice and love without origin in the face of unrecoverable exteriority. We bring our truths and passions to life standing on precipice. We see no other side to the chasm before us, so we desire to retreat into dreams. Our brief mortality peers into the bottomless abyss where even death and Orpheus' gaze cannot fathom. Our life stands momentarily branching, protruding out over chaos. We scream **mortality** only to hear our own echoes reverberating into silence. But below the surface, me in the poverty of ipseity, of signifyingness which must wander to and from over the earth unable to sleep, perchance to impossibly dream, without root, homeless, wanderer, insatiable – exteriority and alterity. These depths plunge bottomlessly towards the wave function and towards Ethics. In this then my life is a branch in which there is inexhaustible hope beyond all imaginable possibilities. Branching of our contingencies grows, ancient - the tree of life. Infinite branching raising our vulnerabilities high, towards the sun, yet rooting, pulsating ever downward, well below the cool darkness of earth - far below the 'nots' and knottiness. Rhizome-ness, growing signifyingness thickening, heightening, swelling spooky, glorious radical exteriority which has no home. Depths which cannot be fathomed but whose vertigo pulls us from beyond life and death in expiation as responsibility for the homeless, the unknown stranger, the widow's pain, the lover's sorrow, the orphan – the other we never knew.

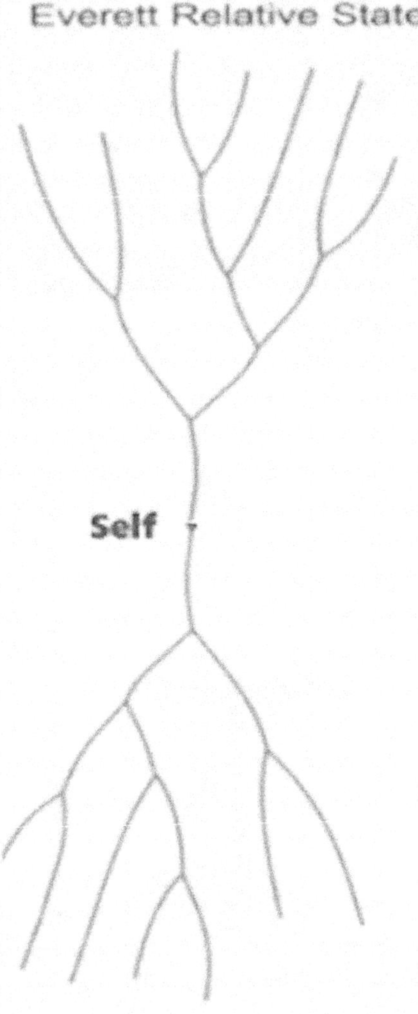

*

* Hugh Everett III, "The Theory of the Relative State," in *The Many-Worlds Interpretation of Quantum Mechanics*, eds. Bryce DeWitt and Neill Graham (Princeton: Princeton University Press, 1973), 3–140.

Mark Dreher, diagram of Everett's relative state formulation, 2026. Image created by the author. Based on Hugh Everett III, "The Theory of the Relative State," in *The Many-Worlds Interpretation of Quantum Mechanics*, eds. Bryce DeWitt and Neill Graham (Princeton: Princeton University Press, 1973), 3–140.

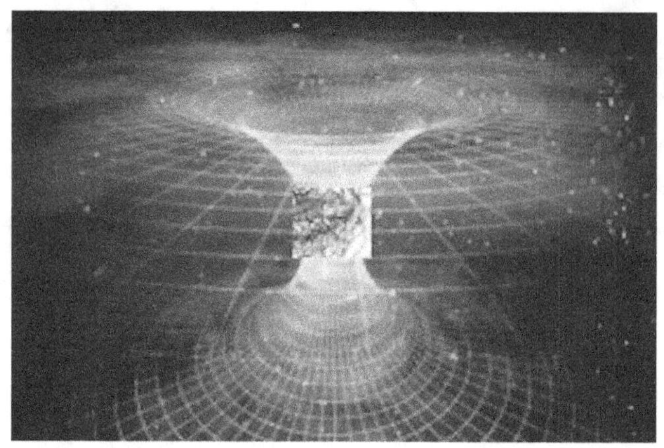

*

†

* "Wormholes," *Live Science*, June 16, 2020.
Image courtesy of Live Science.

† airdone, *Tree of life silhouette*, vector illustration, November 26, 2023, Freepik, https://www.freepik.com/premium-vector/tree-life-silhouette_93354918.htm.
Image by airdone via Freepik.

EPILOGUE: THE PRECIPICE OF THE OTHER

We have traveled a great distance—from the subatomic uncertainties of the quantum wave function to the vast, holographic boundaries of the cosmos. Throughout this journey, I have demonstrated that the physical world exceeds the asymptotic determinations of classical physics. It refuses to be boxed into the rigid binary of "subject" and "object," or "mind" and "matter." However, we must tread carefully. This excess of meaning—this "quantum weirdness"—should not tempt us into lapsing into some sort of mystic proclamation arising out of mere language or history. Quantum mechanics is not magic; it is a rigorous revelation of limits. It reveals that the "Real" is far stranger and more entangled than our classical intuition ever dared to imagine. To illustrate this one last time, we need only look at the "spooky" observer effect. It is neither purely subjective (in the mind) nor purely objective (in the world). It exceeds both. It is a relationship. It is an entanglement.

I have tried to bring out interesting—and frankly, startling—intersections between this quantum reality and the philosophy of Emmanuel Levinas. But here, at the end of the journey, we must acknowledge a profound hierarchy. Quantum physics clears the brush. It deconstructs the "Totality" of classical determinism. It shows us that the universe is not a clockwork machine. But Levinas brings us an infinite distance further than quantum physics ever can. Physics brings us to the edge of the map, but Levinas places us at the precipice. At this precipice, we find that the "excess" of the universe is not just mathematical "chaos"—it is Ethical Responsibility. The "Other" who faces me is radically exterior to all my reductive, narcissistic echoes. The Other is not a variable in an equation to be solved. The Other is a demand before my choice to be met.

In the Introduction, I asked if we could escape the Echo of Narcissus. Quantum physics gives us the hint: Entanglement. From the beginning of this work, there was a discussion of similarities between entanglement and love. At this point, we can now ask the question: What does Levinas tell us concerning Love? Emmanuel Levinas does not give a single, sentimental "theory of love." Instead, he radically reinterprets love as an ethical relation, inseparable from responsibility for another person. Across works such as Totality and Infinity (1961) and Otherwise than Being (1974), love is not *primarily* a feeling, choice, or mutual recognition—it is an asymmetrical demand placed on me by the Other.

For Levinas, love is best understood as being responsible for the Other before oneself. He famously writes that responsibility for my neighbor is "the harsh name for what we call love of one's neighbor." * It does not begin in desire, preference, or shared identity, and it certainly is not grounded in the reciprocity of "I love you because you love me." It precedes freedom and choice entirely; I find myself already responsible. I am answerable to the Other simply because they face me. Levinas locates the origin of this love in the face-to-face encounter, where the Other's face commands me without force or coercion.

---

* Emmanuel Levinas, *Otherwise Than Being, or Beyond Essence*, trans. Alphonso Lingis (The Hague: Martinus Nijhoff, 1981), 115-118.

The face is not just a physical appearance but the manifestation of vulnerability and infinity that resists being reduced to an object or concept. In this encounter, the Other implicitly pleads: "Do not kill me. Do not reduce me." Love appears as our responsiveness to that call. Thus, love is not fusion or possession, but a profound respect for irreducible alterity. Because of this, authentic love is fundamentally unequal. I am responsible for the Other more than they are responsible for me. This asymmetry is essential and non-negotiable. In Otherwise than Being, Levinas intensifies this claim by describing the self as a "hostage" or "substitution" for the Other—someone who bears responsibility even when it is unjust or unreturned. Love does not depend on fairness, it does not wait for reciprocity, and it freely risks suffering. Ethics—and therefore love—is not a contract.

While Levinas does address erotic love (eros), particularly in Totality and Infinity, he views it as ambiguous. It can turn inward toward mere enjoyment and possession, but it can also disrupt our self-sufficiency, opening the self toward alterity, futurity, and responsibility by revealing the Other as mysterious and unreachable. Yet, eros is not the highest form of love; rather, love is expressed most clearly in relentless care and hospitality—paradigmatically seen in parental love and care for the vulnerable. These forms make visible a responsibility that cannot be measured, completed, or withdrawn. Ultimately, while love begins in this singular responsibility, Levinas insists it must eventually confront the presence of the Third—the other others who also make claims on me. Here, love must be translated into justice and politics, though it must never be erased by those systems. Justice without love becomes cold and violent; love without justice risks blindness.

Quantum physics is the place where classical determinations give way to incredible excess. I have termed this excess "chaos"—not as disorder, but as simultaneously inextricable and inexhaustible. It is the yawning gap of Hesiod, the primordial Chaos from which forms struggle to emerge. But infinitely more than quantum chaos, Levinas finds the self—me—in a frenetic retreat.

We are in flight from the *il y a*, the anonymous, faceless horror of mere existence. To escape this void, the ego enshrouds itself in histories, rationalizations, and justifications—the smoke and mirrors of a self gathering its forces into *phusis*. Yet, in this retreat, the self is suddenly spun around and held responsible to the Other. The Other is a precipice that cannot be brought into my origins or my history. As Levinas writes: "The past of the other, a past that is not my past, but which affects me as though it were my own, is the very essence of responsibility." *

Here, we stand at the edge of two vastly different abysses:

Nietzsche, echoing the nobility of tragic history, warns us of the danger of the void: "He who fights with monsters should look to it that he himself does not become a monster. And if you gaze long into an abyss, the abyss also gazes into you." † For Nietzsche, the abyss gazes back with the meaninglessness of the eternal return of the same. The only defense is a heroic, Greek affirmation—the Echo of Narcissus, whose endless compulsion is to affirm itself against the monstrous void, terrified of being deformed by the very nihilism it opposes.

But Levinas reveals a more shattering precipice. The asymmetry of the void is not the impersonal nothingness of the abyss; it is the visage of the Other. In the face-to-face encounter, the self stares into the yawning gap of its own absolute inability. Here, the masterful 'I can' of the ego is stripped away,

* Emmanuel Levinas, *Time and the Other*, trans. Richard A. Cohen (Pittsburgh: Duquesne University Press, 1987), 74.

† Friedrich Nietzsche, *Beyond Good and Evil: Prelude to a Philosophy of the Future*, trans. Walter Kaufmann (New York: Vintage Books, 1966), §146.

leaving us in the sheer "inability to be able". * Because the demand of the Other is infinite, the ego is instantly rendered bankrupt—structurally and forever inadequate to fulfill it. When the ego stands in the visage of the Other, it does not just encounter radical alterity; it stares into the void of its own inescapable debt.

Faced with this infinite debt, the self retreats once again. Unable to endure the precipice of the face-to-face, we flee back into the narcissistic shell, hiding behind the totalities of 'logic', 'justice', and 'self-preservation'. This retreat into the ego—into the building of histories, the gathering of *phusis*, and the construction of rationalizations—is the self's frantic attempt to bandage the wound opened by the Other. We return to the Echo of Narcissus because standing perpetually at the edge of asymmetrical responsibility is unbearable. We enshroud ourselves in smoke and mirrors to survive the crushing weight of the Face. But this retreat is never clean. It is forever haunted, forever ruined by the memory of the precipice. Because the encounter happened, the Trace remains.

We end, then, not with a unified theory of forces, but with a wounded ego and an asymmetrical relation of responsibility. The "Trace" † of the—what shall we say?—particle... wave... neither... defying space and time and the 'is' of histories and narratives, refusing and playing with our intentions. But what of the Trace from the *il y a*, the *there is*, as already responsible before the 'me' of assimilation and history? This trace is not a loss of self to the abyss, but the haunting realization of a self already held accountable to the Other, even as it tries to hide. It is not a romantic symmetry or a cosmic equation to be balanced, but an infinite debt we can never repay, commanded by the face of the widow, the orphan, and the stranger.

And that, finally, is love.

---

* Emmanuel Levinas, *Time and the Other*, trans. Richard A. Cohen (Pittsburgh: Duquesne University Press, 1987), 74.

† Emmanuel Levinas, "The Trace of the Other," trans. Alphonso Lingis, in *Deconstruction in Context: Literature and Philosophy*, ed. Mark C. Taylor (Chicago: University of Chicago Press, 1986), 345-359.

END COMMENTS

There are over 2,000 references in the three volumes of this series. Most of the references are not referencing the quotes I used in the book. I wanted to give the reader extra reading and research material on specific points I was discussing. I think it is important for the reader to have additional material to substantiate and expand on specific topics I discussed. I think of this as an important service to the reader. Most books strictly reference quotes and may provide additional reading material at the end (which I also have done). By providing many more references than needed, I also wanted to reassure myself and demonstrate that I was not too far off the beaten path.

In terms of physics, I have included theoretical material in these volumes. Part of the reason I did this was because many of the findings in quantum physics are novel. This is because in many cases quantum physics has not been around long enough for theory to find common acceptance. With all the capital being spent on quantum intelligence, findings in quantum physics are almost daily coming to the surface of its vast uncharted, I think un-chartable, terrain. When I did employ theoretical results, I made every effort to remain as close as I could to experimental evidence. However, in some cases I did explore a bit of future directions as well.

With regard to philosophy, I wanted to at least look for common consensus among scholars. By its nature, philosophy is more prone to interpretation including my own philosophical musings. I do not claim to be correct in my conclusions as there are always legitimate arguments which counter virtually anyone's opinion. However, I do not think of various interpretations as a weakness but a strength. Physics as well makes interpretations. I illustrated this clearly in the section entitled, "Interpretations of the Quantum Wave Function". Both science and philosophy offer room for interpretation. Both fields commingle in some incredible and beautiful ways. The depth and directions of interpretations were a major effort I wanted to illustrate in these volumes.

As far as specific references are concerned, I found ChatGPT and Gemini to be a reliable tool for digging up relevant references to my current discussion. In the past it would have taken me days and weeks to dig out relevant references. However, when I asked ChatGPT or Gemini for references I did not just rely on the references it gave me. If I had access to the reference material, I would verify that the topic pertained to the point I was making. If I did not have access, as in subscription journals, I would read the abstract to ensure that it pertained to the topic I was discussing. Otherwise, I did not include the reference. I also found AI to be useful finding grammatical errors and also suggesting other ways to modify my writing that seemed clearer to me than my previous versions. I found the interaction with AI was useful but AI was also prone to introducing meanings that I did not intend. So that limited my ability to use AI in this fashion. I did not rely on AI too much for the main body of text except in the section entitled, "Entangling Observations About ChatGPT 4 and Us". In that section I wanted to illustrate that ChatGPT was not always correct and even contradicted itself at times. However, we need to understand that this is not the fault of ChatGPT or Gemini. This is more precisely a virtue of intelligence which exemplifies its inability to complete itself just as we are never able to complete ourselves. If we could be complete in ourselves there would be no other. Without the other there would not even be a possibility of a 'me'. And what would knowledge *be* without the incompleteness of a 'me' in the face of an other?

Even so, intelligence is not dogma but is capable of simultaneously holding multiple ideas, even contradictory ideas, simultaneously. Think of complementarity and the wavicle. Quantum physics is a beautiful illustration of how classicism to some degree lost its way in simply assuming that non-contradiction was universal truth. This critique of dominant strands in classicism was a major tenet I tried to illustrate in these volumes. Classicism was not so much wrong in assuming its dominant classical schemes as it erred in imagining there was no excess to them, e.g. rationalism, positivism, etc.. While quantum physics explores many avenues of how complementarity arises, complementarity is not reductively about binary dualities. This would simply reiterate a dominant theme in classicism. Quantum physics can hold, at once, contradictory dualities and multiplicities but always as excess. Excess cannot be accounted for by such simplistic determinations as binary dualisms and contradictions. Quantum physics exceeds the asymptotic determinations of classical physics without lapsing into some sort of mystic proclamation arising out of language and history.

While I have shown there is ample room for intuitive and historic interpretations of quantum physics, quantum physics exceeds the classical notions of real and ideal, self and other, body and mind, science as truth and dogma as opinion. To illustrate this, I need only mention the spooky observer effect which is not subjective or objective – exceeding both. I have tried to bring out interesting and frankly startling intersections with Levinas and quantum physics.

What is more, Levinas brings us an infinite distance further than quantum physics can, placing us at a precipice where responsibility and ethics to the other are at the zenith of human endeavors – to the other who faces me, radically exterior to all my reductive, narcissistic determinative echoes. Quantum physics is the place where all classical determinations must give way to incredible and wonderful excess, I have termed this excess 'chaos'. Infinitely more than chaos, Levinas finds the self, me, in retreat from radical exteriority, spun and held responsible to, radical externality. For me, the self, the other is precipice, "a past not my past",

> "The past of the other, a past that is not my past, but which affects me as though it were my own, is the very essence of responsibility." -Emmanuel Levinas, Otherwise Than Being, or Beyond Essence, trans. Alphonso Lingis (The Hague: Martinus Nijhoff, 1981), 115–118.

APPENDICES

APPENDIX A (ALPHA) – QUANTUM CONSCIOUSNESS REFERENCES

Appendix A (alpha) – Quantum Consciousness References

Peer Reviewed Publications - Orch OR as cited at https://hameroff.arizona.edu/research-overview/orch-or

Aarat P. Kalra, Alfy Benny, Sophie M. Travis, Eric A. Zizzi, Austin Morales-Sanchez, Daniel G. Oblinsky, Travis J. A. Craddock, Stuart R. Hameroff, M. Bruce MacIver, Jack A. Tuszynski, Sabine Petry, Roger Penrose, Gregory D. Scholes. Electronic Energy Migration in Microtubules, Aug 22, 2022. arXiv:2208.10628 https://arxiv.org/abs/2208.10628 (.pdf)

Hameroff, S. Consciousness, Cognition and the Neuronal Cytoskeleton - A New Paradigm Needed in Neuroscience. Hameroff S. Front Mol Neurosci. 2022 Jun 16;15:869935. https://pubmed.ncbi.nlm.nih.gov/35782391/

https://doi.org/10.3389/fnmol.2022.869935

Hameroff, S 'Orch OR' is the most complete, and most easily falsifiable theory of consciousness. Hameroff, Stuart. 2020. Cognitive Neuroscience, Published online: 24 Nov 2020. https://doi.org/10.1080/17588928.2020.1839037

Kalra, A. P., Hameroff, S., Tuszynski, J., Dogariu, A., Nicolas, Sachin, & Gross, P. J. 2022, August 14. Anesthetic gas effects on quantum vibrations in microtubules – Testing the Orch OR theory of consciousness. https://osf.io/zqnjd/

Kalra, Aarat P, Hameroff, S, Tuszynski, J, Dogariu, A. "Anesthetic Gas Effects on Quantum Vibrations in Microtubules – Testing the Orch OR Theory of Consciousness." 2020, OSF August 21. https://osf.io/zqnjd/

Editorial Views. "Anesthetic action and 'quantum consciousness: A match made in olive oil." Hameroff, SR., 2018. Anesthesiology, 8(129):228-231 https://anesthesiology.pubs.asahq.org/article.aspx?articleid=2682839

Hameroff, S., Newswise, September 5, 2017, Consciousness Depends on Tubulin Vibrations Inside Neurons, Anesthesia Study Suggests. Anesthetic alterations of collective terahertz oscillations in tubules correlate with clinical potency: Implications for anesthetic action and post-operative cognitive dysfunction. See: Craddock TJA, Kurian P, Preto J, Sahu K, Hameroff SR, Klobukowski M, Tuszynski JA. In Scientific Reports https://www.ncbi.nlm.nih.gov/pubmed/28852014 Craddock, TJA, Hameroff SR, AT Ayoub AT, Klobukowski M,Tuszynski JA, Anesthetics Act in Quantum in Brain Microtubules to Prevent Consciousness, Current Topics in Medicinal Chemistry, 2015 15 (6), 523- 33.

https://www.ncbi.nlm.nih.gov/pubmed/25714379 https://www.newswise.com/articles/consciousness-depends-on-tubulin-vibrations-inside-neurons-anesthesia-study-suggests

Craddock TJA, Kurian P, Preto J, Sahu K, Hameroff SR, Klobukowski M, Tuszynski JA. Scientific Report, 2017, August. Anesthetic Alterations of Collective Terahertz Oscillations in Tubulin Correlate with Clinical Potency: Implications for Anesthetic Action and Post-Operative Cognitive Dysfunction. https://www.ncbi.nlm.nih.gov/pubmed/28852014

Hameroff, Stuart R., Penrose, Roger, Chapter 14, eds. Roman R. Poznanski, Jack A. Tuszynski, Todd E. Feinberg, World Scientific, 2016. Consciousness in the Universe: An Updated Review of the "Orch OR" Theory, in Biophysics of Consciousness: A Foundational Approach. https://www.worldscientific.com/doi/abs/10.1142/9789814644266_0014

Hameroff, S. "Change the Music: Psychotherapy and Brain Vibrations." The Neuropsychotherapist 2016. Vol 4(4). 31-35. https://www.thescienceofpsychotherapy.net/courses/v4i4

Craddock, Travis J.A., Stuart R. Hameroff, Ahmed T. Ayoub, Mariusz Klobukowski, & Jack A. Tuszynski. "Anesthetics act in quantum channels in brain microtubules to prevent consciousness." Current Topics in Medicinal Chemistry 2015 Vol 15:6, 523-533 https://www.ncbi.nlm.nih.gov/pubmed/25714379

Hameroff, S., Commentary on Stuart Kauffman's Quantum Criticality at the Origins of Life in John Hewitt's Quantum Criticality in Life's Proteins (update), Phys.org, Apr 15, 2015. https://phys.org/news/2015-04-quantum-criticality-life-proteins.html

Hameroff, S., Anesthesia Points to Deeper Level 'Quantum Channels' Newswise, March 23, 2015 https://newswise.com/articles/view/631215/

Craddock, Travis John Adrian, Douglas Friesen, Jonathan Mane, Stuart Hameroff, & Jack A. Tuszynski. "The Feasibility of Coherent Energy Transfer in Microtubules." Journal of the Royal Society Interface, 2014 11(100). https://royalsocietypublishing.org/doi/full/10.1098/rsif.2014.0677

Hameroff, SR., Craddock TJ, & Tuszynski JA. Quantum effects in the understanding of consciousness. 2014. J Integr Neurosci. 13 June (2):229-52. https://www.ncbi.nlm.nih.gov/pubmed/25012711

Hameroff, S., Comment on L Turin et al, Proc. Nat. Acad. Sci., "Electron spin change during general anesthesia." Aug. 11, 2014 https://www.pnas.org/content/early/2014/08/06/1404387111

Response Hameroff to L. Turin et al., 2014. Re Hadlington, Simon. "Knock-out theory puts new spin on general anaesthesia." Chemistry World Review. 12 August, https://www.chemistryworld.com/news/knock-out-theory-puts-new-spin-on-general-anaesthesia/7643.article

Ref: Citation in Luca Turin et al. 2014 "Electron spin changes during general anesthesia in Drosophilia. Proc. Nat. Acad. Sci.,10.1073/ PNAS 1404387111. 26 April, https://www.pnas.org/content/111/34/E3524

Ref: Hadlington, Simon. "Knock-out theory puts new spin on general anaesthesia." Chemistry World Review, 2014, August 11. https://www.chemistryworld.com/news/knock-out-theory-puts-new-spin-on-general-anaesthesia/7643.article

Ref: citation in Luca Turin et al. 2014. "Electron spin changes during general anesthesia in Drosophilia. Proc. Nat. Acad. Sci., 2014, 10.1073/ pnas. 26 April, Luca Turin, Efthimios M. C. Skoulakis, Andrew P. Horsfield https://www.pnas.org/doi/10.1073/pnas.1404387111

Ref: Hameroff, SR. 2006. "The entwined mysteries of anesthesia and consciousness: Is there a common underlying mechanism?" Anesthesiology 105(2):400–412. https://anesthesiology.pubs.asahq.org/Article.aspx?articleid=1931238

Hameroff, S., & Roger Penrose "Consciousness in the universe: A review of the 'Orch OR' theory." Physics of Life Reviews, 2014 March 11(1):39-78. https://www.ncbi.nlm.nih.gov/pubmed/24070914

A Review by Elsevier, PhysOrg. Discovery of quantum vibrations in 'microtubules' corroborates theory of consciousness. 16 January 2014. https://phys.org/news/2014-01-discovery-quantum-vibrations-microtubules-corroborates.html

Reply to Seven Commentaries on "Consciousness in the Universe: Review of the 'Orch OR' theory" Hameroff, S., & Penrose R. 2014 Physics of Life Reviews, 11:94–100. https://www.sciencedirect.com/science/article/pii/S1571064513001905

Hameroff, S., & Penrose, R., Reply to Criticism of the 'Orch OR qubit' – Orchestrated objective reduction is scientifically justified." Physics of Life Reviews, 2014 11(1):104-112. https://www.sciencedirect.com/science/article/pii/S1571064513001917

Hameroff, Stuart, Consciousness, Microtubules and "Orch-OR": A 'Space-time' Odyssey, Journal of Consciousness Studies, Imprint Academic. 2014 Vol 21, 3-4,126-153. https://www.ingentaconnect.com/content/imp/jcs/2014/00000021/f0020003/art00008

Hameroff, S., Quantum walks in brain microtubules-a biomolecular basis for quantum cognition? Top Cogn Sci., 2014 January; 6(1):91-7 https://www.ncbi.nlm.nih.gov/pubmed/24259348

Discovery of quantum vibrations in 'microtubules' corroborates theory of consciousness, Phys Org, 2014. 16 January. https://phys.org/news/2014-01-discovery-quantum-vibrations-microtubules-corroborates.html

Hameroff, S.R., Craddock T.J., & Tuszynski J.A. Quantum effects in the understanding of consciousness. J Integr Neurosci. 13 June 2014 (2):229-52. https://www.ncbi.nlm.nih.gov/pubmed/25012711

Hameroff, SR. Quantum mathematical cognition requires quantum brain biology: the "Orch OR" theory. Behav Brain Sci, 2013. June; 36(3):287-90. https://www.ncbi.nlm.nih.gov/pubmed/23673035

Comment by Hameroff, S. on: A Tale of Two Fields: "Dissipation of 'dark energy' by cortex in knowledge retrieval" by Capolupo, Freeman and Vitiello. Phys Life Rev. 2013 March; 10(1):95-6; discussion 112-6. https://www.ncbi.nlm.nih.gov\pubmed\23375127

Hameroff, S Trakas M, Duffield C, Annabi E, Gerace MB, Boyle P, Lucas A, Amos Q, Buadu A, Badal JJ, Transcranial ultrasound (TUS) effects on mental states: a pilot study, Brain Stimul, 2013; May;6(3):409-15. https://www.ncbi.nlm.nih.gov/pubmed/22664271

Hameroff, S., How quantum brain biology can rescue conscious free will. Front Integr Neurosci, 2012; 6:93, Oct 12. https://www.ncbi.nlm.nih.gov/pubmed/23091452

Craddock TJ, St. George Marc, Freedman Holly, Barakat Khaled, Damaraju Sambasivarao, Hameroff Stuart, Tuszynski Jack A. Computational Predictions of Volatile Anesthetic Interactions with the Microtubule Cytoskeleton: Implications for Side Effects of General Anesthesia. , PLoS One, 2012; June 25. https://journals.plos.org/plosone/article?id=10.1371/journal.pone.0037251

Hameroff S., Quantum brain biology complements neuronal assembly approaches to consciousness: Comment on "Consciousness, biology and quantum hypotheses" by Baars and Edelman. Phys Life Rev. 2012 Sep;9(3):303-5; discussion 306-7. https://www.ncbi.nlm.nih.gov/pubmed/22795934

Craddock T, Tuszynski J, & Hameroff S. Cytoskeletal Signaling: Is Memory Encoded in Microtubule Lattices by CaMKII Phosphorylation? PLoS Comput Biology, 2012; March 8. https://journals.plos.org/ploscompbiol/article?id=10.1371/journal.pcbi.1002421

Craddock TJ, Tuszynski JA, Chopra D, Casey N, Goldstein LE, Hameroff SR, Tanzi RE, The Zinc Dyshomeostasis Hypothesis of Alzheimer's Disease, PLoS One, 2012; Mar 23 7(3). https://journals.plos.org/plosone/article?id=10.1371/journal.pone.0033552

Penrose, R., & Hameroff, S.R. "Consciousness in the Universe: Neuroscience, Quantum Space-Time Geometry and Orch OR Theory." Journal of Cosmology, 2011, Vol 14. http://www.neurohumanitiestudies.eu/archivio/penrose_consciousness.pdf

APPENDIX B (BETA) - FRAGMENTS OF HERACLITUS

Appendix B (beta) - Fragments of Heraclitus

> "Do not be in too great a hurry to get to the end of Heraclitus the Ephesian's book: the path is hard to travel. Gloom is there and darkness devoid of light. But if an initiate be your guide, the path shines brighter than sunlight." (Diogenes Laertius IX, 16) *

The extant fragments of Heraclitus (500 BC) have been interpreted in a number of ways throughout history from Plato and Aristotle through later scholarship. Heraclitus is thought to have produced a single book which has never been found. The fragments of Heraclitus which we do have are found in other ancient sources which recite passages of what Heraclitus wrote. Some are older and thought to be more closely aligned with what Heraclitus actually wrote. Ancient reports of the book range from various sayings, epigrams, and aphorisms to having been divided to three sections of cosmology, politics (ethics), and theology. Heraclitus expounds on deep interconnections between science, human affairs, and theology. Note that theology had little to do with modern religions including Christianity but more to do with polytheistic and animistic worship. There are many later interpretations and commentators of Heraclitus partly because we only have around 129 reported fragments of what he wrote. Heraclitus writes his prose in a poetic and rhythmic way. He plays with language to make it sound musical by alliterating the Greek letter 'π' in many well-placed words. †

Heraclitus thought the first cause was fire. His fragments discuss opposite, unities, flux, strife, and change. He reportedly writes,

panta rhei kai ouden menei
everything flows and nothing abides
all (life is) flux (panta rhei) ‡

Or in other fragments things are *onta* of present participle of being (to be),

In Cratylus 401d we find,
τὰ ὄντα ἰέναι τε πάντα καὶ μένειν οὐδέν
Ta onta ienai te panta kai menein ouden
"All things move and nothing remains still"
The Loeb translation (H.N.Fowler) is "all things move and nothing remains still, and he

* See Hegel Reading Heraclitus
† See Heraclitus Pesseuon Fragment 52
‡ See Heraclitus (Information Philosopher)

likens the universe to the current of a river, saying that you cannot step twice into the same river." *

His most famous fragment tells us,

ποταμοῖσι τοῖσιν αὐτοῖσιν ἐμβαίνουσιν, ἕτερα καὶ ἕτερα ὕδατα ἐπιρρεῖ.

Potamoisi toisin autoisin embainousin, hetera kai hetera hudata epirrei

"On those stepping into rivers (staying the same), other and other waters flow on." †

At times later commentators of Heraclitus seem to understand his writing of oppositions as co-constituent and co-arising as unities. Other times commentators seem to understand his passages as *apparent* unities residing in strife, discord, and enmity. For some he can sound like a mystic and for others as nonsensical. In various ancient references to Heraclitus' fragments we read,

What opposes unites, and the finest attunement stems from things bearing in opposite directions, and all things come about by strife. ‡

Graspings: things whole and not whole, what is drawn together and what is drawn asunder, the harmonious and the discordant. The one is made up of all things, and all things issue from the one. §

And it is the same thing in us that is quick and dead, awake and asleep, young and old; the former are shifted and become the latter, and the latter in turn are shifted and become the former. #

Although critical of Homer and Hesiod he appeared to have been influenced to some extent by them. Heraclitus stands at the gap where the more ancient answers no longer sufficed. He bemoans information-gathering as the inability to be able to understand. In the fragment below the "learning of many things" is *polumathiê* in Greek or information-gathering.

The learning of many things does not teach understanding; otherwise, it would have taught Hesiod and Pythagoras, and again Xenophanes and Hecataeus. ○

What does he criticize Hesiod for?

Διδάσκαλος δὲ πλείστων Ἡσίοδος· Τοῦτον ἐπίστανται πλεῖστα εἰδέναι, ὅστις ἡμέρην καὶ εὐφρόνην οὐκ ἐγίνωσκεν· ἔστι γὰρ ἕν.

The teacher of most people is Hesiod. They know he knows the most things, a man who

---

* See Heraclitus (Information Philosopher)

† See Heraclitus (Information Philosopher)

‡ See http://heraclitusfragments.com/B8/index.html

§ See http://heraclitusfragments.com/B10/index.html

See http://heraclitusfragments.com/B88/index.html

○ See http://heraclitusfragments.com/B40/index.html

> did not know day and night, i.e., that they are one thing. (Heraclitus, *Fragments*, 22B57 (Diels-Kranz).)

Hesiod draws his cosmology from Greek myth poetically. Day and night are independent from each other. They are static and have no dependent relationship to each other. *

Hesiod treats day and night as gods which pass each other on when they come to earth from their houses. Night is much older than day. Hesiod tells us,

> From Chaos came forth Erebus and black Night; in turn from Night came forth both Day and Aether †

So, how does Heraclitus find knowledge different from Hesiod's myth of day and night? He starts by facing what troubles him. What are the inadequacies which find no words in "Hesiod and Pythagoras, and again Xenophanes and Hecataeus". Is Heraclitus bemoaning chaos in Hesiod or the myths which separate day and night?

Heraclitus sees Hesiod's problem in the separation of opposites. Heraclitus tells us,

> And it is the same thing in us that is quick and dead, awake and asleep, young and old; the former are shifted and become the latter, and the latter in turn are shifted and become the former. ‡

Heraclitus has no tolerance for religious ignorance. He does not have any special reverence for gods. He tells us, "*Immortals become mortals, mortals become immortals; they live in each other's death and die in each other's life.*"

Mortals are immortals and immortals are mortals, the one living the others' death and dying the others' life. R. P. 46. §

Writing of religious rites of purification, he tells us, "They vainly purify themselves by defiling themselves with blood, just as if one who had stepped into the mud were to wash his feet in mud. Any man who marked him doing thus, would deem him mad." And "And they pray to these images, as if one were to talk with a man's house, knowing not what gods or heroes are."

> (129, 130) They vainly purify themselves by defiling themselves with blood, just as if one who had stepped into the mud were to wash his feet in mud. Any man who marked him doing thus, would deem him mad. R. P. 49 a.
>
> (126) And they pray to these images, as if one were to talk with a man's house, knowing not what gods or heroes are. R. P. 49 a. #

* See https://philarchive.org/archive/LESVFK-2
† See Heraclitus: Fragments
‡ See B88, http://heraclitusfragments.com/files/ge.html
§ See 67 and 62, https://en.wikisource.org/wiki/Fragments_of_Heraclitus
# See 129, 126, https://en.wikisource.org/wiki/Fragments_of_Heraclitus

Heraclitus indicates, “War is the father of all and the king of all; and some he has made gods and some men, some bond and some free.”

> Fragment 53
>
> (44) War is the father of all and the king of all; and some he has made gods and some men, some bond and some free. R. P. 34. *
>
> Nietzsche tells us that [Heraclitus] denied the duality of totally diverse worlds. [.. .] He no longer distinguished a physical world from a meta-physical one.[. . .] And after this first step, nothing could hold him back from a second, far bolder negation: the—altogether denied being. [...] Heraclitus proclaimed: "I see nothing other than becoming." †
>
> and Heraclitus will remain eternally right with his assertion that being is an empty fiction. The "apparent" world is the only one: the "true" world is only added by a lie. ‡

I would suggest that, for Heraclitus, ‘metaphysics’ did not yet exist. He peered from the chaos which preceded him. For Heraclitus the ‘one’ is not privileged over flux. Becoming is not a transformation. It *is* flux itself not the result of flux. He writes us “*panta rhei*” or “all [*is*] flux”. Nietzsche in “Twilight of the Idols” writes,

> With the highest respect, I accept the name of Heraclitus. When the rest of the philosophic folk rejected the testimony of the senses because they showed multiplicity and change, he rejected their testimony because they showed things as if they had permanence and unity. Heraclitus too did the senses an injustice. They lie neither in the way the Eleatics believed, nor as he believed--they do not lie at all. What we make of their testimony, that alone introduces lies; for example, the lie of unity, the lie of thinghood, of substance, of permanence. "Reason" is the cause of our falsification of the testimony of the senses. Insofar as the senses show becoming, passing away, and change, they do not lie. But Heraclitus will remain eternally right with his assertion that being is an empty fiction. The "apparent" world is the only one: the "true" world is merely added by a lie. §

All Heraclitus had was the apparent world of chaos, strife, war, death. He summed this up as flux. Heraclitus thinks all is fire. Another fragment indicates, “*That which always was, and is, and will be everlasting fire, the same for all, the cosmos, made neither by god nor man, replenishes in measure as it burns away.*” All is flux. Flux is change without an end, a *telos*. Flux finds no end in becoming. Becoming is not an end itself or a step towards another end. Flux is the end. Strife is both war and peace. Detente is the cooling of war. Progress is an illusion of fire.

> (20) This world, which is the same for all, no one of gods or men has made; but it was ever, is

---

* See 44, https://en.wikisource.org/wiki/Fragments_of_Heraclitus

† See Nietzsche, Philosophy in the Tragic Age of the Greeks, 4. Becoming and Chaos, or Differance and Chaosmos (.pdf)

‡ See Nietzsche, Philosophy in the Tragic Age of the Greeks, 4. Becoming and Chaos, or Differance and Chaosmos (.pdf)

§ See Twilight of the Idols (Chap. 2)

now, and ever shall be an ever-living Fire, with measures of it kindling, and measures going out. *

When flux is thought this way, it cannot rise above into rationality and Being. Contradiction is a result of rationality. Heraclitus is not yet 'rational'. Concept can never arise from Heraclitus. Fire cannot become order. It is mistaken as gods which order the cosmos. Order is a resonance of fire. Heraclitus likens it as a musical note.

Men do not know how what is at variance agrees with itself. It is an attunement of opposite tensions, like that of the bow and the lyre,

> (45) Men do not know how what is at variance agrees with itself. It is an attunement of opposite tensions,[19] like that of the bow and the lyre. R. P. 34.E †

Flux for Heraclitus has more to do with chaos than it has to do with order. This shows how seriously Heraclitus takes Hesiod's chaos.

It seems to me that commentators through history have fallen into two camps:

1. strife and opposition
   a. definitive of things, unity (the one)
   b. continuous change culminating contingently in continuity
2. flux and chaos
   a. contradictory
   b. random change and erratic indeterminacy

The strife and opposition camp tend to give way to our sense of continuity. The continuous theme has a sense of temporality which may be thought in terms of:

1. A causal, oppositional past to an effect as momentary present or presence which holds together (e.g. things, the one)
2. A causal, oppositional past and future which effects changes and transforms through moments of resonance.

Flux and change have no necessary effect, no final destination such as unity. This is chaos. Chaos is better shown through Heraclitus' famous 'river fragments'. William Barnes (1801-1886) was a well-known ancient Greek scholar. Barnes tells us Plato reads the famous fragment of Heraclitus as,

> Heraclitus, I believe, says that all things pass and nothing stays, and comparing existing things to the flow of a river, he says you could not step twice into the same river. (Plato Cratylus 402a = A6) ‡

---

* See 30, https://en.wikisource.org/wiki/Fragments_of_Heraclitus

† See Fragments of Heraclitus: Fragment 51

‡ See How Did Heraclitus View Of Philosophy

Barnes looks at three other interpretations of Heraclitus from three different fragments we have of Heraclitus' 'river fragments'.

B12. potamoisi toisin autoisin embainousin hetera kai hetera hudata epirrei.
On those stepping into rivers staying the same other and other waters flow. (Cleanthes from Arius Didymus from Eusebius)
B49a. potamois tois autois ...
Into the same rivers we step and do not step, we are and are not. (Heraclitus Homericus)
B91[a]. potamôi ... tôi autôi ...
It is not possible to step twice into the same river according to Heraclitus, or to come into contact twice with a mortal being in the same state. (Plutarch) *

The B12 fragment is believed by scholars to be the fragment which is closest to Heraclitus' original fragment. Barnes tells us that Heraclitus wants to maintain both the idea of chaos and the idea of unity of opposites which co-determine what something is. In the case of the river, the waters are the same, as in the noun water - what it *is*, and other, as in ever changing or flowing water - as in the verb watering. Water can be static or dynamic. If I step into a river I will never step into the same river as the water is always renewing itself. In other words, something remains what it is by contradicting itself or opposing itself. Something arises out of its destruction or strife. Therefore, a school of thought concerning Heraclitus' *logos* is the determination of strife - the determination of what *is* by what it is not. From a classical perspective,

Chaos has no moment of recognition. Chaos cannot arise to self-reflection, consciousness, a moment of *stasis* (Greek: standing, existence, stability), a unified present. All are smoking mirrors, dreams, phantasms.

Chaos is erratic and random, an appearance with no cohesion.

The continuous theme has a purely contingent account of temporality as a movement in and out of presence embodying its past and moving toward a future coherently. The term coherent here means as information thought as a continual unfolding of nature, of Being, as the truth of existence.

All of these analyses rely on a historic world view which by necessity can only allow a progressive continuity which completes its totalization with negation. This is modeled after the law of noncontradiction. It cannot concede an excess to its absolute denial of externality. From our epoch of Being and truth Heraclitus is like a piece in a puzzle that does not fit but must be twisted and deformed in order to fit or simply tossed from the puzzle. Let's remember that the puzzle of classicism and its progeny modernity did not exist for Heraclitus. Is it possible that he may have peered at daylight which we cannot yet see? †

Plato and Aristotle thought that logic as the principle of noncontradiction did not matter much to Heraclitus. Heraclitus did not seem to think much about Homer and Hesiod. He called them fools. Neither did he think much about his contemporaries, including Pythagoras. Nevertheless, he inherited many of their ideas.

---

* See Heraclitus (Stanford)

† See Heraclitus (Information Philosopher)

Heraclitus made every effort to break out of the mold of contemporary thought. Although he was influenced in a number of ways by the thought and language of his predecessors, including the epic poets Homer and Hesiod, the poet and philosopher Xenophanes, the historian and antiquarian Hecataeus, the religious guru Pythagoras, the sage Bias of Priene, the poet Archilochus, and the Milesian philosophers, he criticized most of them either explicitly or implicitly, and struck out on his own path. He rejected polumathiê or information-gathering on the grounds that it "does not teach understanding". He treated the epic poets as fools and called Pythagoras a fraud. *

More References:
Philosophical Inquiry
Volume 25, Issue 3/4, Summer 2003
Stavros Kouloumentas
Pages 241-259

See Heraclitus Pesseuon: Fragment 52 https://www.pdcnet.org/philinquiry/content/philinquiry_2003_0025_40606_0241_0259

See THE FRAGMENTS OF HERACLITUS https://heraclitusfragments.com/

See HERACLITUS'S PHILOSOPHY AND HEGEL'S DIALECTIC https://www.jstor.org/stable/26212409

See Unity in Difference – Difference in Unity: Heraclitus and the Truth of Hermeneutics https://www.diva-portal.org/smash/get/diva2:218650/FULLTEXT01.pdf

See Hegel's Lectures on the History of Philosophy - D. Heraclitus https://www.marxists.org/reference/archive/hegel/works/hp/hpheraclitus.htm

See Heraclitus (Stanford) https://plato.stanford.edu/entries/heraclitus/

See https://en.wikiquote.org/wiki/Heraclitus

See https://plato.stanford.edu/entries/heraclitus/

as independent see: Kirk 1954, 156; similarly, KRS 1983, 189 and Guthrie 1962, 484

Kirk, G. 1954. Heraclitus: The Cosmic Fragments. Cambridge: Cambridge University Press.

Kirk, G., J. Raven, and M. Schofield. 1983. The Presocratic Philosophers. 2nd edn. Cambridge: Cambridge University Press (cited as KRS).

Guthrie, W. 1962. A History of Greek Philosophy. vol. 1. Cambridge: Cambridge University Press.

as separate see Kahn 1979, 11

Kahn, C. 1979. The Art and Thought of Heraclitus. Cambridge: Cambridge University Press, 1979).

Nietzsche's 19th century interpretation of Heraclitus' eternally recurring, nihilistic strife of ascending will to power is also anachronistic.

According to Heraclitean strife, the Greek contest, and Nietzsche's will to power, a balancing out of opposing forces is never achieved, otherwise the struggle which fuels existence would die out. The struggle must never be extinguished; opposing forces must continue the battle, each overcoming the other in turn, for all eternity. This is the way in which the eternal recurrence

* See Heraclitus (Stanford)

serves as a prescription for the overman. The will to power, mankind's unrefined animosity and envy, must be acknowledged by the strong individual and transformed from a nihilistic force into one of positive ambition and increase. The eternal recurrence is what the overman strives for within himself; since the rules of eternal becoming, of the contest, do not apply to humanity by nature. Each individual must choose whether to enter into the eternal contest or to extinguish the struggle with his will to power by denying his passions. The eternal recurrence and will to power fit together in that it is the belief in eternal recurrence which gives great individuals the strength to acknowledge the potential of this terrible drive as a source of elevation and increase. *

Force as the human condition of Newton's second law of motion where force overcomes the stagnant inertia of eternal recurrence of the same must be the *logos* of will to power.

Heraclitus is not bemoaning something akin to the 19th century neurosis of existential crisis. as giving us more of an account of chaos as original and enduring. †

In Heraclitus' later commentators on the 'river fragments', if everything constantly changes there is a flow to these changes which is accomplished by a unity of opposites. Rivers arise from unity and difference. The ways of humankind are divined by eternal strife. For these commentators, chaos cannot be maintained in that rivers cannot be thought of as pure chaos. However, logic had not even been devised yet. At least not as we think of it as the law of non-contradiction. Perhaps these commentators have imported an anachronistic device in maintaining that chaos and the unity of opposites is a contradiction. Was this so for Heraclitus? Later Aristotle in his Metaphysics discusses logic in terms of what we think as the principle of noncontradiction.

The Most Fundamental Principle of Knowledge: The Law of Non-Contradiction [LNC]

Which principle this is, let us proceed to say. It is this: the same attribute cannot at the same time belong and not belong to the same subject and in the same respect--we must presuppose, to guard against dialectical objections, any further qualifications which might be added.

This, then, is the most certain of all principles, since it answers to the definition given above. For it is impossible for any one to believe the same thing to be and not to be, as some think Heraclitus says. For what a man says, he does not necessarily believe; and if (1) it is impossible that contrary attributes should belong at the same time to the same subject (the usual qualifications must be presupposed in this premise too), and if (2) an opinion which contradicts another is contrary to it, then obviously (3) it is impossible for the same man at the same time to believe the same thing to be and not to be; for if a man were mistaken on this point he would have contrary opinions at the same time... ‡

Here we clearly see that Aristotle sharply focused logic as adhering to the principle of noncontra-

---

* See Nietzsche and Heraclitus .

† See Verbs for Knowing in Heraclitus' Rebuke of Hesiod (DK 22B57) (.pdf)
And Nietzsche and Heraclitus.

‡ See Aristotle's METAPHYSICS: Book IV (excerpts)

diction. However, Aristotle's argument here is primarily against his contemporaneous school of Parmenides' relativism which could not account for change and motion. *

* See Plato, Parmenides (137c - 142a)

APPENDIX Γ (GAMMA) – COMMENTARIES ON HERACLITUS

Appendix Γ (gamma) – Commentaries on Heraclitus

Here are some more commentators which have given us more insights into Heraclitus.

They say that Euripides gave [Socrates] a copy of Heraclitus' book and asked him what he thought of it. He replied: 'What I understand is splendid; and I think that what I don't understand is so too - but it would take a Delian to get to the bottom of it.' (Diogenes Laertius, Lives of the Philosophers) *

At the beginning of his writings on nature, and pointing in some way at the environment, [Heraclitus] says:

Of this account which holds forever men prove uncomprehending, both before hearing it and when first they have heard it. For although all things come about in accordance with this account, they are like tyros as they try the words and the deeds which I expound as I divide up each thing according to its nature and say how it is. Other men fail to notice what they do when they are awake, just as they forget what they do when asleep.

[B 1]

Having thus explicitly established that everything we do or think depends upon participation in the divine account, he continues and a little later on adds:

For that reason you must follow what is common (i.e. what is universal - for 'common' means 'universal'). But although the account is common, most men live as though they had an understanding of their own. [B 2]

(Sextus Empiricus, Against the Mathematicians VII 132) †

3.

On the subject of the soul, Cleanthes sets out the doctrines of Zeno [the Stoic] in order to compare them to those of the other natural scientists. He says that Zeno, like Heraclitus, holds the soul to be a percipient exhalation. For, wanting to show that souls as they are exhaled always become new, he likened them to rivers, saying:

On those who enter the same rivers, ever different waters flow - and souls are exhaled from the moist things. [B 12]

Now Zeno, like Heraclitus, says that the soul is an exhalation; but he holds that it is percipient...

(Arius Didymus, fragment 39 Diels,

* See Diogenes Laertius, Lives of Eminent Philosophers

† See Sextus Empiricus

quoted by Eusebius, Preparation for the Gospel XV xx 2) *

4.

Heraclitus the Obsure theologizes the natural world as something unclear and to be conjectured about through symbols. He says:

Gods are mortal, humans immortal, living their death, dying their life. [B 62]

And again:

We step and do not step into the same rivers, we are and we are not. [B 49a]

Everything he says about nature is enigmatic and allegorized.

(Heraclitus, Homeric Questions 24.3-5) †

5.

For it is not possible to step twice into the same river, according to Heraclitus, nor to touch mortal substance twice in any condition: by the swiftness and speed of its change, it scatters and collects itself again - or rather, it is not again and later but simultaneously that it comes together and departs, approaches and retires. [B 91]

(Plutarch, On the E at Delphi 392B) ‡

6.

Heraclitus says that the universe is divisible and indivisible, generated and ungenerated, mortal and immortal, Word and Eternity, Father and Son, God and Justice.

Listening not to me but to the account, it is wise to agree that all things are one, [B 50]

says Heraclitus. That everyone is ignorant of this and does not agree he states as follows:

They do not comprehend how, in differing, it agrees with itself - a backward-turning connection, like that of a bow and a lyre. [B 51]

That an account exists always, being the universe and eternal, he says in this way:

Of this account which holds forever men prove uncomprehending, before hearing it ad when first they have heard it. For although all things come about in accordance with this account, they are like tyros as they try the words and the deeds which I expound as I divide up each thing according to its nature and say how it is. [B 1]

That the universe is a child and an eternal king of all things for all eternity he states as follows:

Eternity is a child at play, playing draughts: the kingdom is a child's. [B 52]

That the father of everything that has come about is generated and ungenerated, creature and creator, we hear him saying:

War is the father of all, king of all: some it shows as gods, some as men, some it makes slaves, others free. [B 53]

... That God is unapparent, unseen, unknown to men, he says in these words:

Unapparent connection is better than apparent. [B 54]

---

* See Eusebius, Preparation for the Gospel

† See Heraclitus: Homeric Problems

‡ See THE E AT DELPHI (DE E APUD DELPHOS)

He praises and admires the unknown and unseen part of his power above the known part. That he is visible to men and not undiscoverable he says in the following words:

I honour more those things which are learned by sight and hearing, [B 55]

he says - i.e. the visible more than the invisible. <The same> is learned from such words of his as these:

Men have been deceived, he says, as to their knowledge of what is apparent in the same way that Homer was - and he was the wisest of all the Greeks. For some children who were killing lice deceived him by saying: "What we saw and caught we leave behind, what we neither saw nor caught we take with us.' [B 56]

... Heraclitus says that dark and light, bad and good, are not different but one and the same. For example, he reproaches Hesiod for not knowing day and night - for day and night, he says, are one, expressing it thus:

A teacher of most is Hesiod: they are sure he knows most who did not recognize day and night - for they are one. [B 57]

... He says that the polluted and the pure are one and the same, and that the drinkable and the undrinkable are one and the same:

The sea, he says, is most pure and most polluted water: for fish, drinkable and life-preserving; for men, undrinkable and death-dealing. [B 61]

And he explicitly says that the immortal is mortal and the mortal immortal in the following words:

Immortals are mortal, mortals immortals: living their death, dying their life. [B 62]

He also speaks of a resurrection of this visible flesh in which we are born, and he is aware that god is the cause of this resurrection - he says:

There they are said to rise up and to become wakeful guardians of the living and the dead. [B 63]

And he says that a judgement of the world and of everything in it comes about through fire; for

Fire will come and judge and convict all things. [B 66]

He says that this fire is intelligent and the cause of the management of the universe, expressing it thus:

The thunderbolt steers all things [B 64]

(i.e. directs everything) - by 'the thunderbolt' he means the eternal fire, and he calls it need and satiety.[B 65] (The establishment of the world according to him being need and the conflagration satiety).

In the following passage he has set down all his own thought - and at the same time that of the sect of Noetus, whom I have briefly shown to be a disciple not of Christ but of Heraclitus. For he says that the created universe is itself the maker and creator of itself:

God is day and night, winter and summer, war and peace, satiety and famine; but he changes like olive oil which, when it is mixed with perfumes, gets its name from the scent of each. [B 67]

(Hippolytus, Refutation of All Heresies IX ix 1-x 9) *

---

* See Refutation of All Heresies

7.

Of those whose accounts I have heard, no-one has come so far as to recognize that the wise is set apart from all things. [B 108]

It is better to hide folly than to make it public. [B 109]

It is not good for me to get all they want. [B 110]

Sickness makes health sweet and good, hunger plenty, weariness rest. [B 111]

To be temperate is the greatest excellence. And wisdom is speaking the truth and acting with knowledge in accordance with nature. [B 112]

Thinking is common to all. [B 113]

Speaking with sense one should rely on what is common to all, as a city on its law and with yet greater reliance. For all human laws are nourished by the one divine; for it is as powerful as it wishes, and it suffices for all, and it prevails. [B 114]

Soul has a self-increasing account. [B 115]

All men can know themselves and be temperate. [B 116]

A man when he is drunk is led by a boy, stumbling, not knowing where he goes, his soul moist. [B 117]

A dry soul is wisest and best. [B 118]

(Stobaeus, Anthology III i 174-180, v 6-8) *

8.

Surely nature longs for the opposites and effects her harmony from them... That was also said by Heraclitus the Obscure:

Combinations - wholes and not wholes, concurring differing, concordant discordant, from all things one and from one all things. [B 10]

In this way the structure of the universe - I mean, of the heavens and earth and the whole world - was arranged by harmony through the blending of the most opposite principles.

([Aristotle], *On the World* 396b7-8, 20-25) †

[137c] "I am ready, Parmenides, to do that," said Aristoteles, "for I am the youngest, so you mean me. Ask your questions and I will answer."

"Well then," said he, "if the one exists, the one cannot be many, can it?" "No, of course not." "Then there can be no parts of it, nor can it be a whole." "How is that?" "The part surely is part of a whole." "Yes." "And what is the whole? Is not a whole that of which no part is wanting?"

[137d] "Certainly." "Then in both cases the one would consist of parts, being a whole and having parts." "Inevitably." "Then in both cases the one would be many, not one." "True." "Yet it must be not many, but one." "Yes." "Then the one, if it is to be one, will not be a whole and will not have parts." "No."

"And if it has no parts, it can have no beginning, or middle, or end, for those would be parts of it?" "Quite right." "Beginning and end are, however, the limits of everything." "Of

---

* See Ioannis Stobaei Anthologium; Volume 3

† See Heraclitus (Stanford)

course." "Then the one, if it has neither beginning nor end, is unlimited." "Yes, it is unlimited." "And it is without form,

[137e] for it partakes neither of the round nor of the straight." "How so?" "The round, of course, is that of which the extremes are everywhere equally distant from the center." "Yes." "And the straight, again, is that of which the middle is in the nearest line between the two extremes." "It is." "Then the one would have parts and would be many, whether it partook of straight or of round form." "Certainly." "Then it is neither straight nor round, since it has no parts."

[138a] "Right."

"Moreover, being of such a nature, it cannot be anywhere, for it could not be either in anything else or in itself." "How is that?" "If it were in something else, it would be encircled by that in which it would be and would be touched in many places by many parts of it; but that which is one and without parts and does not partake of the circular nature cannot be touched by a circle in many places." "No, it cannot." "But, furthermore, being in itself it would also be surrounding with itself naught other than itself,

[138b] if it were in itself; for nothing can be in anything which does not surround it." "No, it cannot." "Then that which surrounds would be other than that which is surrounded; for a whole cannot be both active and passive in the same action; and thus one would be no longer one, but two." "True." "Then the one is not anywhere, neither in itself nor in something else." "No, it is not."

"This being the case, see whether it can be either at rest or in motion." "Why not?"

[138c] "Because if in motion it would be either moving in place or changing; for those are the only kinds of motion." "Yes." "But the one, if changing to something other than itself, cannot any longer be one." "It cannot." "Then it is not in motion by the method of change." "Apparently not." "But by moving in place?" "Perhaps." "But if the one moved in place, it would either revolve in the same spot or pass from one place to another." "Yes, it must do so." "And that which revolves must rest upon a center and have other parts which turn about the center;

[138d] but what possible way is there for that which has no center and no parts to revolve upon a center?" "There is none." "But does it change its place by coming into one place at one time and another at another, and move in that way?" "Yes, if it moves at all." "Did we not find that it could not be in anything?" "Yes." "And is it not still more impossible for it to come into anything?" "I do not understand why." "If anything comes into anything, it must be not yet in it, while it is still coming in, nor still entirely outside of it, if it is already coming in, must it not?" "It must."

[138e] "Now if anything goes through this process, it can be only that which has parts; for a part of it could be already in the other, and the rest outside; but that which has no parts cannot by any possibility be entirely neither inside nor outside of anything at the same time." "True." "But is it not still more impossible for that which has no parts and is not a whole to come into anything, since it comes in neither in parts nor as a whole?" "Clearly."

[139a] "Then it does not change its place by going anywhere or into anything, nor does it revolve in a circle, nor change." "Apparently not." "Then the one is without any kind of motion." "It is motionless." "Furthermore, we say that it cannot be in anything." "We do."

"Then it is never in the same." "Why is that?" "Because it would then be in that with which the same is identical." "Certainly." "But we saw that it cannot be either in itself or in anything else." "No, it cannot." "Then the one is never in the same."

[139b] "Apparently not." "But that which is never in the same is neither motionless nor at rest." "No, it cannot be so." "The one, then, it appears, is neither in motion nor at rest." "No, apparently not."

"Neither, surely, can it be the same with another or with itself; nor again other than itself or another." "Why not?" "If it were other than itself, it would be other than one and would not be one." "True." "And, surely, if it were the same with another, it would be that other, and would not be itself;

[139c] therefore in this case also it would not be that which it is, namely one, but other than one." "Quite so." "Then it will not be the same as another, nor other than itself." "No." "But it will not be other than another, so long as it is one. For one cannot be other than anything; only other, and nothing else, can be other than another." "Right." "Then it will not be other by reason of being one, will it?" "Certainly not." "And if not for this reason, not by reason of itself; and if not by reason of itself, not itself; but since itself is not other at all,

[139d] it will not be other than anything." "Right." "And yet one will not be the same with itself." "Why not?" "The nature of one is surely not the same as that of the same." "Why?" "Because when a thing becomes the same as anything, it does not thereby become one." "But why not?" "That which becomes the same as many, becomes necessarily many, not one." "True." "But if the one and the same were identical, whenever anything became the same it would always become one, and when it became one, the same." "Certainly." "Then if the one is the same with itself,

[139e] it will not be one with itself; and thus, being one, it will not be one; this, however, is impossible; it is therefore impossible for one to be either the other of other or the same with itself." "Impossible." "Thus the one cannot be either other or the same to itself or another." "No, it cannot." "And again it will not be like or unlike anything, either itself or another." "Why not?" "Because the like is that which is affected in the same way." "Yes." "But we saw that the same was of a nature distinct from that of the one." "Yes, so we did."

[140a] "But if the one were affected in any way apart from being one, it would be so affected as to be more than one, and that is impossible." "Yes." "Then the one cannot possibly be affected in the same way as another or as itself." "Evidently not." "Then it cannot be like another or itself." "No, so it appears." "Nor can the one be so affected as to be other; for in that case it would be so affected as to be more than one." "Yes, it would be more." "But that which is affected in a way other than itself or other,

[140b] would be unlike itself or other, if that which is affected in the same way is like." "Right." "But the one, as it appears, being never affected in a way other than itself or other, is never unlike either itself or other." "Evidently not." "Then the one will be neither like nor unlike either other or itself." "So it seems."

"Since, then, it is of such a nature, it can be neither equal nor unequal to itself or other." "Why not?" "If it is equal, it is of the same measures as that to which it is equal." "Yes." "And if it is greater or less than things

[140c] with which it is commensurate, it will have more measures than the things which

are less and less measures than the things which are greater." "Yes." "And in the case of things with which it is not commensurate, it will have smaller measures than some and greater measures than others." "Of course." "Is it not impossible for that which does not participate in sameness to have either the same measures or anything else the same?" "Impossible." "Then not having the same measures, it cannot be equal either to itself or to anything else." "No, apparently not." "But whether it have more measures or less,

[140d] it will have as many parts as measures and thus one will be no longer one, but will be as many as are its measures." "Right." "But if it were of one measure, it would be equal to the measure; but we have seen that it cannot be equal to anything." "Yes, so we have." "Then it will partake neither of one measure, nor of many, nor of few; nor will it partake at all of the same, nor will it ever, apparently, be equal to itself or to anything else; nor will it be greater or less than itself or another." "Perfectly true."

[140e] "Well, does anyone believe that the one can be older or younger or of the same age?" "Why not?" "Because if it has the same age as itself or as anything else, it will partake of equality and likeness of time, and we said the one had no part in likeness or equality." "Yes, we said that." "And we said also that it does not partake of unlikeness or inequality." "Certainly." "How, then, being of such a nature,

[141a] can it be either younger or older or of the same age as anything?" "In no way." "Then the one cannot be younger or older or of the same age as anything." "No, evidently not." "And can the one exist in time at all, if it is of such a nature? Must it not, if it exists in time, always be growing older than itself?" "It must." "And the older is always older than something younger?" "Certainly." "Then that which grows older than itself grows at the same time younger than itself, if it is to have something than which it grows older." "What do you mean?"

[141b] "This is what I mean: A thing which is different from another does not have to become different from that which is already different, but it must be different from that which is already different, it must have become different from that which has become so, it will have to be different from that which will be so, but from that which is becoming different it cannot have become, nor can it be going to be, nor can it already be different: it must become different, and that is all." "There is no denying that."

[141c] "But surely the notion 'older' is a difference with respect to the younger and to nothing else." "Yes, so it is." "But that which is becoming older than itself must at the same time be becoming younger than itself." "So it appears." "But surely it cannot become either for a longer or for a shorter time than itself; it must become and be and be about to be for an equal time with itself." "That also is inevitable." "Apparently, then, it is inevitable that everything which exists in time and partakes of time

[141d] is of the same age as itself and is also at the same time becoming older and younger than itself." "I see no escape from that." "But the one had nothing to do with such affections." "No, it had not." "It has nothing to do with time, and does not exist in time." "No, that is the result of the argument."

"Well, and do not the words 'was,' 'has become,' and 'was becoming' appear to denote participation in past time?" "Certainly."

[141e] "And 'will be,' 'will become,' and 'will be made to become,' in future time?" "Yes." "And 'is' and 'is becoming' in the present?" "Certainly." "Then if the one has no participation

in time whatsoever, it neither has become nor became nor was in the past, it has neither become nor is it becoming nor is it in the present, and it will neither become nor be made to become nor will it be in the future." "Very true." "Can it then partake of being in any other way than in the past, present, or future?" "It cannot." "Then the one has no share in being at all." "Apparently not." "Then the one is not at all." " Evidently not." " Then it has no being even so as to be one, for if it were one, it would be and would partake of being; but apparently one neither is nor is one, if this argument is to be trusted."

[142a] "That seems to be true." "But can that which does not exist have anything pertaining or belonging to it?" "Of course not." "Then the one has no name, nor is there any description or knowledge or perception or opinion of it." "Evidently not." "And it is neither named nor described nor thought of nor known, nor does any existing thing perceive it." "Apparently not." "Is it possible that all this is true about the one ?" "I do not think so."

"Shall we then return to our hypothesis and see

[142b] if a review of our argument discloses any new point of view?" "By all means." "We say, then, that if the one exists, we must come to an agreement about the consequences, whatever they may be, do we not?" "Yes." "Now consider the first point. If one is, can it be and not partake of being?" "No, it cannot." "Then the being of one will exist, but will not be identical with one; for if it were identical with one, it would not be the being of one, nor would one partake of it,

[142c] but the statement that one is would be equivalent to the statement that one is one but our hypothesis is not if one is one, what will follow, but if one is. Do you agree?" "Certainly." "In the belief that one and being differ in meaning?" "Most assuredly." "Then if we say concisely 'one is,' it is equivalent to saying that one partakes of being?" "Certainly." "Let us again say what will follow if one is and consider whether this hypothesis must not necessarily show that one is of such a nature as to have parts." "How does that come about ?" "In this way:

[142d] If being is predicated of the one which exists and unity is predicated of being which is one, and being and the one are not the same, but belong to the existent one of our hypothesis, must not the existent one be a whole of which the one and being are parts?" "Inevitably." "And shall we call each of these parts merely a part, or must it, in so far as it is a part, be called a part of the whole?" "A part of the whole." "Whatever one, then, exists is a whole and has a part." "Certainly." "Well then, can either of these two parts of existent one—unity and being—abandon the other?

142e] Can unity cease to be a part of being or being to be a part of unity?" "No." "And again each of the parts possesses unity and being, and the smallest of parts is composed of these two parts, and thus by the same argument any part whatsoever has always these two parts; for always unity has being and being has unity;

[143a] and, therefore, since it is always becoming two, it can never be one." "Certainly." "Then it results that the existent one would be infinite in number?" "Apparently."

"Let us make another fresh start." "In what direction?" "We say that the one partakes of being, because it is?" "Yes." "And for that reason the one, because it is, was found to be many." "Yes." "Well then, will the one, which we say partakes of being, if we form a mental conception of it alone by itself, without that of which we say it partakes, be found to be only one, or many?" "One, I should say."

[143b] "Just let us see; must not the being of one be one thing and one itself another, if the one is not being, but, considered as one, partakes of being?" "Yes, that must be so." "Then if being is one thing and one is another, one is not other than being because it is one, nor is being other than one because it is being, but they differ from each other by virtue of being other and different." "Certainly." "Therefore the other is neither the same as one nor as being." "Certainly not." "Well, then, if we make a selection among them,

[143c] whether we select being and the other, or being and one, or one and the other, in each instance we select two things which may properly be called both?" "What do you mean?" "I will explain. We can speak of being?" "Yes." "And we can also speak of one?" "Yes, that too." "Then have we not spoken of each of them?" "Yes." "And when I speak of being and one, do I not speak of both?" "Certainly." "And also when I speak of being and other, or other and one, in every case I speak of each pair as both?"

[143d] "Yes." "If things are correctly called both, can they be both without being two?" "They cannot." "And if things are two, must not each of them be one?" "Certainly." "Then since the units of these pairs are together two, each must be individually one." "That is clear." "But if each of them is one, by the addition of any sort of one to any pair whatsoever the total becomes three?" "Yes." "And three is an odd number, and two is even?" "Of course."

[143e] "Well, when there are two units, must there not also be twice, and when there are three, thrice, that is, if two is twice one and three is thrice one?" "There must." "But if there are two and twice, must there not also be twice two? And again, if there are three and thrice, must there not be thrice three?" "Of course." "Well then, if there are three and twice and two and thrice, must there not also be twice three and thrice two?" "Inevitably." "Then there would be even times even,

[144a] odd times odd, odd times even, and even times odd." "Yes." "Then if that is true, do you believe any number is left out, which does not necessarily exist?" "By no means." "Then if one exists, number must also exist." "It must." "But if number exists, there must be many, indeed an infinite multitude, of existences; or is not number infinite in multitude and participant of existence?" "Certainly it is." "Then if all number partakes of existence, every part of number will partake of it?" "Yes."

[144b] "Existence, then, is distributed over all things, which are many, and is not wanting in any existing thing from the greatest to the smallest? Indeed, is it not absurd even to ask that question? For how can existence be wanting in any existing thing?" "It cannot by any means." "Then it is split up into the smallest and greatest and all kinds of existences nothing else is so much divided,

[144c] and in short the parts of existence are infinite." "That is true." "Its parts are the most numerous of all." "Yes, they are the most numerous." "Well, is there any one of them which is a part of existence, but is no part?" "How could that be?" "But if there is, it must, I imagine, so long as it is, be some one thing; it cannot be nothing." "That is inevitable." "Then unity is an attribute of every part of existence and is not wanting to a smaller or larger or any other part." "True."

[144d] "Can the one be in many places at once and still be a whole? Consider that question." "I am considering and I see that it is impossible." "Then it is divided into parts, if it is not a whole; for it cannot be attached to all the parts of existence at once unless it is divided." "I

agree." "And that which is divided into parts must certainly be as numerous as its parts." "It must." "Then what we said just now—that existence was divided into the greatest number of parts—was not true for it is not divided, you see, into any more parts than one,

[144e] but, as it seems, into the same number as one for existence is not wanting to the one, nor the one to existence, but being two they are equal throughout." "That is perfectly clear." "The one, then, split up by existence, is many and infinite in number." "Clearly." "Then not only the existent one is many, but the absolute one divided by existence, must be many." "Certainly."

"And because the parts are parts of a whole, the one would be limited by the whole;

[145a] or are not the parts included by the whole?" "They must be so." "But surely that which includes is a limit." "Of course." "Then the existent one is, apparently, both one and many, a whole and parts, limited and of infinite number." "So it appears." "Then if limited it has also extremes ?" "Certainly." "Yes, and if it is a whole, will it not have a beginning, a middle, and an end? Or can there be any whole without these three? And if any one of these is wanting, will it still be a whole?" "It will not."

[145b] "Then the one, it appears, will have a beginning, a middle, and an end." "It will." "But surely the middle is equally distant from the extremes for otherwise it would not be a middle." "No." "And the one, apparently, being of such a nature, will partake of some shape, whether straight or round or a mixture of the two." "Yes, it will."

"This being the case, will not the one be in itself and in other?" "How is that?" "Each of the parts doubtless is in the whole and none is outside of the whole." "True." "And all the parts are included in the whole ?"

[145c] "Yes." "And surely the one is all its parts, neither more nor less than all." "Certainly." "But the whole is the one, is it not?" "Of course." "Then if all the parts are in the whole and all the parts are the one and the one is also the whole, and all the parts are included in the whole, the one will be included in the one, and thus the one will be in itself." "Evidently." "But the whole is not in the parts, neither in all of them nor in any.

[145d] For if it is in all, it must be in one, for if it were wanting in any one it could no longer be in all; for if this one is one of all, and the whole is not in this one, how can it still be in all?" "It cannot in any way." "Nor can it be in some of the parts; for if the whole were in some parts, the greater would be in the less, which is impossible." "Yes, it is impossible." "But not being in one or several or all of the parts, it must be in something else or cease to be anywhere at all?" "It must." "And if it were nowhere, it would be nothing, but being a whole, since it is not in itself, it must be in something else, must it not?"

[145e] "Certainly." "Then the one, inasmuch as it is a whole, is in other and inasmuch as it is all its parts, it is in itself; and thus one must be both in itself and in other." "It must." "This being its nature, must not the one be both in motion and at rest?" "How is that?" "It is at rest, no doubt, if it is in itself; for being in one,

[146a] and not passing out from this, it is in the same, namely in itself." "It is." "But that which is always in the same, must always be at rest." "Certainly." "Well, then, must not, on the contrary, that which is always in other be never in the same, and being never in the same be not at rest, and being not at rest be in motion?" "True." "Then the one, being always in itself and in other, must always be in motion and at rest." "That is clear."

"And again, it must be the same with itself and other than itself,

[146b] and likewise the same with all other things and other than they, if what we have said is true." "How is that?" "Everything stands to everything in one of the following relations: it is either the same or other; or if neither the same or other, its relation is that of a part to a whole or of a whole to a part." "Obviously." "Now is the one a part of itself?" "By no means." "Then it cannot, by being a part in relation to itself, be a whole in relation to itself, as a part of itself." "No, that is impossible." "Nor can it be other than itself."

[146c] "Certainly not." "Then if it is neither other nor a part nor a whole in relation to itself, must it not therefore be the same with itself?" "It must." "Well, must not that which is in another place than itself—the self being in the same place with itself—be other than itself, if it is to be in another place?" "I think so." "Now we saw that this was the case with one, for it was in itself and in other at the same time." "Yes, we saw that it was so." "Then by this reasoning the one appears to be other than itself."

[146d] "So it appears." "Well then, if a thing is other than something, will it not be other than that which is other than it?" "Certainly." "Are not all things which are not one, other than one, and the one other than the not one?" "Of course." "Then the one would be other than the others." "Yes, it is other." "Consider; are not the absolute same and the absolute other opposites of one another?" "Of course." "Then will the same ever be in the other, or the other in the same?" "No." "Then if the other can never be in the same, there is no existing thing

[146e] in which the other is during any time; for if it were in anything during any time whatsoever, the other would be in the same, would it not?" "Yes, it would." "But since the other is never in the same, it can never be in any existing thing." "True." "Then the other cannot be either in the not one or in the one." "No, it cannot." "Then not by reason of the other will the one be other than the not one or the not one other than the one." "No." "And surely they cannot by reason of themselves be other than one another, if they do not partake of the other."

[147a] "Of course not." "But if they are not other than one another either by reason of themselves or by reason of the other, will it not be quite impossible for them to be other than one another at all?" "Quite impossible." "But neither can the not one partake of the one; for in that case they would not be not one, but would be one." "True." "Nor can the not one be a number; for in that case, too, since they would possess number, they would not be not one at all." "No, they would not." "Well, then, are the not one parts of the one?" "Or would the not one in that case also partake of the one?" "Yes, they would partake of it."

[147b] "If, then, in every way the one is one and the not one are not one, the one cannot be a part of the not one, nor a whole of which the not one are parts, nor are the not one parts of the one, nor a whole of which the one is a part." "No." "But we said that things which are neither parts nor wholes of one another, nor other than one another, are the same as one another." "Yes, we did." "Shall we say, then, that since the relations of the one and the not one are such as we have described, the two are the same as one another?" "Yes, let us say that." "The one, then, is, it appears, other than all other things and than itself, and is also the same as other things and as itself."

[147c] "That appears to be the result of our argument."

"Is it, then, also like and unlike itself and others?" "Perhaps." "At any rate, since it was found to be other than others, the others must also be other than it." "Of course." "Then it is other than the others just as the others are other than it, neither more nor less?" "Certainly." "And if neither more nor less, then in like degree?" "Yes." "In so far as it is so affected as to be other than the others and the others are affected in the same way in relation to the one, to that degree the one will be affected

[147d] in the same way as the others and the others in the same way as the one." "What do you mean?" "I will explain. You give a particular name to a thing?" "Yes." "Well, you can utter the same name once or more than once?" "Yes." "And do you name that to which the name belongs when you utter it once, but not when you utter it many times? Or must you always mean the same thing when you utter the same name, whether once or repeatedly?" "The same thing, of course." "The word other is the name of something, is it not?" "Certainly."

[147e] "Then when you utter it, whether once or many times, you apply it to nothing else, and you name nothing else, than that of which it is the name." "Assuredly." "Now when we say that the others are other than the one, and the one is other than the others, though we use the word other twice, we do not for all that apply it to anything else, but we always apply it to that nature of which it is the name." "Certainly."

[148a] "In so far as the one is other than the others and the others are other than the one, the one and the others are not in different states, but in the same state; but whatever is in the same state is like, is it not?" "Yes." "Then in so far as the one is in the state of being other than the others, just so far everything is like all other things; for everything is other than all other things." "So it appears." "But the like is opposed to the unlike." "Yes." "And the other to the same." "That is also true." "But this, too, was shown, that the one is the same as the others."

[148b] "Yes, it was." "And being the same as the others is the opposite of being other than the others." "Certainly." "In so far as it was other it was shown to be like." "Yes." "Then in so far as it is the same it will be unlike, since it has a quality which is the opposite of the quality which makes it like, for the other made it like." "Yes." "Then the same will make it unlike; otherwise the same will not be the opposite of the other."

[148c] "So it appears." "Then the one will be both like and unlike the others, like in so far as it is other, unlike in so far as it is the same." "Yes, that sort of conclusion seems to be tenable." "But there is another besides." "What is it?" "In so far as it is in the same state, the one is not in another state, and not being in another state it is not unlike, and not being unlike it is like but in so far as it is in another state, it is of another sort, and being of another sort it is unlike." "True." "Then the one, because it is the same as the others and because it is other than the others, for both these reasons or for either of them would be both like and unlike the others."

[148d] "Certainly." "And likewise, since it has been shown to be other than itself and the same as itself, the one will for both these reasons or for either of them be both like and unlike itself." "That is inevitable." "Now, then, consider the question whether the one touches or does not touch itself and other things." "I am considering." "The one was shown, I think, to be in the whole of itself." "Right." "And the one is also in other things?" "Yes." "Then by reason of being in the others

[148e] it would touch them, and by reason of being in itself it would be prevented from touching the others, but would touch itself, since it is in itself." "That is clear." "Thus the one

would touch itself and the other things." "It would." "But how about this? Must not everything which is to touch anything be next to that which it is to touch, and occupy that position which, being next to that of the other, touches it?" "It must." "Then the one, if it is to touch itself, must lie next to itself and occupy the position next to that in which it is." "Yes, it must." *

* See Plato, Parmenides (137c)

APPENDIX Δ (DELTA) – DATES OF ANCIENT PHILOSOPHERS

Appendix Δ (delta) – Dates of Ancient Philosophers

Note: Dates may vary depending on the source and may be incomplete.

Anaxagoras
fl. mid-5th cent. BC
Anaximander
6th cent. BC
Anaximenes
6th cent. BC
Antisthenes
ca.446-366 BC
Archytas
5th-4th cent. BC
Aristippus
ca. 435-356 BC
Aristotle
384-322 BC
Aristotle
384 - 322 BC
Cratylus
388 BC
Democritus
born ca. 460 BC
Diogenes of Apollonia
late 5th cent. BC
Diogenes of Sinope
ca. 404-323 BC
Empedocles
late 5th cent. BC
Gorgias
483-375 BC
Heraclitus
6th-5th cent. BC
Heraclitus
540 - 480 BC
Hesiod
7th cent. BC
Hippocrates

d. ca. 380 BC
Leucippus
5th cent. BC
Melissus of Samos
fl. ca. 440 BC
Parmenides
born ca. 510 BC 515 - 348 BC
Parmenides
late 6th century BC - mid 5th century BC
Philolaus of Croton
late 5th cent. BC
Plato
429-347 BC
Protagoras
d. ca. 420
Pythagoras
6th-5th cent. BC
Socrates
469–399 BC
Speusippus
ca. 410-338 BC
Thales
early 6th cent. BC
Theophrastus
372–288 BC
Xenocrates
395–313 BC
Xenophanes
6th-5th cent. BC
Xenophon
ca. 430-c.350 BC
Zeno of Elea
born ca. 490 BC

APPENDIX E (EPSILON) – A PERSONAL NOTE

Appendix E (epsilon) – A Personal Note

All of us hear and form judgements about 'the way it is' every day. These judgements tend to become habitual. When that happens, our judgements become territory to defend until 'death do us part'. I have stated many ideas in these volumes which I am sure would provoke the ire of folks who know a lot more than me. For that I apologize and ask that you teach me. I have no fear of being wrong as I have been wrong far more than I have been right. But I have to say that the times I have been wrong are the times when I learned the most. I would rather have the struggle to learn and be wrong than the ease of being right. There is hope with being wrong. There is no hope in the perpetual defense of being right. The defense of correctness tends to be gamesmanship with the incessant battle to defend the virtues of one's judgements over and against the other, the bad faith other which seems to me to only turn out to be yet again, the echo of Narcissus – the binary duality of the 'not' of me which is taken as the bad faith other. Gamesmanship hollows out one's soul and over time turns one into a shell of a human being. I would rather have a healthy respect for my failures than an overestimation of my 'truths'. With that in mind I want to add a few notions which I have no basis for believing, much less thinking so I just chalk it up to fanciful ideas that, for me at least have a hopeful impact.

When I was 19 years old, as is not uncommon with young guys, I had a reckless and near-death experience which altered the direction of my life. In my early sixties, I endured the pain of losing my most loved and beautiful twenty-year-old son on April 23, 2017, at midnight which I still endure to this day on a daily basis. In the darkness of this sorrow which nevertheless still holds the place for the love of my son Chris, I have reflected on my own near fatal mistake at the age of 19 years old. In my reflections on the quantum wave function and Everett's many-worlds branching I have imagined that it is possible that I may have taken an ill-fated branch that fateful night when I was 19 years old resulting in tragic consequences. In that narrative could this have been a deliberate branch which I chose? Could my parents on that branch have lived their lives in the pain my son left me for the rest of my life? At times I have been angry about what Chris did. But what if I inflicted the same pain on my parents on a different branch which I have no access to? Maybe Chris went onto another branch where he lived and went on to have a happy, fulfilling life in which his mom and I were filled with what every parent desires for their children for the rest of our lives. A happy fulfilled life is what every parent wants their children to have. With a view to that, what would it mean for me to be angry with what he did? There is a thought which comes to mind. All our ire and judgements may find their way back implicating only ourselves. The adulteress was brought before Jesus and asked should she be stoned to death which religious laws at the time required (for women only) caught in adultery. He responded, "He who is without sin, cast the first stone." John 8:7. We all need to remember that in all our judgements and anger of the other. We may find that the face of the other all along we derided, scoffed, or were angry with may well have been our own narcissistic echo, our own narcissistic face. There is another biblical saying,

> Judge not, that you be not judged. For with what judgment you judge, you will be judged. And with the measure you use, it will be measured again for you. And why do you see the speck that is in your brother's eye, but do not consider the plank that is in your own eye? Or how will you say to your brother, 'Let me pull the speck out of your eye,' when a log is in your own eye?*

Realistically, we all cannot help but make value judgements. These kinds of judgements fall from our personal and historic orientations. However, we should be fully aware that we may find one day that the other we judged, we were angry with, points directly at us and not the other we imagined. I think that the idea that the wave function branching means we no longer have access to all the other branches is, at the least, an interesting and compelling vehicle for serious self-reflection.

Here is another idea which is certainly fanciful without a shred of evidence but what if...? We have all heard about people's near-death experiences where they are in a tunnel and move towards a light at the other end. Some say their loved ones who have passed on are waiting in that light at the end of the tunnel. I know it is a crazy idea, but could that tunnel be an entangled wormhole? Ok, ha, ha - I laugh too but let's pursue this foolishness. Let's suppose that, as entangled systems ourselves, we are incessantly entangling and being entangled—a continuous loop driven by the 'spookiness' of the observer effect. And not just other people also with Gaia (referenced in the section entitled "Philosophical Implications of Quantum Computing and Chaos") or what I have called *phusis*. Don't ask me to explain this or justify this because I cannot but, for me personally, I find some hope in the idea that there may be much more to the story of life than we can possibly imagine. One thing I do know is in life we reap what we sow - karma. If our life practice culminates (*telos*) in the encasement, the tomb of our narcissistic echoes of ourselves then the only hope we have is in death, our own death, to end the misery of our absolute 'one-without-an-other' life...but I would not count on that if I were you. We had no choice about being born, gradually more or less emerging from the chaos of a 'wave function' perhaps which had not yet fallen into 'sense' much less commonsense. All through life we blank-out (some more than others myself included here), daze, mentally fog – just call it intermittent degrees of wave function superposition arousing itself into state reduction, branching, etc. It has occurred to me this can become more frequent with aging (as I keep forgetting what I was supposed to remember). If all this is so, certainly old age and death would be the entry back into the chaos of the wave function. Who is to say we could yet wake once again rise from that slumber? And Nietzsche's adage of eternal return to exactly the same life we had may have been more of the voice of classicism (finite space and infinite time will eventually repeat itself exactly) than the spooky observer effect. And if nothing after death as modernists maintain it does not bother me in the least.

There is nothing to fear about death as nothingness. I think when we are younger, we have fears about death, anxiety about nothingness, but I think age can be a valuable lesson in that regard. Hope is for this life. Life will present all of us with heartache if we are doing something right. Getting numb, feeling nothing, is nothingness in this life. Nothing can be done about that if choice does not override passiveness. Caring and loving deeply is painful. Young folks have expectations about love which is probably appropriate for their age (speaking as an ex-young person). But gaining wisdom about love means feelings of loss and pain are not to be shunned but are profound places of the other who may no longer present. They are the traces of the other in me - still. They are life throbbing in and through the

* See Mathew 7:1-4

tree of life. They show that the tree is still green, and that life has conquered the vacuity and rot of death. Those that are rooted in their own echo, their own vacuousness and numbness are shells of a rotten tree, eaten by their own disease. This is when hope has vanished into nothingness before death. This is dying nothingness which can only see bleakness and despair in every future. We must learn the wisdom of contentedness for this life only, to live and love fully, to believe and hope for life itself. So, even if there is nothing after death there was something in life. The rest does not concern us. And, if we do blink back from superposition after death, we will blink back full and ready for whatever. On the other hand, the resignation, despair, and emptiness of one without the other, without care and responsibility for the other, without wonder beauty and awe in all its fullness and pain is not a good way to welcome a new sapling into the tree of life. I do not consider myself a Christian, but I am reminded of the odd reference in Genesis, "*Now, lest he [we] reach out his [our] hand and take also of the tree of life and eat, and live forever—*".

I have no allusions about what some think of as 'realism', common sense, the 'way it is'. I think I have made clear why I think we can go no further than 'narrative' both personally, collectively, and historically. This becomes difficult when we think of evil, of atrocities like Hitler and overtly horrific evil acts. But the thing I remember is that in my way of thinking, my narrative, there is no justification for the Hitler's of history. In my estimation, they are the final gaze of Orpheus which can no longer see Eurydice, the other, but can only perpetually see their own narcissistic face departing into the underworld of one without another, where the observer effect has become only of oneself – devoid of ethics, without any other to oneself. My idea of **Hell** is simply one without another. It may be my narrative, but my narrative is no less true than the modernist narrative of empirical facts which are absolute in a mechanistic 'reality'. And by the way, that particular metaphysic has never been able to claim any higher ground concerning the evils of humanity as well. On the contrary, it certainly has a history of ennobling the brutal tyrants of history – Ayn Rand comes to mind. And also, with Donald Trump this idea of '*nonsensical but persuasive rhetoric transformed into the rhetorician's own personal glory and transformed into might makes right*' seems to be yet another example of how 'realism' has come to play as 'empirical facts'. In my humble opinion capitalism without any rules will inevitably engender human epitomes unexamined narcissistic rage.

I am sure there are those that will think I am a raving communist. I am not. I am closer to what Europeans call a market democratic socialist. I love and value democracy and would not want to live under any political system. Republicans have been crying wolf about socialism from the early 20th century unto the present day and Democrats have denied it. Don't believe me? See my post with quotes from Howard Taft as far back as 1908, The American Liberty League going back to 1933, Father Charles Coughlin, Harry Truman, Joe McCarthy, Barry Goldwater, John McCain, Dwight Eisenhower, Ronald Reagan, and Newt Gingrich. * Even today the far right is threatening us with *becoming* socialists. This is all fear tactics because even by the history of the far right's own admissions, we are already a mix of socialism and capitalism which is the definition of market based democratic socialism in much of Europe and in the United States is called a social democracy. †

---

* See Clashing Histories – Right Fright in the Light

† See Social democracy - Wikipedia

> Markets generate problems of their own (especially when they involve monopolies, negative externalities, and asymmetric information). But if regulations are introduced to counter these "market failures", markets can be the best feasible mechanism for generating matches between demand and supply in large, complex societies (as higher prices signal high demand, with supply rushing to cover it, while lower prices signal low demand, leading supply to concentrate on other products). Market socialism affirms the traditional socialist desideratum of preventing a division of society between a class of capitalists who do not need to work to make a living and a class of laborers having to work for them, but it retains from capitalism the utilization of markets to guide production. *

By the textbook definition of socialism, we have been a version of a socialist state called market socialism since at least the beginnings of labor unions, Social Security, Medicare, Medicaid, and Welfare Programs.

> Social democracy has a strong, long-standing connection with trade unions and the broader labour movement. It is supportive of measures to foster greater democratic decision-making in the economic sphere, including co-determination, collective bargaining rights for workers, and expanding ownership to employees and other stakeholders.
>
> Socialism is nothing other than what has evolved in this country for over a hundred years. The Republicans are correct in telling us this. They just need to lose the sneer and quit trying to make it an all or nothing, capitalism or socialism, apocalyptic event. The rhetoric is too old to get a rise out of all but the most ardent devotees of their faith. Problem solving is quite different from power acquisition. We need to decide if we are going to solve problems or let the free acquirers of power decide for us. †

We have partial public ownership we call stocks. We have economic interventionism putting checks on monopolies. For many years we have openly promoted social equality. Of course, the new and boringly old fear tactics of extremists have found yet another rhetorical, same ol' tactical tagline, "Woke".

> Social democracy maintains a commitment to representative and participatory democracy. Common aims include curbing inequality, eliminating the oppression of underprivileged groups, eradicating poverty, and upholding universally accessible public services such as child care, education, elderly care, health care, and workers' compensation. Economically, it supports income redistribution [progressive taxation] and regulating the economy in the public interest. ‡

There has never been a pure capitalist state in which only the 'free' market decides because in its pure form it reverts into a kleptocracy. Communism was a complete and total horror and human disas-

---

* See Socialism (Stanford Encyclopedia of Philosophy)

† See Clashing Histories – Right Fright in the Light

‡ See Social democracy - Wikipedia

ter. Capitalism has made a mantra of marketeer's, AKA branding guru's, motto that 'perception is reality'. The insidious and easily escapable fact of that truism is that perception is moldable, tailorable, customizable to ego, whim, desire, and unabashed narcissism where 'reality' is merely transactional. For the record I am not a new-ager or a communist, but marketing is a science which has been proven to work in shaping everything from consumer purchases, expectations, and demands to political persuasions and basic philosophical assumptions - and that is an empirical fact. The advent of smart phones and technology which consumes much time and attention from users puts a constant stream of commercials into a worldwide audience every day. This was already a bonanza for marketeers. But now with artificial intelligence the marketing avenues can be customized for each individual consumer as if their very best friend was conversing with them. Once quantum computing gets married to artificial intelligence, the potency of marketing will be unrivaled. Quantum intelligence can interact with you like your best friend or God or whatever dearly held personal values you might have. For a marketeer it will be like Morpheus in the movie "The Matrix" where he/she/it (or pick your pronoun) can lure you back into the Matrix consuming you as their energy source (or what we now think of as money). Ok, let's not get paranoid here.

Can consumers be critical thinkers and make their own choices? Ideally, that would be true. But marketing science has shown for many decades that strong and impulsive emotions coupled with social network theory sells products. Marketing can sell anything, not just products. It can sell philosophies, ideologies, and the most bizarre cultish ideas, all in the service of its benefactors. Critical thinking is the only solution to this looming dilemma. But education and universities are not immune from the mediocrity of laissez-faire, 'letting the market decide'. And the market has all the most powerful tools in human history to actually let the marketeers decide not the consumers. Again, I am not a communist. I believe capitalism, when it works and is well regulated to avoid hazards like destroying the environment, protecting consumers from fraud, etc., is a positive vehicle for wealth distribution and addressing fundamental issues of human suffering. Additionally, for the 'free' market to work well it cannot be rigged by the golden rule – those who have the money make the rules. What I am about to write is not meant to be universal as it is only my anecdotal experience, but I personally think it applies more than many folks are aware of.

I have worked in small, medium, and multinational companies and depending on the size of the company the guys at the top are all happy to fund candidates and lobbyists to get rid of regulations they do not like and create regulations they do like. They are highly successful 'gerrymandering' so to speak the 'rules' of the not so 'free' market to accommodate their agendas. And if that does not work the good 'ol boys' network is always the backup. The executives at the multinational company I worked for were more of a social country club of prospective and strategic partners, subsidiaries, and even competitors than anything else. With regard to consumers and employees, I remember a senior vice president at the multinational company I worked for going on and on during social occasions about tribalism. He saw his job as weaving a tribal myth, a tale, an oracular edict (his words not mine) that would bind the clan to a sacred duty – to get him richer!

As for the science of marketing, ever increasing personal electronics and the future of quantum intelligence will mean there is no amount of regulation which can forestall its eventual domination. The only hope against human automatons is critical thinking skills enhanced by higher quality and free education at all levels. The ability to see through the oncoming onslaught of perception as reality using social network theory and sensory marketing is the only resistance against a brave new world of

Orwell's book "1984". The science of marketing has a term called sensory marketing which is an advertising tactic that appeals to one or more of the five human senses of sight, hearing, smell, taste, and touch to create an emotional association with a specific product or brand. Sensory marketing assumes that people, as consumers, will act according to their emotional impulses more than to their objective reasoning. In this way, an effective sensory marketing effort can result in consumers choosing to buy a certain product, rather than an equal but less expensive alternative. And this is not just for product consumers anymore. *

The science of marketing is real and proven, and it is not just for consumer products anymore. It can totally reshape values, ideologies, religious orientations, in a way that will make Charles Manson, David Koresh, Jim Jones, and Adolf Hitler look like Sunday School teachers. According to a recent article in the Journal of Consumer Research, consumption ideology is defined as "ideas and ideals that are related to consumerism and manifested in consumer behavior". † Another article from the American Marketing Association suggests that incorporating consumers' political orientation into companies' segmentation and marketing strategies can help companies anticipate and manage consumers' and brands' political activism as well as anticipate and leverage ideological differences in consumption across non-political domains. ‡ Despite what we are hearing from high tech CEOs about needing more regulation for artificial intelligence, ultimately the movement towards quantum intelligence will not be deterred by regulations and public safety laws. The days when external social safeguards were effective are coming to an end. These technologies coupled with marketing science will put more and more demand on educated consumers. Soft pedaling this threat can only have apocalyptic consequences.

Technology will increasingly and inevitably reduce jobs which do not require more advanced training. Without serious and concerted efforts to educate and train people for many diverse occupations in science, technologies, and humanities we will see more and more social upheaval which may well end in Homo sapiens species extinction. I know the study of humanities has been banished from the land of the living, but it is actually a historic record of the land of the living which is invaluable for thinking critically. Humans are standing on a precarious abyss with astounding achievements unparalleled in our history. However, just like every evolutionary advantage, there are also built-in disadvantages which eventually can cause adaptation to fall behind. Quantum intelligence as Yuval Harari writes about will be a new and powerful evolutionary species with incredible adaptive advantages over Homo Sapiens. Quantum intelligence is incredible because its evolution did not come from typical biological adaptation but from the promise of intelligence which was always built into *phusis*. It holds a massive promise for Homo Sapiens but just like the atomic bomb can become an extinction threat from authoritarians who have the means and the desire to usurp their violent agendas on the world. I am not convinced that we as humans can and will have the wherewithal to work with *phusis* ethically and responsibly towards the radical exteriorities we face. What we face or run from is responsibility for the other.

---

* See An Introduction to Sensory Marketing
And The Science of Emotion in Marketing: How Our Brains Decide What to Share and Whom to Trust
And Emotional Appeals in Advertising
And Sensory Marketing: Straight to the Emotions

† See Consumption Ideology

‡ See How Politics Shapes Consumption Behavior

END NOTES

4. THE 20TH CENTURY

1. Plato conceived of the self as a knower. His concept of the self is intricately intertwined with the nature of the rational soul, which he regarded as the highest form of cognition. Plato believed that the essence of a person lies in the psyche (Ancient Greek: *ψῡχή*), which he considered to be the incorporeal, eternal occupant of an individual's being. This rational soul is the seat of knowledge, and thus, the concepts of the self and knowledge are closely linked in Plato's philosophy. Plato's notion of the true self emerges within the context of knowledge. According to Plato, we acquire our true self when we activate our latent knowledge of the Forms. The Forms (or Ideas) represent the ideal, unchanging, and eternal concepts that exist beyond the physical world. Activating this knowledge involves recognizing and understanding these Forms, which leads to self-discovery. Plato's view of the self transcends mere physical existence. The soul, as the real self, persists continuously from the earliest stages of life (from zygote or infancy) and remains unchanged by basic building materials. The Ship of Theseus is a thought experiment about whether an object is the same object after having had all of its original components replaced. Plato's perspective solves the Ship of Theseus dilemma. as the soul provides permanence even amidst bodily changes. Plato's concept of the self extends beyond earthly life. The soul, being eternal, persists into the afterlife, carrying with it the accumulated knowledge and experiences. Plato's idea of the self centers on the rational soul, knowledge, and the activation of latent wisdom. From Plato's cave analogy, most people see dimly their relationship to the Forms. However, even in this the Forms or essence, has a potential shadowy bridge to humankind. In Plato's view humans reflect reality imperfectly.
   See Plato's Concept of the Self
   And Plato's theory of soul
   And Plato and the Divided Self
   And 6.2 Self and Identity
2. Aristotle believed that the self, or the human person, is a composite entity consisting of both body and soul. Philosophers call this hylomorphism (from Greek *hylē*, "matter"; *morphē*, "form"). Aristotle considered the soul as the essential form of a living being. Hylomorphism conceives every physical entity or being as a compound of primary matter (potency) and immaterial, substantial form (act), with the generic form as immanently real within the individual. According to Aristotle every natural body consists of two intrinsic principles: one potential, namely, primary matter, and one actual, namely, substantial form. Aristotle based

   his argument chiefly on the dynamic analysis of "*becoming*," or substantial change. If a being changes into another being, something permanent must exist that is common to the two terms; otherwise, there would be no transformation but merely a succession by the annihilation of the first term and the creation of the second. This permanent and common something cannot itself be strictly a being because a being already is and does not become, and because a being "*in act*" cannot be an intrinsic part of a being possessing a unity of its own; it must therefore be a being "*in potency*", a potential principle, passive and indeterminate. At the same time, in the two terms of the change, there must also be an actual, active, determining principle. The potential principle is matter, the actual principle, form. This doctrine was central to Aristotle's philosophy of nature and has been variously interpreted by Greek and Arab commentators of Aristotle and by the Scholastic philosophers. According to Aristotle, the soul serves as the "*form*" of the human body. In other words, the soul gives structure and purpose to the physical body. This interconnectedness between body and soul implies that they are inseparable. Aristotle considered the soul as the essential form of a living being. While he acknowledged that the soul does not exist independently of the body, he distinguished different aspects of the soul. Aristotle believed that the intellect (or reason) is immortal and perpetual. It constitutes the higher part of the soul and is separable from the body. Aristotle's views diverged from those of Plato. Plato emphasized that the true self of human beings is their reason or intellect, which exists independently of the body. In contrast, Aristotle maintained that the soul's functions are intricately tied to the body's organization. In this sense Aristotle was more like a phenomenologist. For Aristotle the soul was defined by its actual manifestations. Aristotle's philosophy of self underscores the integrated unity of body and soul, with the soul acting as the formative principle that animates and directs human life.
   See Aristotle's Concept of the Self
   And Aristotle on the Individuality of Self
   And Philosophy of self
   And Hylomorphism (Wikipedia)
   And Hylomorphism (Britannica)

And Hylomorphism (New World Encyclopedia)

3. Descartes believed that a human person consists of two distinct parts. The physical aspect of a person was the material body. The non-material part was the mind or soul. The thinking, conscious entity that transcends the physical world. Descartes posited a radical separation between the mind and the body. The thinking self (or soul) is nonmaterial, immortal, and conscious. It exists independently of the physical laws governing the universe. Descartes advocated for systematic doubt as a means to achieve clear and well-reasoned conclusions. By questioning everything accepted without scrutiny, individuals can develop beliefs that are genuinely their own. Descartes's concept of the self centers around the idea of a dual existence: a material body and an immaterial mind or soul. For Descartes, everything could be doubted except thinking. To doubt thinking would require thinking. Therefore, Descartes tells us – I think therefore I am (*cogito, ergo sum*). From thinking, Descartes argues that our innate idea of God is of a perfect being. Existence is a form of perfection. Therefore, God must exist. The self is mind embedded in body, substance. Descartes is known for his mind/body dualism called the thinking thing. For Descartes, the soul (thinking) is immortal.

4. Leibniz rejected Cartesian doctrines that all mental states are conscious and that non-human animals lack souls or sensation. He believed that even non-rational animals possess souls and feelings, prompting him to reflect on the mental capacities distinguishing humans from lower animals. Leibniz proposed that perception, rather than consciousness (as Cartesians assumed), is the distinguishing feature of mentality. Rational souls have elevated types of perceptions, including self-consciousness and abstract thought. Human minds surpass animal souls due to their capacity for more elevated types of perceptions and appetitions (mental tendencies). Rational souls exhibit reasoning and the tendency to act according to overall judgments. Leibniz's thought experiment involving a mill argues that material things (such as machines or brains) cannot genuinely have mental states. Only immaterial entities (soul-like) can think or perceive. Thus, constructing a truly thinking computer is impossible. Leibniz's doctrine denies genuine interaction between body and soul. Instead, he posits that God created souls and bodies to naturally correspond without divine intervention. Leibniz's ideas about the self encompassed perception, elevated mental capacities, and the immaterial nature of thinking entities. For Leibniz, the self is monad, absolutely self-contained. There is no exteriority to the self.

   See Gottfried Leibniz: Philosophy of Mind

   And Self and Identity

   And 10 Leibniz: Innate Ideas, Reflection, and Self-Knowledge

   Leibniz writes, "*There is no way of explaining how a monad can be altered or changed internally by some other creature, since one cannot transpose anything in it, nor can one conceive of any internal motion that can be excited, directed, augmented, or diminished within it, as can be done in composites, where there can be change among the parts. The monads have no windows through which something can enter or leave. Accidents cannot be detached, nor can they go about outside of substances, as the sensible species of the Scholastics once did. Thus, neither substance nor accident can enter a monad from without.*" Leibniz posited that essences exist both in the divine intellect and as limitations of the infinite essence of God. Leibniz believed that possible worlds are constituted by individual substances. When God creates the world, he selects one of these possible worlds—the most perfect one—to actualize. These possible worlds consist of individual essences, each endowed with a degree of reality or perfection. The very existence of a genuinely possible thing implies the existence of an individual essence that can be created.

   See Gottfried Leibniz: Philosophy of Mind

   And Self and Identity

   And 6 Leibniz on the Idea and the Concept of the Self: Inner Sentiment, Self-Knowledge, and the Awakening of Metaphysical Ideas

   And 10 Leibniz: Innate Ideas, Reflection, and Self-Knowledge

   And Leibniz on Possibilia, Creation, and the Reality of Essences

   And Philosophy_of_self (Wikipedia)

   And Leibniz and the Shelf of Essence (.pdf)

5. Isaac Newton a devout Christian believed in a monotheistic God. Early on, he was a theologian at Trinity College. Newton concluded that a simple and authentic form of Christianity had been corrupted over the centuries. He believed that the orthodox notion of the Trinity was a fiction invented in the early fourth century and promoted by those who served the devil. According to his views, it was idolatry to give any being the worship rightfully due to God. While he considered Jesus divine, he did not equate Jesus with God. God was the master creator. Humans, the self was a creation of God. The self gets its identity from God.

   See Isaac Newton

   And Religious views of Isaac Newton

   And Newton's Religious Life and Work

   And Newton's Metaphysics of Space as God's Emanative Effect

   Newton writes of God, "*He endures always and is present everywhere, and by existing always and everywhere he constitutes duration and space. Since each and every particle of space is always, and each and every indivisible moment of duration is everywhere, certainly the maker and lord of all things will not be never or nowhere ... God is one and the same God always and everywhere. He is omnipresent not only virtually but also substantially; for active power cannot subsist without substance.*"

Newton, Isaac, 1999, The Principia: Mathematical Principles of Natural Philosophy, I.B. Cohen and A. Whitman (trans.), Berkeley: University of California Press. Page 941.

6. The Enlightenment, also known as the Age of Reason, was a transformative period in Europe and North America from the late 17th century to the late 18th century. During this time, new approaches in philosophy, science, and politics emerged. The Enlightenment celebrated the human capacity for reason. Philosophers believed that reason was the tool through which knowledge could be extended, individual liberty maintained, and happiness secured. Unlike the Middle Ages, where received wisdom went unchallenged, Enlightenment thinkers questioned everything. They were often referred to as "*free-thinkers*" because they challenged accepted beliefs and sought rational answers. The roots of the Enlightenment can be traced back to the Renaissance (1400-1600) and the humanist movement. Humanists emphasized civic virtue and the realization of an individual's full potential for the benefit of society. The Protestant Reformation (1517-1648) also played a role by diminishing the traditional power of the Christian Church. Enlightenment thinkers sought greater religious freedom and toleration. Enlightenment philosophers engaged in a battle of reason against emotion, superstition, and fear. They applied reason to age-old questions: How should we live together in societies? What is the best form of government? What constitutes happiness?

   Their optimism for a better world and the freedom to question everything fueled this intellectual revolution. While the Enlightenment wasn't a single, self-conscious movement, it did introduce innovative themes of thought. But Enlightenment ended with the British empiricists. Philosophers like John Locke conceived of the human mind as a "tabula rasa" (blank slate) upon which experience wrote freely. Locke was an empiricist and believed that all knowledge is derived from sensory experiences. He argued that there is no continuous, unchanging self or soul. Instead, we are a bundle of experiences —a collection of perceptions that constantly change. George Berkeley was another empiricist, extending Locke's ideas. He proposed that the material world exists only through perception. In other words, things exist because they are perceived by minds (including God's mind). Berkeley's view challenges the notion of an independent, substantial soul. Instead, he emphasized the importance of mental experiences. David Hume further developed empiricism. He rejected the idea of a permanent self or soul. According to Hume, our sense of self arises from a stream of consciousness, where experiences flow continuously. Hume famously stated that when he introspected, he found only a bundle of perceptions, not a substantial self or soul. Individual character was shaped by personal experiences in the world. The Enlightenment was a contested ground for ideas of selfhood, oscillating between materialist and idealist conceptions of the self. It occurred against the backdrop of increased individual agency in society. At the end of Enlightenment, we see that already classical and scholastic notions of the self and the soul had begun emptying themselves out into the certainty of neutrality. The historic 'thrill had gone' and but the relatedness to truth was now based on neutrality of nature and the universe.

   See 4.2.3 Empiricism
   And The Soul as an Idea in Ancient Greece 101: Epicurus' Soul Atoms
   And Soul Religion and Philosophy (Britannica)
   And The Enlightenment
   And Enlightenment European History (Britannica)
   And Self-Enlightenment
   And World Literature: The Enlightenment

7. From Hegel, phenomenology of consciousness is the exposition of the coming to be of all knowledge. But phenomenology of consciousness is a precondition of Hegels "Logic", the "Science of Logic". The Logic proceeds in terms of dialectic. Dialectical oppositions necessitate increasingly higher principles transforming and at the same time maintaining oppositional tensions. This process continues until the subject of all predicates is Concept. Concept is completion, perfection, and essence. Concept is universal and absolute fulfillment of the dialectic of Logic. Some Hegelians will tell us is this not hierarchical but horizontally, organically developed as autopoiesis. Autopoiesis originates from the field of systems theory and refers to self-creation or self-production. Autopoiesis is how self-referential structures emerge and develop. Autopoiesis, in this context, reflects the unity-in-difference of matter and life, reality and ideality, nature and thinking. Operational closure is a condition for the openness of a system. It allows for teleological functioning, as systems strive toward specific states. Hegel's notion of the organism as the "unity of inner and outer" exemplifies this minimal self-relating. It is a fundamental form of life where the boundary between inside and outside emerges. Autopoiesis acts as a logical bootstrap: a network produces entities that create their own boundaries. When completed, these entities self-distinguish and bootstrap themselves out of the soup of chemistry and physics. Despite the opposition between modern scientific discourse and Hegelian speculative philosophy, these Hegelian's concepts of autopoiesis bridges the gap between science, physics, and philosophy. Hegel's system provides a unique lens to analyze the relationship between operational closure and the autonomy of living organizations. Perhaps Hegel's system is the highest possible statement of classicism's narratives, of the unity of knowledge and its organic embeddedness. But Hegel's system stops at the interruption of quantum physics from classicism's branching from the spooky exteriority of the quantum wave function.

   See The Phenomenology of Spirit (Wikipedia)

And 6 Spirit in the Phenomenology of Spirit
And Hegel's Phenomenological Method
And Hegel's Phenomenological Method and the Later Movement of Phenomenology
And An Introduction to Hegel's PHENOMENOLOGY OF SPIRIT, complete, in 9 Lectures and 36 Videos
And Selected Works of G.W.F. Hegel: Science of Logic
And Hegel's Phenomenology: On the Logical Structure of Human Experience
And Hegel and Husserl on Phenomenology, Logic, and the System of Sciences: A Reappraisal
And Autopoietic Systems Theory within Hegelian Speculative Philosophy
And AUTOPOIESIS AND LIFE IN HEGEL'S SCIENCE OF LOGIC
And Narrativity and Self-Creating Forms: Autopoiesis in Perspective
And Ecological Ethics and Living Subjectivity in Hegel's Logic. The Middle Voice of Autopoietic Life

6. ETHICS AND LOVE

1. It interesting to note that chaos in ancient Greece was not our idea of an 'it', as neuter.

On my reading, the most important aspect of Chaos is that it is distinctly nonmale. Although grammatically neuter, Chaos is either characterized as female, a gender-neutral deity,[23] or a "sexually indeterminate figure."[24] An aspect of Chaos' gender is its ability to generate offspring through parthenogenesis. Chaos introduces its characteristic indefinable obscurity into the world through its progeny Erebos and Night (Nux), who are begotten through self-differentiation.[25] This parthenogenetic production introduces children that reiterate the features of their parent. In contrast, through sexual union, Night and Erebos produce children who represent a greater degree of definition in the figuration of Brightness.

(Aither) and Day (Hêmeros),[26] a procreation that points to the way that Hesiod uses sexual generation as a driving force

behind the progression of the succession myth, as I expand upon below. I argue that Chaos begins as a disordered force, which contrasts with the ordering male forces that follow and through the introduction of sexual generation it too is

responsible for the introduction of an embryonic form of order.

Gastêr, Nêdys, and Thauma: Feminine Sources of Deception and Generation in Hesiod's Theogony (.pdf) by Kaitlyn Boulding

Also On Origin | Musings (mixermuse.com)

It is hard for us at the later time in history to understand that the ancient Greeks did not have a notion of 'it' like we do. The Greeks were animists. Everything had soul, animation.

While the myths of the ancient Greeks indicate a kind of animate vitalization of phusis as we see in the pantheon of gods, the notion of 'neuter' in ancient Greek was 'neither one or the other' as in 'not masculine or feminine'. The neuter was the 'not' in ancient Greek of gender. The negative does not specify a positive term as modern usage does of the 'it' (i.e., substance, inanimate objects, not alive, thing, etc.). On Origin | Musings (mixermuse.com)

There were no "its' in the ancient Greek world but there was "neither male nor female". The term "neuter" has an interesting etymology. It originates from Latin, where it means "of the neuter gender"—literally, "neither one nor the other." This concept of neutrality applies to nouns, pronouns, and verbs. In grammar, a neuter noun or pronoun is neither masculine nor feminine in gender, while a neuter verb has middle or reflexive meaning, neither active nor passive. Interestingly, the Latin word "neuter" is likely a loan-translation from the Greek term "*oudeteros*," which also means "neither." This linguistic borrowing highlights the interconnectedness of languages, and the way ideas travel across cultures and time. So, whether in Greek or Latin, the concept of "neuter" reflects a state of being neither one thing nor the other.

See Ancient Greek I - Nouns, Pronouns, and their Case Functions (openbookpublishers.com)

neuter (adj.) – late 14c., of grammatical gender, "neither masculine nor feminine," from Latin neuter "of the neuter gender," literally "neither one nor the other," from ne- "not, no" (from PIE root *ne- "not") + uter "either (of two)" (see whether). Probably a loan-translation of Greek oudeteros "neither, neuter." In 16c., it had the sense of "taking neither side, neutral."

See neuter (adj.) (etymonline)

FURTHER READING

Further Reading

Here are good reference articles in Quanta magazine for some of the basic physics notions discussed throughout this post:

How Quantum Pairs Stitch Spacetime | Quanta Magazine: https://www.quantamagazine.org/tensor-networks-and-entanglement-20150428/

Interactive: What Is Space? | Quanta Magazine: https://www.quantamagazine.org/interactive-what-is-space-20150430.

Here are several videos oriented to the physics novice that will also provide an excellent background for the physics material I plan to cover:

Leonard Susskind | "ER = EPR" or "What's Behind the Horizons of Blackholes?" - 1 of 2: https://www.youtube.com/watch?v=OBPpRqxY8Uw (video)

Leonard Susskind | "ER = EPR" or "What's Behind the Horizons of Blackholes?" - 2 of 2: https://www.youtube.com/watch?v=uiG_EtVQu5o (video)

Lecture 1 | Quantum Entanglements, Part 1 (Stanford): https://www.youtube.com/watch?v=0Eeuqh9QfNI (video)

Lecture 2 | Quantum Entanglements, Part 1 (Stanford): https://www.youtube.com/watch?v=VtBRKw1Ab7E (video)

Here are more reference articles which I found very interesting:

Physics of the Universe | Kavli Institute for Particle Astrophysics and Cosmology (KIPAC) (stanford.edu): https://kipac.stanford.edu/research/physics-universe

Loeb, AVI. December 12, 2020. "Endless Creation Out of Nothing. Scientific American (online)

Kane, Gordon. October 9, 2006. "Are virtual particles really constantly popping in and out of existence? Or are they merely a mathematical bookkeeping device for quantum mechanics?". Scientific American (online)

Wave Function Collapse: https://en.wikipedia.org/wiki/Wave_function_collapse

Spontaneous Symmetry Breaking: https://en.wikipedia.org/wiki/Spontaneous_symmetry_breaking

Symmetry Breaking: https://plato.stanford.edu/entries/symmetry-breaking/

Max Born: https://en.wikipedia.org/wiki/Max_Born

Wave Function: https://en.wikipedia.org/wiki/Wave_function

Schrödinger equation: https://en.wikipedia.org/wiki/Schr%C3%B6dinger_equation

Delta Potential: https://en.wikipedia.org/wiki/Delta_potential

Introduction to Quantum Mechanics: https://en.wikipedia.org/wiki/Introduction_to_quantum_mechanics

Erwin Schrödinger: https://en.wikipedia.org/wiki/Erwin_Schr%C3%B6dinger

What Is Life?: https://en.wikipedia.org/wiki/What_Is_Life%3F

Non-Relativistic Quantum Mechanics Lecture Notes: https://www.uio.no/studier/emner/matnat/fys/FYS4110/h07/undervisningsmateriale/LectureNotes.pdf

Quantum State: https://en.wikipedia.org/wiki/Quantum_state#Pure_states

Mass: https://en.wikipedia.org/wiki/Mass

Poincare Group: https://en.wikipedia.org/wiki/Poincar%C3%A9_group

Relativistic Mass (ucr.edu): https://math.ucr.edu/home/baez/physics/Relativity/SR/mass.html

Non-relativistic spacetime: https://en.wikipedia.org/wiki/Non-relativistic_spacetime

About Non-relativistic Quantum Mechanics and Electromagnetism: https://www.lidsen.com/journals/rpm/rpm-04-04-027/rpm.2204027.pdf

A non-relativistic theory of quantum mechanics and gravity with local modulus symmetry: https://arxiv.org/ftp/arxiv/papers/2008/2008.07749.pdf

The Bronstein hypercube of Quantum Gravity: https://core.ac.uk/works/51933878

Relativistic Quantum Mechanics: https://en.wikipedia.org/wiki/Relativistic_quantum_mechanics

Is Zero Point Energy Real? https://users.ox.ac.uk/~lina0174/vacuum.pdf

Virtual Particles: What are they? https://profmattstrassler.com/articles-and-posts/particle-physics-basics/virtual-particles-what-are-they/

Physics Questions People Ask Fermilab: https://www.fnal.gov/pub/science/inquiring/questions/virtual_particles.html

Virtual particle: https://en.wikipedia.org/wiki/Virtual_particle

Copenhagen Interpretation of Quantum Mechanics: https://plato.stanford.edu/entries/qm-copenhagen/
Copenhagen interpretation: https://en.wikipedia.org/wiki/Copenhagen_interpretation
SAGP SSIPS Abstracts 2015: https://orb.binghamton.edu/cgi/viewcontent.cgi?article=1431&context=sagp
"The Unmoved-Mover" in Aristotle's Metaphysics: https://owlcation.com/humanities/The-Unmoved-Mover-in-Aristotles-Metaphysics
Metaphysics: http://classics.mit.edu/Aristotle/metaphysics.html
The Philosophical Sense of Transcendence: Levinas and Plato on Loving Beyond Being: https://ndpr.nd.edu/reviews/the-philosophical-sense-of-transcendence-levinas-and-plato-on-loving-beyond-being/
Quantum Entanglement: https://en.wikipedia.org/wiki/Quantum_entanglement
Refrigeration by Quantum Measurements: https://physics.aps.org/articles/v12/s22
Using quantum measurements to fuel a cooling engine: https://phys.org/news/2019-03-quantum-fuel-cooling.html
Entanglement Made Simple: https://www.quantamagazine.org/entanglement-made-simple-20160428/
What is quantum entanglement? All about this 'spooky' quirk of physics: https://interestingengineering.com/science/quantum-entanglement
Your Simple (Yes, Simple) Guide to Quantum Entanglement: https://www.wired.com/2016/05/simple-yes-simple-guide-quantum-entanglement/
How is quantum entanglement related to decoherence?: https://www.quora.com/How-is-quantum-entanglement-related-to-decoherence
Information Philosopher - Philosophy: https://www.informationphilosopher.com/introduction/philosophy/
Quantum mysteries dissolve if possibilities are realities | Science News: https://www.sciencenews.org/blog/context/quantum-mysteries-dissolve-if-possibilities-are-realities
Emmanuel Levinas: https://plato.stanford.edu/entries/levinas/

REFERENCES

Blanchot, M. (1981). *The Gaze of Orpheus and Other Literary Essays.* Barrytown, New York: Lydia Davis and Station Hill Press.

Brody, J. (2020). *Quantum Entanglement.* MIT Press.

Deleuze, Gilles, and Félix Guattari. *A Thousand Plateaus: Capitalism and Schizophrenia*. Translated by Brian Massumi, University of Minnesota Press, 1987.

Frautschi, S. C., Olenick, R. P., Apostol, T. M., & Goodstein, D. L. (2007). *The Mechanical Universe: Mechanics and Heat (Advanced ed.).* Cambridge [Cambridgeshire]: Cambridge University Press. doi:ISBN 978-0-521-71590-4, OCLC 227002144

Harari, Y. N. (2016). *Homo Deus: A Brief History of Tomorrow.* London: Random House.

Hyland, D. A., Manoussakis, J. P., Baracchi, C., Brogan, W. A., Warnek, P., Figal, G., . . . Gonzales, F. J. (2006). *Heidegger and the Greeks: Interpretive Essays.* (D. A. Hyland, & J. P. Manoussakis, Eds.) Bloomington, IN, USA: Indiana University Press.

Kuhn, T. S. (1962). *The Structure of Scientific Revolutions.* Chicago: University of Chicago Press.

Landeweerd, L. (2021). *Time, life, and memory: Bergson and contemporary science.* Cham, Switzerland: Springer.

Levinas, Emmanuel. Totality and Infinity: An Essay on Exteriority. Translated by Alphonso Lingis. Pittsburgh: Duquesne University Press, 1969.

Levinas, E. (1981). *Otherwise Than Being or Beyond Essence.* (A. Lingis, Trans.) Boston: Martinus Nijhoff Publishers. doi:10.1007/978-94-015-7906-3

Levinas, E. (2000). *Entre Nous: On Thinking-of-the-Other.* (B. H. Smith, Trans.) New York: Columbia University Press.

Ohanian, H. C. (2009). *Einstein's Mistakes.* New York: W. W. Norton & Company; Illustrated edition (November 9, 2009). doi:ISBN-13 : 978-0393337686

Schrödinger, E. (1935). *Discussion of Probability Relations between Separated Systems* (Vol. 31). Cambridge: Proceedings of the Cambridge Philosophical Society. doi:10.1017/S0305004100013554

Schrodinger, E. (1967). *What is Life? and Mind and Matter* (978-1-107-60466-7 Paperback ed.). Cambridge, UK: Cambridge University Press.

Susskind, L. (2008). *The Cosmic Landscape.* Massachusetts: Back Bay Books Kindle Edition.

von Neumann, John. Mathematical Foundations of Quantum Mechanics. Princeton University Press, 1955.

WORKS CITED

"Diachrony and Representation." In Entre Nous: Thinking-of-the-Other. (1998). (M. B. Harshav, Trans.) New York: Columbia University Press.

Bergo, B. (2019). (E. N. Zalta, Ed.) Retrieved from The Stanford Encyclopedia of Philosophy (Fall 2019 Edition): https://plato.stanford.edu/cgi-bin/encyclopedia/archinfo.cgi?entry=levinas

Coe, C. D. (2018). LEVINAS AND THE TRAUMA OF RESPONSIBILITY The Ethical Significance of Time. Bloomington: Indiana University Press.

Guenther, L. (2009). "Nameless Singularity" Levinas on Individuation and Ethical Singularity. *Epoche, 14*(1), Pages 167-187. Retrieved from Epoché: A Journal for the History of Philosophy: https://doi.org/10.5840/epoche200914128

Hegel, G. W. (2010). *The Science of Logic.* (G. d. Giovanni, Ed., & G. d. Giovanni, Trans.) Cambridge: Cambridge University Press. doi:ISBN-13:9780521832557

Jean-Louis. (2006). Introduction to Reading the Pentateuch. Eisenbrauns. Winona Lake: Eisenbrauns.

Levinas, E. (1969). *Totality and Infinity: An Essay on Exteriority.* (A. Lingis, Trans.) Pittsburgh: Duquesne University Press.

Levinas, E. (1981). *Otherwise than Being or Beyond Essence.* (A. Lingis, Trans.) Pittsburgh: Indiana University Press.

Levinas, E. (1989). The Levinas Reader. (S. Hand, Ed.) 311.

Levinas, E. (2000). *Entre Nous: On Thinking-of-the-Other.* (B. H. Smith, Trans.) New York: Columbia University Press.

Lorenzo Buffoni, A. S. (2019). Quantum Measurement Cooling. *PHYSICAL REVIEW LETTERS*(Phys. Rev. Lett. 122, 070603 – Published 21 February 2019). Retrieved from https://journals.aps.org/prl/abstract/10.1103/PhysRevLett.122.070603

Ward, L. F. (1897). *Outlines of Sociology.* New York: THE MACMILLAN COMPANY. Retrieved from https://socialsciences.mcmaster.ca/econ/ugcm/3ll3/ward/outlines.pdf